Wavelets

Wavelets

The Key to Intermittent Information?

EDITED BY

B. W. Silverman

School of Mathematics
University of Bristol

and

J. C. Vassilicos

Department of Applied Mathematics and Theoretical Physics
University of Cambridge

Originating from contributions to a Discussion Meeting
of The Royal Society

OXFORD
UNIVERSITY PRESS

OXFORD

UNIVERSITY PRESS

Great Clarendon Street, Oxford OX2 6DP

Oxford University Press is a department of the University of Oxford.
It furthers the University's objective of excellence in research, scholarship,
and education by publishing worldwide in

Oxford New York

Athens Auckland Bangkok Bogotá Buenos Aires Calcutta
Cape Town Chennai Dar es Salaam Delhi Florence Hong Kong Istanbul
Karachi Kuala Lumpur Madrid Melbourne Mexico City Mumbai
Nairobi Paris São Paulo Singapore Taipei Tokyo Toronto Warsaw

with associated companies in Berlin Ibadan

Oxford is a registered trade mark of Oxford University Press
in the UK and in certain other countries

Published in the United States
by Oxford University Press Inc., New York

First published in the Philosophical Transactions of The Royal Society 1999
First published by Oxford University Press 2000

A catalogue record for this book is available from the British Library

Library of Congress Cataloging-in-Publication Data

ISBN 0 19 850716 X

Typeset by Focal Image Ltd, London
Printed in Great Britain
on acid-free paper by
Biddles Ltd, Guildford and King's Lynn

Preface

In recent years there has been an explosion of interest in wavelets, in a wide range of fields in science and engineering and beyond. Our aim in organizing the Royal Society Discussion Meeting on which this issue is based was to bring together researchers in wavelets from disparate fields, both in order to provide a showcase for a wider audience, and to encourage cross-fertilization of ideas. The meeting, held on 24 and 25 February 1999, attracted a large and enthusiastic audience. Apart from the main papers collected here, there was lively discussion as well as some very interesting contributed posters, many of which will be published elsewhere in the scientific literature.

Of course, many of the ideas behind wavelets are by no means new. What has been achieved in the past 10–15 years is the common mathematical framework into which these have been incorporated; the understanding of the connection between discrete filter banks and wavelet bases for functions in continuous time; the development of software for wavelets and many related methods; and a broadening perspective showing just how widely wavelets can be used.

The basic ideas are simply stated. In broad terms, a wavelet decomposition provides a way of analysing a signal both in time and in frequency. If f is a function defined on the whole real line, then, for a suitably chosen *mother wavelet* function ψ, we can expand f as

$$f(t) = \sum_{j=-\infty}^{\infty} \sum_{k=-\infty}^{\infty} w_{jk} 2^{j/2} \psi(2^j t - k),$$

where the functions $\psi(2^j t - k)$ are all orthogonal to one another. The coefficient w_{jk} conveys information about the behaviour of the function f concentrating on effects of scale around 2^{-j} near time $k \times 2^{-j}$. This wavelet decomposition of a function is closely related to a similar decomposition (the discrete wavelet transform (DWT)) of a signal observed at discrete points in time.

The DWT can be calculated extremely quickly, and has the property of being very good at *compressing* a wide range of signals actually observed in practice— a very large proportion of the coefficients of the transform can be set to zero without appreciable loss of information, even for signals that contain occasional abrupt changes of level or other behaviour. It is this ability to deal with heterogeneous and intermittent behaviour that makes wavelets so attractive. Classical methods of signal processing depend on an underlying notion of stationarity, for which methods such as Fourier analysis are very well adapted. If one moves away from stationary behaviour, particularly towards the intermittent behaviour with sharp individual events found in many or most real systems, then methods like wavelets are likely to come into their own. Non-stationarity and intermittency, or sparsity of information, can appear in many different guises, but one can also find in the wavelet tool-kit a wide variety of mother wavelets and wavelet-like transforms, each one suited to a specific class of problems. Examples discussed in some length in this issue are ridgelets, harmonic wavelets, complex wavelets and the Mexican hat wavelet.

It has already become clear that in many cases the standard wavelet methodology has to be extended and modified in order to be useful. The paper by Daubechies *et al.* demonstrates how wavelet ideas can be applied to data obtained on irregular spatial grids, by making use of the ideas of the lifting scheme. They retain the basic principle in wavelet analysis of splitting the data into 'detail' and 'signal', but the concept of 'level' in the transformation is defined by reference to the local geometry of the surface. An extension in a different direction is explored by Candès & Donoho, who deal with certain multi-dimensional surfaces by using *ridgelets*, which are only localized in one direction in space. They investigate the properties of ridgelets, and in particular demonstrate that they may have much better compression performance than standard multi-dimensional wavelets on images observed in practice, especially those with discontinuities along straight edges.

Kingsbury provides a description of a particular implementation of complex wavelets to image processing. He shows how complex wavelet filter banks overcome some of the limitations of classical wavelet filter banks and are both shift invariant and directional. Nicolleau & Vassilicos use the Mexican hat wavelet to define a new fractal dimension, the eddy capacity, and they discuss in what sense this eddy capacity is a direct geometrical measure of scale-invariant intermittency. Of course, wavelets are not a panacea, and Prandoni & Vetterli investigate ways that wavelets are not the best approach to the compression of piecewise polynomial signals.

Why are wavelets useful in statistical applications? This question was approached both from a general and a specific point of view. Johnstone explains and explores the way in which sparsity of representation allows for improvement in the statistical performance of estimation methods. Combined with other insights into the way that wavelets and related families represent phenomena of genuine interest, this provides a unified approach to understanding the merits of wavelet techniques. Silverman gives some directions in which wavelet methodology, especially in statistics, can be extended beyond the standard assumptions, and among other things considers ways in which wavelets can be used to estimate curves and functions observed with correlated noise and on irregular grids. The time–frequency properties of wavelet analysis make it appropriate for wavelets to be used in the analysis of time-series. Nason & von Sachs explore this area, drawing examples from anaesthesiology and other fields.

Several authors brought to the meeting their expertise in specific areas of application. Ramsey gives a wide-ranging discussion of the relevance of wavelets to economics. One of his examples concerns the notion that transactions of different sizes have effects over different time-scales; the capacity of wavelets to analyse processes in both scale and time is important here. Pen's paper gives a practical illustration of the way that a wavelet approach gives a good denoising of images obtained in astronomy. Newland demonstrates that harmonic wavelets are a very efficient tool for the time–frequency analysis of transient and intermittent data in vibrations and acoustics. Arneodo *et al.* use the continuous wavelet transform to educe an underlying multiplicative structure from velocity data of homogeneous isotropic turbulence of high Reynolds number. This structure is found to be intermittent and dependent on Reynolds number, and may be the kinematic reflection of a cascading process in the turbulence. Moving from the physical to the biological sciences, Field explains that wavelet-like structures may be involved in the mammalian visual system because of their efficiency in capturing the sparse statistical structure of natural scenes.

The genuine interdisciplinary nature of wavelet research is demonstrated by the breadth of contributions to the meeting and to this issue. It is perhaps too early to answer definitely the question posed in the title of the meeting: whether wavelets are indeed *the* key to understanding and modelling heterogeneous and intermittent phenomena. However, it is already clear that they have a very substantial role and that the Discussion Meeting has played an important part in developing this. We are extremely grateful to our fellow organizers, Ingrid Daubechies and Julian Hunt, for their suggestions in constructing the programme, to Irene Moroz and Guy Nason for organizing the poster sessions of contributed papers, to the Novartis Foundation for their support, and to the Royal Society for sponsoring and hosting the meeting.

Bristol   B. W. S.
Cambridge   J. C. V.

Contents

3 Wavelets for the study of intermittency and its topology   49
F. Nicolleau and J. C. Vassilicos

4 Wavelets in statistics: beyond the standard assumptions   71
Bernard W. Silverman

Contributors

A. Arneodo	Centre de Recherche Paul Pascal, Pessac, France
Emmanuel J. Candès	Stanford University, USA
Ingrid Daubechies	Princeton University, USA
David L. Donoho	Stanford University, USA
D. J. Field	Cornell University, USA
Igor Guskov	Princeton University, USA
Iain M. Johnstone	Stanford University, USA
Nick Kingsbury	University of Cambridge, UK
S. Manneville	Laboratoire Ondes et Acoustique, ESPCI, Paris, France
J. F. Muzy	Centre de Recherche Paul Pascal, Pessac, France
Guy P. Nason	University of Bristol, UK
David E. Newland	University of Cambridge, UK
F. Nicolleau	University of Sheffield, UK
Ue-Li Pen	University of Toronto, Canada
Paolo Prandoni	Swiss Federal Institute of Technology, Lausanne, Switzerland
James B. Ramsey	New York University, USA
S. G. Roux	NASA, Climate and Radiation Branch, Greenbelt, USA
Rainer von Sachs	Catholic University of Louvain, Belgium
Peter Schröder	California Institute of Technology, Pasadena, USA
Bernard W. Silverman	University of Bristol, UK
Wim Sweldens	Lucent Technologies, Murray Hill, USA
J. C. Vassilicos	University of Cambridge, UK
Martin Vetterli	University of California, Berkeley, USA and Swiss Federal Institute of Technology, Lausanne, Switzerland

1

Wavelets on irregular point sets

Ingrid Daubechies and Igor Guskov[1]
*Program for Applied and Computational Mathematics, Princeton University,
Fine Hall, Washington Road, Princeton, NJ 08544-1000, USA*

Peter Schröder
*Department of Computer Science, California Institute of Technology,
Pasadena, CA 91125, USA*

Wim Sweldens
*Lucent Technologies, Bell Laboratories, 600 Mountain Avenue,
Murray Hill, NJ 07974, USA*

Abstract

In this article we review techniques for building and analysing wavelets on irregular point sets in one and two dimensions. We discuss current results on both the practical and theoretical sides. In particular, we focus on subdivision schemes and commutation rules. Several examples are included.

Keywords: multiresolution; wavelets; irregular grid; subdivision; mesh; commutation

1 Introduction

Wavelets are a versatile tool for representing general functions and datasets, and they enjoy widespread use in areas as diverse as signal processing, image compression, finite-element methods and statistical analysis (among many others). In essence we may think of wavelets as *building blocks* with which to represent data and functions. The particular appeal of wavelets derives from their representational and computational *efficiency*: most datasets exhibit correlation both in time (space) and frequency, as well as other types of structure. These can be modelled with high accuracy through sparse combinations of wavelets. Wavelet representations can also be computed *fast*, because they can be built using multiresolution analysis and subdivision.

Traditionally, wavelet functions $\psi_{j,m}$ are defined as translates and dilates of one particular function, the *mother* wavelet ψ. We refer to these as *first-generation* wavelets. This paper is concerned with a more general setting in which wavelets need not—and, in fact, cannot—be translates and dilates of one or a few templates. Generalizations of this type were called *second-generation* wavelets in Sweldens (1997); they make it possible to reap the benefit of wavelet algorithms in settings with

---

[1]Department of Computer Science, California Institute of Technology, Pasadena, CA 91125, USA.

irregularly spaced samples, or on 2-manifolds which cannot be globally parametrized to the plane. In generalizing wavelet analysis to these more general settings one would like to preserve many of the properties enjoyed by first-generation wavelets. In particular, they should still be associated with fast algorithms and have appropriate smoothness and localization properties. In addition, they should be able to characterize various functional spaces of interest. In this paper we shall be mostly concerned with fast algorithms, localization and smoothness; we will not address function-space characterizations. Note though that the smoothness of the wavelets is related to their ability to form unconditional bases for certain function spaces (Dahmen, 1996; Donoho, 1992).

The key to generalizing wavelet constructions to these non-traditional settings is the use of generalized *subdivision schemes*. The first-generation setting is already connected with subdivision schemes, but they become even more important in the construction of second-generation wavelets. Subdivision schemes provide fast algorithms, create a natural multiresolution structure and yield the underlying scaling functions and wavelets we seek.

Subdivision is a technique originally intended for building smooth functions starting from a coarse description. In this setting there is no need for irregular grids, as one is free to choose the finer grid to be regular. However, we intend to use subdivision as part of an entire multiresolution analysis which starts from a finest irregular grid. This finest grid is gradually 'coarsified'; subdivision then gives an approximation of the original data by extrapolating the reduced data on the coarser grid back to the original finest grid. In such a setting the geometry of the grids is fixed by the finest irregular grid and the coarsification procedure; thus subdivision on irregular grids is called for.

Remark 1.1 Another approach would be to resample the original finest level data on a regular grid and use first-generation wavelets. Resampling, however, can be costly, introduce artefacts and is generally impossible in the surface setting. Therefore we choose to work on the original grid.

1.1   One-dimensional subdivision

The main idea behind subdivision is the iteration of upsampling and local averaging to build functions and intricate geometrical shapes. Originally, such schemes were studied in computer-aided geometric design in the context of corner cutting (de Rham, 1956; Chaikin, 1974) and the construction of piecewise polynomial curves, e.g. the de Casteljau algorithm for Bernstein–Bézier curves (de Casteljau, 1959) and algorithms for the iterative generation of splines (Lane & Riesenfeld, 1980; De Boor, 1978). Later subdivision was studied independently of spline functions (Dyn *et al.*, 1987; Dubuc, 1986; Deslauriers & Dubuc, 1987; Cavaretta *et al.*, 1991; Cavaretta & Micchelli, 1987, 1989) and the connection to wavelets was made (Mallat, 1989; Daubechies, 1988).

For example, figure 1 demonstrates the application of the four-point scheme. New points are defined as local averages of two old points to the left and two old

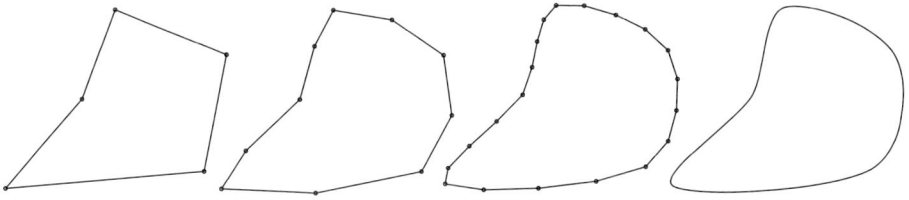

Fig. 1: Subdivision is used to generate a smooth curve starting from a coarse description.

Fig. 2: Regular, semiregular and irregular grid hierarchies in one dimension.

points to the right with weights $\frac{1}{16}(-1, 9, 9, -1)$.

In the case of spline functions, smoothness follows from simple algebraic conditions on the polynomial segments at the knots. However, in the general setting convergence and smoothness of the limit function are harder to prove. Various approaches have been explored to find the Hölder exponent of the limit function, or to determine its Sobolev class. Early references in this context are Dubuc (1986), Dyn *et al.* (1987, 1990*b*), Deslauriers & Dubuc (1987), Micchelli & Prautzsch (1987), Daubechies & Lagarias (1991), Cavaretta *et al.* (1991), Rioul (1992), Villemoes (1994) and Eirola (1992). These studies and their results all rely on regular, i.e. equispaced, grids. The analysis uses tools such as the Fourier transform, spectral analysis and the commutation formula.

In this paper we focus on irregular point sets. To describe the settings we are interested in, we distinguish three types of refinement grids: regular, semiregular and irregular (see figure 2). A *regular* grid has equidistant points on each level, and, each time a new point is inserted, it is placed exactly between two old points. For example, the curve shown in figure 1 is parametrized over a regular grid. A *semiregular* grid (middle, figure 2) starts with an irregular coarse grid and adds new points at parameter locations midway between successive old points. Thus the finer grids are locally regular except around the original coarsest level points. In *irregular* grids (right) parameter locations of new points need not be midway between successive old points. Note that regular grids are translation and dilation invariant, while semiregular grids are locally dilation invariant around coarsest level vertices and irregular grids generally possess no invariance property.

Similarly the weights used in subdivision schemes come in three categories: uniform, semi-uniform and non-uniform. *Uniform* schemes like the four-point scheme of figure 1 correspond to first-generation wavelets and use the same subdivision

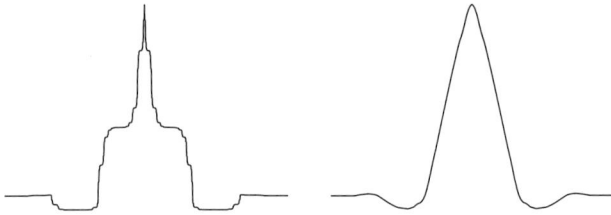

Fig. 3: An example why non-uniform subdivision is needed. The limit function with uniform (left), non-uniform (right) subdivision. The same irregular grid is used in both figures.

weights within a level and across all levels; they are typically used on regular grids or grids which can be smoothly remapped to a regular grid. *Semi-uniform* schemes are used on semiregular grids; they vary the weights within each level (special weights are used in the neighbourhood of the coarsest level points), but the same weights are used across levels. Such schemes are sometimes referred to as *stationary*. Wavelets and subdivision schemes on an interval also fall in this category. *Non-uniform* schemes use varying weights within *and* across levels and correspond to the second-generation setting.

Almost all work on smoothness for non-regular grids concerns the semiregular grids with semi-uniform subdivision schemes. Because translation-invariance is lost, Fourier-transform-based arguments can no longer be used. However, since the same weights are used on successive levels, one has dilation invariance around coarsest level points and can reduce the smoothness analysis to the study of spectral properties of certain fixed matrices. In Warren (1995, unpublished research[2]) it is shown that the semi-uniform version of the four-point scheme on a semiregular grid yields a C^1 limit function.

In the irregular case the subdivision scheme must become non-uniform to account for the irregularity of the associated parameter locations. This is illustrated in figure 3, which shows the limit functions of the uniform four-point rule (left) and non-uniform four-point rule (Sweldens & Schröder, 1996) (right); both use the same irregular grid.

The study of irregular subdivision is not only theoretically interesting, but also of great importance in practical applications. For example, in the semiregular setting, one can use adapted weights to better control the shape of a curve (Kobbelt & Schröder, 1997) or surface (Zorin *et al.*, 1996). More importantly, in many practical set-ups we start with samples associated to a very fine, but irregular, grid. Now the main task for subdivision is not further refinement, but rather aid in a multiresolution analysis on *coarser* grids. The wavelet and scaling functions from the coarsest level are generated with a subdivision scheme with new points which are no longer

---

[2]Available from www.cs.rice.edu/~jwarren.

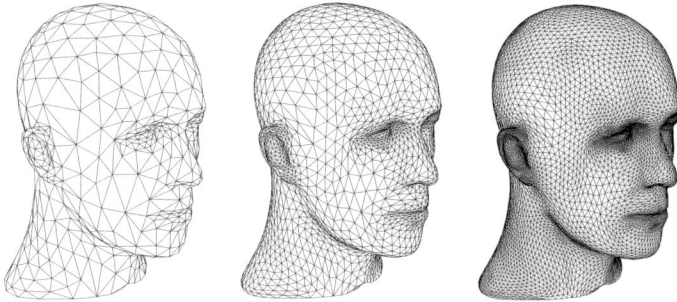

Fig. 4: Two-dimensional loop subdivision is used to generate smooth surfaces from a coarse description.

parametric midpoints, but are dictated by the finest level grid on which the data were originally sampled. Even though the actual number of levels is always finite for any concrete application of these methods, the asymptotic behaviour of irregular subdivision is still relevant as the finest and coarsest level could be arbitrarily far apart.

In these settings smoothness results become much harder to obtain. Because the subdivision weights vary within a level, the Fourier transform can no longer be used, and because they vary across levels, even spectral analysis cannot help. In this paper we discuss some tools that can be used to analyse smoothness; in particular, we demonstrate that the commutation formula still holds and becomes a critical tool for smoothness analysis.

1.2 Two-dimensional subdivision

The two-dimensional setting appears in the context of generating smooth surfaces (see figure 4). Here regular grids are too restrictive. For example, tensor product settings are only applicable for surfaces homeomorphic to a plane, cylinder or torus due to the Euler characteristic. Historically, this challenge was addressed by generalizing traditional spline patch methods to the semiregular biquadratic (Doo & Sabin, 1978), bicubic (Catmull & Clark, 1978) and quartic-box-spline settings (Loop, 1987). Similar to the one-dimensional setting, researchers also developed interpolating constructions (Dyn *et al.*, 1990a; Zorin *et al.*, 1996; Kobbelt, 1996). All these settings (and others since; for an overview see Schröder & Zorin (1998)) proceed by applying quadrisection to an initial mesh consisting of either quadrilaterals or triangles and thus belong to the semiregular setting. The weights used in the subdivision scheme are semi-uniform since they take into account the local neighbourhood structure of a vertex, i.e. how many edge neighbours a given vertex has. As in the one-dimensional semiregular setting, spectral analysis is the key to understanding the smoothness of these constructions. We refer to Reif (1995),

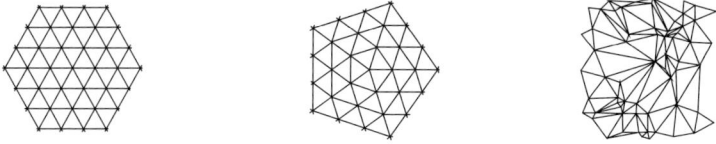

Fig. 5: Sections of regular, semiregular and irregular triangle grids in two dimensions.

Warren (unpublished research), Zorin (1996) and Schweitzer (1996) for more details.

The irregular setting appears in two dimensions just as in the one-dimensional case when some finest irregular level is presented on input and the main task is to build a multiresolution analysis on *coarser* levels. In this case, however, we can no longer define downsampling as simply retaining every other sample. This brings us to the realm of mesh simplification; we postpone the discussion of mesh simplification and the construction of appropriate non-uniform subdivision operators to Section 3.

1.3   Overview

This paper summarizes the results obtained in Daubechies *et al.* (1998, *b*), Guskov (1998) and Guskov *et al.* (1999). We start with the one-dimensional results of Daubechies *et al.* (1998, *b*). We show that even simple subdivision rules, such as cubic Lagrange interpolation, can lead to very intricate subdivision operators. To control these operators, we use commutation: because the subdivision scheme maps the space of cubic polynomial sequences to itself, we can define derived subdivision schemes for the divided difference sequences. These simpler schemes can then be used to prove growth bounds on divided differences of some order, corresponding to smoothness results for the limit function of the original scheme. The commutation formula enables us to control smoothness and is the key to the construction of wavelets associated with the subdivision scheme.

In Guskov (1998), inspiration from the one-dimensional analysis is used to tackle the much more complex two-dimensional case. Again, differences and divided differences are introduced, which can be computed from level to level with their own derived subdivision scheme. Control on the growth of these divided differences then leads to smoothness results. In practice, finding the right ansatz for irregular subdivision in the two-dimensional setting is much harder than in the already difficult one-dimensional case. Finally, we show how irregular subdivision schemes can be used in multiresolution pyramids for two-dimensional meshes embedded in $\mathbb{R}^3$ and review several applications from Guskov *et al.* (1999). The 'wavelets' associated with these schemes are overcomplete and are related to frames rather than bases.

2 The one-dimensional case

2.1 Multilevel grids

Consider grids X_j, which are strictly increasing sequences of points $\{x_{j,k} \in \mathbb{R} \mid k \in \mathbb{Z}\}$, and which are consecutive binary refinements of the initial grid X_0, i.e. $X_j \subset X_{j+1}$ and $x_{j+1,2k} = x_{j,k}$ for all j and k. Thus in every refinement step we insert one odd indexed point $x_{j+1,2k+1}$ between each adjacent pair of 'even' points $x_{j,k} = x_{j+1,2k}$ and $x_{j,k+1} = x_{j+1,2k+2}$, as in figure 2. We define $d_{j,k} := x_{j,k+1} - x_{j,k}$. We shall also use the term *grid size* on level j, for the quantity $d_j := \sup_k d_{j,k}$. As $j \to \infty$ we want the grids to become dense, with 'no holes left'; this translates to the requirement that the d_j be summable.

Remark 2.1

(1) The above multilevel grids are called *two-nested*. One can also consider more general irregular grids such as q-nested grids, where we insert $q-1$ new points in between old points or even non-nested but 'threadable' grids. See Daubechies *et al.* (1998) for more details on this.

(2) In case the ratio between the lengths of any two neighbouring intervals is globally bounded, we call the grid *homogeneous*. An example of an irregular two-nested grid that is not homogeneous is built by $x_{j+1,2k+1} = \beta x_{j,k} + (1 - \beta)x_{j,k+1}$, where β is a fixed parameter satisfying $0 < \beta < 1$. This is an example of a *dyadically* balanced grid: the ratio between the lengths of two 'sibling' intervals $d_{j,2l}$ and $d_{j,2l+1}$ is bounded. However, the ratio between $d_{j,-1} = \beta^j$ and $d_{j,0} = (1 - \beta)^j$ is unbounded.

2.2 Subdivision schemes

Subdivision starts with a set of initial function values $f_0 = \{f_{0,k}\}$, which live on the coarsest grid X_0. The subdivision scheme S is a sequence of linear operators S_j, $j \geq 0$, which iteratively computes values $f_j = \{f_{j,k}\}$ on the finer grids via the formula $f_{j+1} = S_j f_j$, or

$$f_{j+1,l} = \sum_k S_{j,l,k} f_{j,k}.$$

We consider only *local* schemes in the sense that the above summation has a globally bounded number of terms centred around $k = 2l$. Subdivision gives us values defined on the grid points $x_{j,k}$. By connecting these points we can define a piecewise linear function $f_j(x)$ (see figure 6). Our ambition is to synthesize a continuous limit function $\phi(x)$ as the pointwise limit for $j \to \infty$ of $f_j(x)$. We are interested in the existence and smoothness of $\phi(x)$.

The subdivision coefficients $S_{j,k,l}$ will depend on the application one has in mind. We pointed out in the introduction that one cannot simply stick with the coefficients

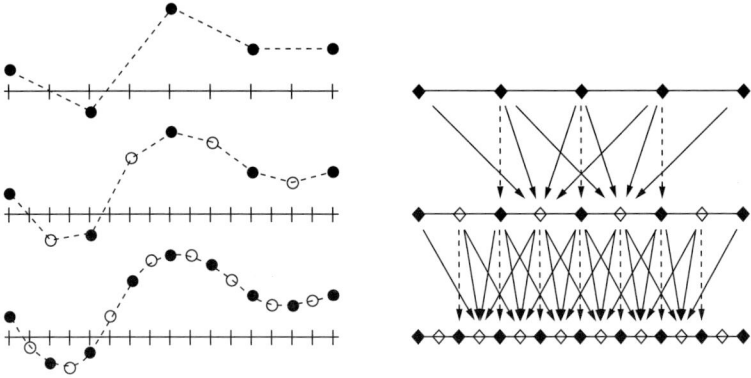

Fig. 6: On the left an interpolating scheme is applied to function values; the new function values are shown as open circles. On the right, arrows show the dependencies in the computation of those values; the vertical dashed arrows indicate that function values which were already assigned are kept unchanged in the subdivision, because this is an interpolating scheme.

from the regular case; typically the coefficients need to be spatially varying, and will be linked to the spatial variation of the grid.

One such subdivision scheme which allows for a spatial interpretation is Lagrangian interpolating subdivision (Dubuc 1986; Dyn *et al.* 1987; Deslauriers & Dubuc 1987, 1989). Here, the value $f_{j+1,2k+1}$ at a new point is found by defining a polynomial which interpolates the points $(x_{j,l}, f_{j,l})$ for l in the neighbourhood of k, and evaluating this polynomial at $x_{j+1,2k+1}$ (see figure 7). In the regular cubic case, this corresponds to the standard four-point scheme, with

$$S_{j,2k+1,k} = S_{j,2k+1,k+1} = \tfrac{9}{16} \text{ and } S_{j,2k+1,k-1} = S_{j,2k+1,k+2} = -\tfrac{1}{16}.$$

In the irregular setting the coefficients are a non-trivial quotient of cubic polynomials in the $x_{j,k}$ (see Daubechies *et al.*, 1999).

Lagrangian subdivision is *interpolating* in the sense that in each subdivision step the values at the even grid points are kept, i.e. $f_{j+1,2k} = f_{j,k}$, and the limiting function thus interpolates the original data $\phi(x_{0,k}) = f_{0,k}$. For non-interpolating or *approximating* schemes, the $f_{j+1,2k}$ can differ from $f_{j,k}$ (see figure 8).

2.3   Smoothness results

To derive smoothness estimates, we use Lemarié's *commutation formula* idea, generalized to the present irregular setting. (Note that this is similar to Rioul (1992) and Dyn *et al.* (1991), who studied the regular case.) For the cubic Lagrange interpolation

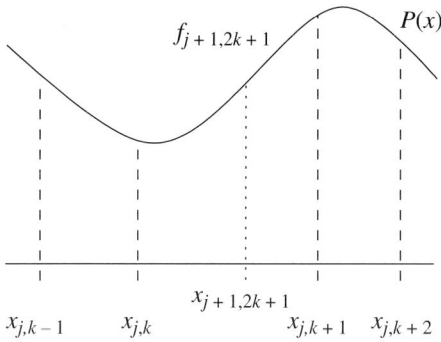

Fig. 7: Cubic Lagrangian interpolation. The value $f_{j+1,2k+1}$ at the odd grid point $x_{j+1,2k+1}$ is obtained by evaluating a cubic polynomial $P(x)$ interpolating values at four neighbouring even grid points $x_{j+1,2k-2} = x_{j,k-1}, \ldots, x_{j+1,2k+4} = x_{j,k+2}$.

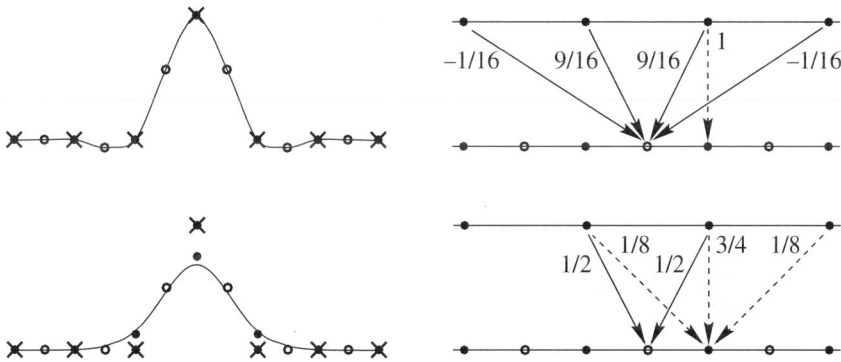

Fig. 8: The top row shows the limit function and weights of the cubic Lagrange interpolation scheme; the bottom row illustrates the non-interpolating subdivision scheme producing cubic B-splines.

example, this amounts to introducing divided difference sequences,

$$f_{j,k}^{[1]} := \frac{f_{j,k+1} - f_{j,k}}{x_{j,k+1} - x_{j,k}},$$

and observing that the $f_{j,k}^{[1]}$ are *also* related by local subdivision, i.e. there exists a local subdivision scheme with entries $S_{j,l,k}^{[1]}$ so that

$$f_{j+1,l}^{[1]} = \sum_k S_{j,l,k}^{[1]} f_{j,k}^{[1]}.$$

The existence of the $S_j^{[1]}$ follows from the fact that every $S_j^{[0]} := S_j$ maps a constant sequence to itself (see Daubechies *et al.*, 1999). (It is clear that if $S^{[0]}$ does not leave constants invariant, then $S^{[1]}$ cannot exist, since it would need to map the zero sequence to a non-zero result.) Moreover, one can show that if the $f_j^{[1]}$ 'converge' to a continuous function ϕ_1, then the $f_j^{[0]} := f_j$ also converge, to a continuously differentiable function ϕ, and that $\phi' = \phi_1$. (For details, see Daubechies *et al.* (1998, *b*).) This is the essence of the commutation idea:

$$
\begin{array}{ccc}
f_j & \xrightarrow{\text{divided difference}} & f_j^{[1]} \\
\downarrow \text{limit} & & \downarrow \text{limit} \\
\phi & \xrightarrow{\text{differentiation}} & \phi_1 = \phi'
\end{array}
$$

It turns out that one can also consider higher-order divided differences; for the cubic Lagrange interpolation case, one can go up to fourth-order differences because $S^{[0]}$ maps cubic polynomials sampled at $x_{j,k}$ to cubic polynomials sampled at $x_{j+1,k}$. These $f_{j,k}^{[4]}$ no longer converge, but we can control their growth, and this helps us prove that ϕ_1 is continuous, and ϕ continuously differentiable. In fact, detailed (and rather technical) estimates in Daubechies *et al.* (1999) show that, for homogeneous grids,

$$
|f_{j,k}^{[4]}| \le C \frac{\lambda^j}{d_{j,k}^3},
$$

where $\lambda < 1$ is determined by the bound on the ratio between neighbouring interval lengths. Once such a bound is known, a general theorem (see theorem 4 in Daubechies *et al.* (1999)) can be used to prove that $\phi \in C^{2-\epsilon}$. This result is optimal in the sense that even in the regular case better smoothness cannot be obtained.

Remark 2.2

(1) 1 This result for cubic Lagrange interpolation on homogeneous grids can be extended to grids that are dyadically balanced only. The analysis becomes much more delicate.

(2) 2 A similar approach can be used for non-interpolating subdivision. In that case it turns out that one has to use appropriately defined divided differences, which are different from the 'standard' definition. See Daubechies *et al.* (1998) for a complete discussion of this situation.

2.4   Wavelets

Wavelets at level j are typically used, in the regular case, as building blocks to represent any function in the multiresolution analysis that 'lives' in the $(j + 1)$st approximation space V_{j+1}, but not in the coarser resolution approximation space

$V_j \subset V_{j+1}$. One can introduce similar wavelets in the present irregular setting. The scaling functions $\phi_{j,k}$ are the limit functions obtained from starting the subdivision scheme at level j, from the 'initial' data $f_{j,l} = \delta_{l,k}$, and refining from there on. Under appropriate assumptions on the subdivision operators S_j, the $\phi_{j,k}$ are independent; V_j is the function space spanned by them. Clearly, $V_j \subset V_{j+1}$. As in the regular case, there are many different reasonable choices for complement spaces W_j (which will be spanned by the wavelets at level j) that satisfy $V_{j+1} = V_j \oplus W_j$.

When the scaling functions are interpolating as in the Lagrangian case, i.e.

$$\phi_{j,k}(x_{j,k'}) = \delta_{k,k'},$$

then a simple choice for a wavelet is given by $\psi_{j,m} = \phi_{j+1,2m+1}$, i.e. the wavelet is simply a finer-scale scaling function at an odd location. This is sometimes called an *interpolating* wavelet. This is in general not a very good wavelet as it does not have any vanishing moments. It can be turned into a wavelet with vanishing moments using the lifting scheme (Sweldens & Schröder, 1996).

Another way to select a complement space W_j is to use commutation between two biorthogonal multiresolution hierarchies, V_j and $\tilde{V}_j$. If both are associated to local subdivision schemes, then the biorthogonality of the $\phi_{j,k}$ and $\tilde{\phi}_{j,l}$ imposes consistency requirements on the S_j and $\tilde{S}_j$. Commutation can be used, as in the regular case, to pass from one dual pair of multiresolution analyses to another, by operations related to differentiating and integrating, respectively. For instance, the above choice of an interpolating wavelet corresponds formally to letting the dual scaling function be a Dirac. Applying the commutation rule each time reduces the order of the scaling functions, but increases the order of the dual scaling function. In particular, the Dirac will become a box and later on a general B-spline. It turns out that there is a natural definition of wavelets $\psi_{j,k}$ and $\tilde{\psi}_{j,k}$ corresponding to the dual multiresolution structures. It is shown in Daubechies *et al.* (1998) that, as in the regular case, the new wavelet after commutation is the derivative of the old wavelet and the new dual wavelet is the integral of the old dual wavelet. By repeatedly applying commutation starting from the Lagrangian setting, one can thus build the entire family of biorthogonal compactly supported irregular B-spline wavelets and their duals (Daubechies *et al.*, 1998).

3   The two-dimensional case

We mentioned in the introduction that the importance of the irregular setting arises from the practical need to coarsify in settings in which the initial input is given as a function over a fine triangulation of the plane (functional setting) or as a triangulation of a 2-manifold (surface setting). In the one-dimensional setting, the downsampling operation to create a coarser level is straightforward as we can simply 'skip' every other sample. In the irregular two-dimensional setting, downsampling is much less straightforward. Before delving into the details of irregular two-dimensional subdivision, we first discuss a number of approaches which can be employed to

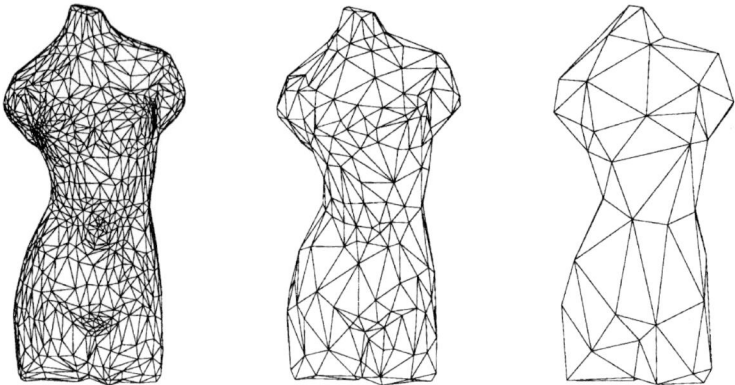

Fig. 9: If an irregular finely detailed mesh is given, the first task in building a multiresolution analysis is coarsification.

define irregular downsampling in the surface setting. This problem has received a lot of attention in computer graphics, where it is generally referred to as polygonal simplification.

3.1 Polygonal simplification

In polygonal mesh simplification, the goal is to simplify a given (triangulated) mesh $\mathcal{M}^L = (\mathcal{P}^L, \mathcal{K}^L)$ into successively coarser, homeomorphic meshes $(\mathcal{P}^l, \mathcal{K}^l)$ with $0 \leq l < L$, where $(\mathcal{P}^0, \mathcal{K}^0)$ is the coarsest or base mesh. Here $\mathcal{P}^l$ is a set of l point positions, while $\mathcal{K}^l$ encodes the topological structure of the mesh and consists of triples $\{i, j, k\}$ (triangles), pairs $\{i, j\}$ (edges) and singletons $\{i\}$ (vertices). The goal now is to allow certain topological operations on $\mathcal{K}^l$ which preserve the manifold property and genus of the mesh. These changes go hand in hand with geometric changes which are typically subject to an approximation quality criterion.

Several approaches for such mesh simplification have been proposed (the interested reader is referred to the excellent survey by Heckbert & Garland (1997) for more details). The most popular methods are the so-called 'progressive meshes' (PM). In a PM construction a sequence of edge collapses is prioritized based on the error it introduces. An edge collapse brings the endpoints of the chosen edge into coincidence, in the process removing two triangles, three edges and one vertex (in the case of interior edges). The point location of the merged vertex can be chosen so as to minimize some error criterion with respect to the original mesh. The error can be measured in various norms such as L^∞ (symmetric Haussdorff distance), L^2 and Sobolev norms.

For our purposes we are using a PM construction based on half-edge collapses, i.e. the point position for the collapsed edge is one of its end points. This results in a mesh hierarchy which is interpolating in the sense that the point position sets

$\mathcal{P}^l$ are nested. There are several possible ways to define levels of a hierarchy. The most flexible way treats a single half-edge collapse operation as defining a level. In contrast to the usual wavelet setting this results in a linear, rather than logarithmic, number of levels.

Before going to the surface case, we first consider the functional setting and then treat the surface setting as three instances of a functional setting.

3.2 Functional setting: multivariate commutation formula

Just as in the one-dimensional case, irregular multivariate subdivision schemes act on sequences whose elements are associated with irregular parameter locations. We introduce levels numbered 0, 1, ... with level 0 corresponding to the coarsest scale. Within each level n, the collection of all parameter locations constitute an irregular grid χ_n.

We can now introduce a subdivision scheme S as a sequence of linear operators $S_n, n \geq 0$, which iteratively compute sequences f_n defined on χ_n, starting from some coarsest level data f_0 via

$$f_{n+1} = S_n f_n.$$

In the one-dimensional setting we analysed the regularity of the functions produced by subdivision through the behaviour of properly defined divided differences. We proceed similarly for the irregular two-dimensional setting. Let $\mathcal{D}_n^{[p]}$ denote the operator which maps the data sequence f_n into the corresponding sequence $f_n^{[p]}$ of divided differences of order p, that is $f_n^{[p]} = \mathcal{D}_n^{[p]} f_n$. We say that there exists a derived subdivision scheme $S^{[p]}$ satisfying the commutation formula if the sequences $f_n^{[p]}$ are related via the relation $f_{n+1}^{[p]} = S_n^{[p]} f_n^{[p]}$, where the $S_n^{[p]}$ constitute a local bounded subdivision scheme. Thus we can write

$$\mathcal{D}_{n+1}^{[p]} S_n = S_n^{[p]} \mathcal{D}_n^{[p]}.$$

We then prove that the bounds on the growth of sequences $f_n^{[p]}$ can be translated into the smoothness estimates for the functions produced by the original subdivision scheme S.

In order to extend this construction to the multivariate case we need to define the multivariate divided differences in such a way that the algebra of the commutation formula works. This is done in Guskov (1998) for a class of polynomial reproducing subdivision schemes. It is also shown there that, for multilevel grids satisfying some natural conditions, the bounds on the growth of these divided differences can be used to analyse the regularity of functions produced by subdivision.

3.3 Constructing a subdivision scheme

In this section we provide a particular example of a subdivision scheme which in the functional setting produces visually smooth functions on irregular triangulations. Our subdivision algorithm relies on minimizing divided differences. Consider a

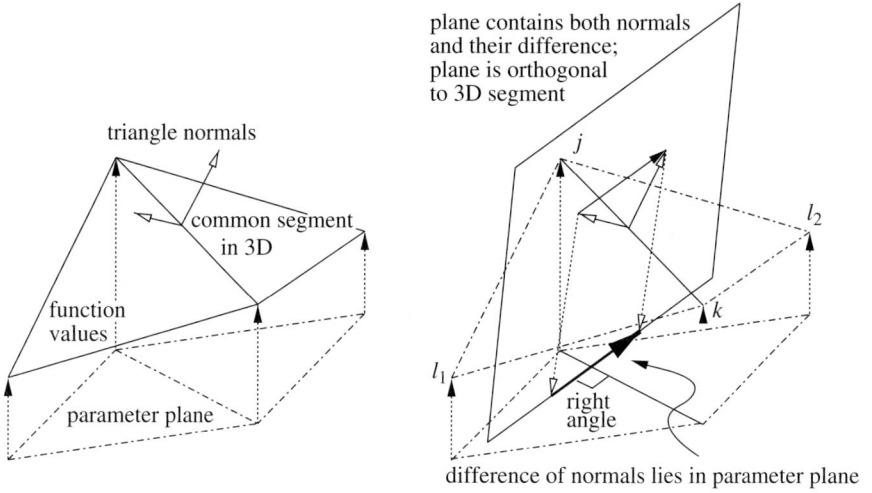

plane contains both normals
and their difference;
plane is orthogonal
to 3D segment

triangle normals

common segment
in 3D

function
values

parameter plane

difference of normals lies in parameter plane

Fig. 10: Second differences are associated with an edge. Since they are the difference of two adjacent triangle normals (first divided differences), one can see that the second differences are orthogonal to the common edge in the parameter plane.

triangle $\{i, j, k\}$ in the parameter plane with corners (x_i, y_i), (x_j, y_j) and (x_k, y_k), and function values f_i, f_j and f_k. These three function values define a plane. The gradient to this plane can be seen as a first-order divided difference corresponding to this triangle. The gradient is zero only if the plane is horizontal ($f_i = f_j = f_k$).

Next, we define the second-order differences. They are computed as the difference between two normals on neighbouring triangles and can be thought of as being associated with the common edge (see figure 10, left). It is easy to see that the difference between gradients of two adjacent triangles is orthogonal to their common edge (see figure 10, right). Thus the component $D_e^2 f$ normal to the edge e can be used for the second-order difference. It depends linearly on the four function values of these two triangles. The coefficients can be found in Guskov (1998) or Guskov *et al.* (1999). The second-order difference operator is zero only if the two triangles lie in the same plane, and one can see that its behaviour is closely related to the dihedral angle.

The central ingredient in the design of our subdivision scheme is the use of a non-uniform relaxation operator which inserts new values in such a manner that second-order differences are minimized. Define a quadratic energy, which is an instance of a discrete fairing functional (Kobbelt, 1997):

$$Rf_i = \arg\min E(f_i) = \sum_{e \in \mathcal{K}} (D_e^2 f)^2.$$

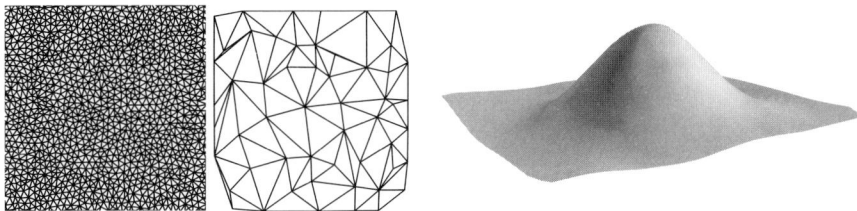

Fig. 11: From left to right: a portion of the fine mesh; the coarse mesh; function produced by the non-interpolating scheme.

Setting $\partial E / \partial f_i = 0$ yields

$$Rf_i = \sum_{j \in \mathcal{V}_2(i)} w_{i,j} f_j \quad \text{with } \mathcal{V}_2(i) = \quad . \tag{3.1}$$

Note that if f is a linear function, i.e. all triangles lie in one plane, the fairing functional E is zero. Consequently, linear functions are invariant under R. In particular, R preserves constants from which we deduce that the $w_{i,j}$ sum to one.

Subdivision is computed one level at a time starting from level n_0 in the PM. Reversing the PM construction back to the finest level adds one vertex (x_n, y_n, f_n) per level; the non-uniform subdivision is computed one vertex at a time. The position of each new vertex n is computed according to 3.1, using areas and lengths of the original finest level mesh. Next, the immediate neighbours of n are relaxed using 3.1 as well. The ambition of our strategy of minimizing $D_e^2 f$ is to obtain C^1 smoothness. However, there is currently no ansatz on the bounds of the divided differences to prove regularity of the limit function. Figure 11 shows an irregular grid of 20 493 triangles (left), simplified down to 86 triangles (middle). Now associate the value $f = 1$ with the centre vertex and 0 with all others. The figure on the right is the result of running the subdivision scheme back to the finest level. Even though the grid is irregular the resulting function appears smooth.

3.4 Functions on surfaces

In order to build a multiresolution structure on meshes, we first need to introduce the relaxation operator acting on functions defined over triangulated surfaces in three dimensions. We shall follow the strategy of the planar case and introduce second differences for such functions. For this we need to specify some locally consistent parametrization over the support of the difference operator. Consider a triangular mesh $\mathcal{P}$ in $\mathbb{R}^3$, and let $f : \mathcal{P} \to \mathbb{R}$. We would like to define the second difference operator $D_e^2 f$ for an edge e from the triangulation $\mathcal{P}$. For this we only need a consistent parametrization (i.e. flattening) for two neighbouring triangles at a time. Let the edge $e = \{i, j\}$ be adjacent to two triangles $\{i, j, k\}$ and $\{j, i, l\}$. We use the 'hinge map'

$$s^{(n-1)}$$

| pre-smooth | subdivision |

$s^{(n)}$

$d^{(n)}$

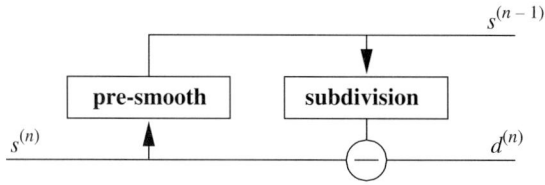

Fig. 12: Burt–Adelson-style pyramid scheme.

to build a pair of adjacent triangles in the plane. These two triangles in the parameter plane have the same angles and edge lengths as the two triangles in $\mathbb{R}^3$. We then define $D_e^2 f$ as described in the previous section. Using these second differences, it is easy to extend the definition of the relaxation operator and the corresponding subdivision scheme to work with functions defined over triangulated surfaces.

3.5   Burt–Adelson pyramid

For meshes we found it more useful to generalize an oversampled Burt–Adelson-type pyramid (Burt & Adelson, 1983) than a critically sampled wavelet pyramid. Let $(\mathcal{P}^n)$ be some fixed PM hierarchy of triangulated surfaces. We start from the function $f_N : \mathcal{P} = \mathcal{P}^N \to \mathbb{R}$, defined on the finest level, and compute a sequence of functions $(f_n)(n_0 \le n \le N)$ as well as oversampled differences $d_i^{(n)}$ between levels.

Like subdivision, the Burt–Adelson pyramid is computed vertex by vertex. Thus the four critical components of a BA pyramid: presmoothing, downsampling, subdivision and detail computation are done for one vertex n at a time (see figure 12). The presmoothing comes down to applying the relaxation operator to the neighbours of n. Downsampling simply removes the vertex n through a half-edge collapse. We perform subdivision as described above and compute details $d^{(n)}$ for all neighbours of n.

In order to see the potential of a mesh pyramid in applications, it is important to understand that the details $d^{(n)}$ can be seen as an approximate frequency spectrum of the mesh. The details $d^{(n)}$ with large n come from edge collapses on the finer levels and thus correspond to small scales and high frequencies, while the details $d^{(n)}$ with small n come from edge collapses on the coarser levels and thus correspond to large scales and low frequencies. Hence, the sequence of $d^{(n)}$ for running n can be seen as an approximate frequency spectrum. Moreover, while the superscript n of an individual detail vector $d_i^{(n)}$ corresponds to its level/frequency, the subscript i corresponds to its location. Thus we actually have a space-frequency decomposition.

It is theoretically possible to build a critically sampled wavelet transform based on the lifting scheme (Sweldens, 1997). The idea is to use an interpolating subdivision scheme which only affects the new vertex and omits the relaxation of the even vertices. Consequently, only one detail per vertex is computed and the sampling is always critical. However, at this point it is not clear how to design updates that make the transform numerically stable. Additionally, interpolating subdivision schemes do not yield very smooth meshes and have unwanted undulations. Therefore, critically sampled wavelet transforms have had limited use in graphics applications.

4 Applications

In the *surface setting* we deal with a triangulated mesh $\mathcal{P}$ of arbitrary topology and connectivity embedded in three dimensions with vertices $p_i = (x_i, y_i, z_i)$. It is important to separate the two capacities the mesh $\mathcal{P}$ fulfils in our analysis. First, the original mesh and its PM representation serve as the source of local parametrization and connectivity information which determines the coefficients of our adaptive relaxation operator.

Second, if our purpose is to process the geometry of the mesh, it is crucial to treat all three coordinates x, y and z as *dependent* variables. In fact, we consider the coordinates of the mesh to be real functions on the current PM vertex set. Initially, before any changes in geometry take place, these functions can be viewed as identities. When the wanted processing operations, such as filtering or editing, are applied to the data, these functions become more meaningful.

As an example of possible application of our scheme we present various manipulations of the scanned Venus's head model. The original mesh has 50 000 vertices. After building a PM hierarchy, we use our BA pyramid scheme to build a multiresolution representation. We can use different manipulations of the detail coefficients in order to achieve various signal processing tasks. Specifically, if all the detail coefficients finer than some level are put to zero, we achieve a smoothing effect (in figure 13*b* all the details on the levels above 1000 were set to zero). The stopband filter effect is achieved by setting to zero some range of coefficients (in figure 13*c* all the details between the levels 1000 and 15 000 were set to zero). One can also enhance certain frequencies (in figure 13*d* all the details between the levels 1000 and 15 000 were multiplied by two).

5 Conclusion

One of the current frontiers in wavelet research and applications is the generalization of multiresolution methods from the regular to the semiregular and, more recently, irregular setting. We have given a brief review of these developments, starting with the one-dimensional setting and moving on to the two-dimensional functional and manifold settings. While there exists an extensive set of tools for the analysis of wavelet constructions in the regular setting, such tools have only recently begun to emerge for the irregular setting. One such tool is the generalization of commutation from the regular to the irregular setting. We have applied these ideas by proposing new irregular subdivision schemes in the manifold setting which are explicitly designed to minimize certain differences. Little is as yet known about the

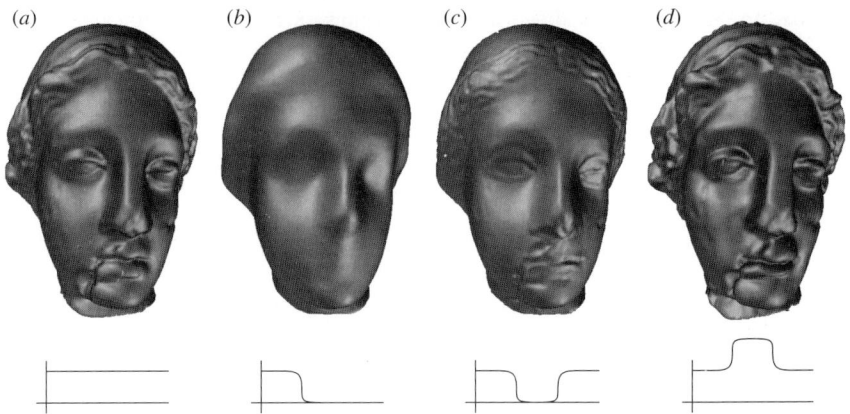

Fig. 13: Smoothing and filtering of the Venus head. (*a*) Original; (*b*) low-pass filter; (*c*) stopband filter; (*d*) enhancement filter.

analytic smoothness properties of the resulting constructions, but numerical evidence suggests that they are quite useful for practical applications.

6 Acknowledgements

The work reviewed in this paper was supported in part by grants from the NSF (ACI-9624957, ACI-9721349, DMS-9874082, DMS-9872890), DOE (W-7405-ENG-48), ONR (N00014-96-1-0367 P00004) and AFOSR (F49620-98-1-0044). I.G. was partly supported by a Harold W. Dodds Fellowship and a Summer Internship at Bell Laboratories, Lucent Technologies.

7 References

Burt, P. J. & Adelson, E. H. 1983 Laplacian pyramid as a compact image code. *IEEE Trans. Commun.* **31**, 532–540.

Catmull, E. & Clark, J. 1978 Recursively generated B-spline surfaces on arbitrary topological meshes. *Computer Aided Design* **10**, 350–355.

Cavaretta, A. S. & Micchelli, C. A. 1987 Computing surfaces invariant under subdivision. *Computer Aided Geom. Design* **4**, 321–328.

Cavaretta, A. S. & Micchelli, C. A. 1989 The desing of curves and surfaces by subdivision algorithms. In *Mathematical aspects of computer aided geometric design* (ed. T. Lyche & L. L. Schumaker). Tampa, FL: Academic.

Cavaretta, A. S., Dahmen, W. & Micchelli, C. A. 1991 Stationary subdivision. *Mem. Am. Math. Soc.* **93**.

Chaikin, G. 1974 An algorithm for high speed curve generation. *Comp. Graphics Image Process.* **3**, 346–349.

Dahmen, W. 1996 Stability of multiscale transformations. *J. Fourier Analysis Appl.* **2**, 341–361.

Daubechies, I. 1988 Orthonormal bases of compactly supported wavelets. *Commun. Pure Appl. Math.* **41**, 909–996.

Daubechies, I. & Lagarias, J. C. 1991 Two-scale difference equations. I. Existence and global regularity of solutions. *SIAM Jl Math. Analysis* **22**, 1388–1410.

Daubechies, I., Guskov, I. & Sweldens, W. 1998 Commutation for irregular subdivision. Technical Report, Lucent Technologies, Bell Laboratories.

Daubechies, I., Guskov, I. & Sweldens, W. 1999 Regularity of irregular subdivision. *Constr. Approx.* **15**, 381–426.

De Boor, C. 1978 *A practical guide to splines*. Applied Mathematical Sciences, no 27. New York: Springer.

de Casteljau, F. 1959 *Outillages méthodes calcul*. Paris: André Citroën Automobiles SA.

de Rham, G. 1956 Sur une courbe plane. *J. Math. Pures Appl.* **39**, 25–42.

Deslauriers, G. & Dubuc, S. 1987 Interpolation dyadique. In *Fractals, dimensions non entières et applications*, pp. 44–55. Paris: Masson.

Deslauriers, G. & Dubuc, S. 1989 Symmetric iterative interpolation processes. *Constr. Approx.* **5**, 49–68.

Donoho, D. L. 1992 Interpolating wavelet transforms. Preprint, Department of Statistics, Stanford University.

Doo, D. & Sabin, M. 1978 Analysis of the behaviour of recursive division surfaces near extraordinary points. *Computer Aided Design* **10**, 356–360.

Dubuc, S. 1986 Interpolation through an iterative scheme. *J. Math. Analyt. Appl.* **114**, 185–204.

Dyn, N., Levin, D. & Gregory, J. 1987 A 4-point interpolatory subdivision scheme for curve design. *Computer Aided Geom. Design* **4**, 257–268.

Dyn, N., Levin, D. & Gregory, J. A. 1990*a* A butterfly subdivision scheme for surface interpolation with tension control. *ACM Trans. Graphics* **9**, 160–169.

Dyn, N., Levin, D. & Micchelli, C. A. 1990*b* Using parameters to increase smoothness of curves and surfaces generated by subdivision. *Computer Aided Geom. Design* **7**, 129–140.

Dyn, N., Gregory, J. & Levin, D. 1991 Analysis of uniform binary subdivision schemes for curve design. *Constr. Approx.* **7**, 127–147.

Eirola, T. 1992 Sobolev characterization of solutions of dilation equations. *SIAM Jl Math. Analysis* **23**, 1015–1030.

Guskov, I. 1998 Multivariate subdivision schemes and divided differences. Technical Report, Department of Mathematics, Princeton University.

Guskov, I., Sweldens, W. & Schröder, P. 1999 Multiresolution signal processing for meshes. Computer Graphics Proceedings (SIGGRAPH 99, Los Angeles), pp. 325–334. ACM SIGGRAPH.

Heckbert, P. S. & Garland, M. 1997 Survey of polygonal surface simplification algorithms. Technical Report, Carnegie Mellon University.

Kobbelt, L. 1996 Interpolatory subdivision on open quadrilateral nets with arbitrary topology. In *Proc. Eurographics 96*, Computer Graphics Forum, pp. 409–420.

Kobbelt, L. 1997 Discrete fairing. In *Proc. 7th IMA Conf. on the Mathematics of Surfaces*, pp. 101–131.

Kobbelt, L. & Schröder, P. 1997 Constructing variationally optimal curves through subdivision. Technical Report CS-TR-97-05, Department of Computer Science, California Institute of Technology.

Lane, J. M. & Riesenfeld, R. F. 1980 A theoretical development for the computer generation of piecewise polynomial surfaces. *IEEE Trans. Patt. Anal. Mach. Intell.* **3**, 35–46.

Loop, C. 1987 Smooth subdivision surfaces based on triangles. Master's thesis, Department of Mathematics, University of Utah.

Mallat, S. G. 1989 Multiresolution approximations and wavelet orthonormal bases of $0L^2(\mathbb{R})$. *Trans. Am. Math. Soc.* **315**, 69–87.

Micchelli, C. A. & Prautzsch, H. 1987 Computing surfaces invariant under subdivision. *Computer Aided Geom. Design* **4**, 321–328.

Reif, U. 1995 A unified approach to subdivision algorithms near extraordinary vertices. *Computer Aided Geom. Design* **12**, 153–174.

Rioul, O. 1992 Simple regularity criteria for subdivision schemes. *SIAM Jl Math. Analysis* **23**, 1544–1576.

Schröder, P. & Zorin, D. (eds) 1998 *Course notes: subdivision for modeling and animation.* ACM SIGGRAPH.

Schweitzer, J. E. 1996 Analysis and application of subdivision surfaces. PhD thesis, University of Washington.

Sweldens, W. 1997 The lifting scheme: a construction of second generation wavelets. *SIAM Jl Math. Analysis* **29**, 511–546.

Sweldens, W. & Schröder, P. 1996 Building your own wavelets at home. In *Wavelets in computer graphics*, pp. 15–87. Course notes ACM SIGGRAPH.

Villemoes, L. F. 1994 Wavelet analysis of refinement equations. *SIAM Jl Math. Analysis* **25**, 1433–1460.

Warren, J. 1995 Binary subdivision schemes for functions over irregular knot sequences. In *Mathematical methods in CAGD III* (ed. M. Daehlen, T. Lyche & L. Schumaker). Academic.

Zorin, D. 1996 C^k continuity of subdivision surfaces. Technical Report, California Institute of Technology.

Zorin, D., Schröder, P. & Sweldens, W. 1996 Interpolating subdivision for meshes with arbitrary topology. In *Computer graphics (SIGGRAPH '96 Proc.)*, pp. 189–192. ACM SIGGRAPH.

2

Revealing a lognormal cascading process in turbulent velocity statistics with wavelet analysis

A. Arneodo
Centre de Recherche Paul Pascal, Avenue Schweitzer, 33600, Pessac, France

S. Manneville
Laboratoire Ondes et Acoustique, ESPCI, 10 rue Vauquelin, 75005 Paris, France

J. F. Muzy
Centre de Recherche Paul Pascal, Avenue Schweitzer, 33600, Pessac, France

S. G. Roux
NASA Goddard Space Flight Center, Climate and Radiation Branch (Code 913), Greenbelt, MD 20771, USA

Abstract

We use the continuous wavelet transform to extract a cascading process from experimental turbulent velocity signals. We mainly investigate various statistical quantities such as the singularity spectrum, the self-similarity kernel and space-scale correlation functions, which together provide information about the possible existence and nature of the underlying multiplicative structure. We show that, at the highest accessible Reynolds numbers, the experimental data do not allow us to distinguish various phenomenological cascade models recently proposed to account for intermittency from their lognormal approximation. In addition, we report evidence that velocity fluctuations are not scale-invariant but possess more complex self-similarity properties, which are likely to depend on the Reynolds number. We comment on the possible asymptotic validity of the multifractal description.

Keywords: turbulence; wavelet analysis; intermittency; self-similarity; cascade models; multifractals

1 Introduction

Since Kolmogorov's founding work (Kolmogorov, 1941) (hereafter called K41), fully developed turbulence has been intensively studied for more than 50 years (Monin & Yaglom, 1975; Frisch & Orzag, 1990; Frisch, 1995). A standard way of analysing a turbulent flow is to look for some universal statistical properties of the fluctuations of the longitudinal velocity increments over a distance l, $\delta v_l = v(x + l) - v(x)$. For instance, investigating the scaling properties of the structure functions,

$$S_p(l) = \langle |\delta v_l|^p \rangle \sim l^{\zeta_p}, \quad p > 0, \tag{1.1}$$

where $\langle \cdots \rangle$ stands for ensemble average, leads to a spectrum of scaling exponents ζ_p, which has been widely used as a statistical characterization of turbulent fields (Monin & Yaglom, 1975; Frisch & Orzag, 1990; Frisch, 1995). Based upon assumptions of statistical homogeneity, isotropy and of constant rate ϵ of energy transfer from large to small scales, K41 theory predicts the existence of an inertial range $\eta \ll l \ll L$ (η and L being, respectively, the dissipative and integral scales), where $S_p(l) \sim \epsilon^{p/3} l^{p/3}$. Although these assumptions are usually considered to be correct, there has been increasing numerical (Briscolini *et al.*, 1994; Vincent & Meneguzzi, 1995) and experimental (Monin & Yaglom, 1975; Anselmet *et al.*, 1984; Gagne, 1987; Frisch & Orzag, 1990; Frisch, 1995; Tabeling & Cardoso, 1995; Arneodo *et al.*, 1996) evidence that ζ_p deviates substantially from the K41 prediction $\zeta_p = \frac{1}{3}p$, at large p. The observed nonlinear behaviour of ζ_p is generally interpreted as a direct consequence of the intermittency phenomenon displayed by the rate of energy transfer (Castaing *et al.*, 1990; Meneveau & Sreenivasan, 1991). Under the so-called Kolmogorov's refined hypothesis (Kolmogorov, 1962), the velocity structure functions can be rewritten as

$$S_p(l) \sim \langle \epsilon_l^{p/3} \rangle l^{p/3} \sim l^{\tau(p/3)+p/3},$$

where ϵ_l is the local rate of energy transfer over a volume of size l. The scaling exponents of S_p are thus related to those of the energy transfer: $\zeta_p = \tau(\frac{1}{3}p) + \frac{1}{3}p$.

Richardson's 1926 cascade pioneering picture is often invoked to account for intermittency: energy is transferred from large eddies (of size of order L) down to small scales (of order η) through a cascade process in which the transfer rate at a given scale is not spatially homogeneous as in the K41 theory but undergoes local intermittent fluctuations. Over the past 30 years, refined models—including the lognormal model of Kolmogorov (1962) and Obukhov (1962) (hereafter called KO62), multiplicative hierarchical cascade models, such as the random β-model, the α-model, the p-model (for a review see Meneveau & Sreenivasan 1991), the log-stable models (Schertzer & Levejoy, 1987; Kida, 1990), and more recently the log-infinitely divisible cascade models (Novikov 1990, 1995; Dubrulle 1994; She & Waymire 1995; Castaing & Dubrulle 1995), together with the rather popular log-Poisson model advocated by She & Leveque (1994)—have appeared in the literature as reasonable models for mimicking the energy cascading process in turbulent flows. Unfortunately, all the existing models appeal to adjustable parameters that are difficult to determine by plausible physical arguments and that generally provide enough freedom to account for the experimental data for the two sets of scaling exponents ζ_p and $\tau(p)$.

The scaling behaviour of the velocity structure functions (equation (1.1)) is at the heart of the multifractal description pioneered by Parisi & Frisch (1985). K41 theory is actually based on the assumption that at each point of the fluid the velocity field has the same scaling behaviour $\delta v_l(x) \sim l^{1/3}$, which yields the well-known $E(k) \sim k^{-5/3}$ energy spectrum. By interpreting the nonlinear behaviour of ζ_p as a direct consequence of the existence of spatial fluctuations in the local regularity of the velocity field, $\delta v_l(x) \sim l^{h(x)}$, Parisi & Frisch (1985) attempt to capture intermittency in a geometrical framework. For each h, let us call $D(h)$ the fractal dimension of the set for which $\delta v_l(x) \sim l^h$. By suitably inserting this local scaling behaviour into equation (1.1), one can bridge the so-called singularity spectrum $D(h)$ and the set of scaling exponents ζ_p by a Legendre transform: $D(h) = \min_p(ph - \zeta_p + 1)$. From the properties of the Legendre transform, a nonlinear ζ_p spectrum is equivalent to the assumption that there is more than a single scaling exponent h. Let us note that from low- to moderate-Reynolds-number turbulence, the inertial scaling range is small and the evaluation of ζ_p is not very accurate. Actually, the existence of scaling laws like equation (1.1) for the structure functions is not clear experimentally (Arneodo *et al.*, 1996; Pedrizetti *et al.*, 1996), even at the highest accessible Reynolds numbers; this observation questions the validity of the multifractal description. Recently, Benzi *et al.* (1993b, c, 1995) have shown that one can remedy the observed departure from scale-invariance by looking at the scaling behaviour of one structure function against another. More precisely, ζ_p can be estimated from the behaviour $S_p(l) \sim S_3(l)^{\zeta_p}$, if one assumes that $\zeta(3) = 1$ (Frisch, 1995). The relevance of the so-called extended self-similarity (ESS) hypothesis is recognized to improve and to further extend the scaling behaviour towards the dissipative range (Benzi *et al.* 1993b, c, 1995; Briscolini *et al.* 1994). From the application of ESS, some experimental consensus has been reached on the definite nonlinear behaviour of ζ_p and its possible universal character, at least as far as isotropic homogeneous turbulence is concerned (Arneodo *et al.*, 1996). But beyond some practical difficulties there exists a more fundamental insufficiency in the determination of ζ_p. From the analogy between the multifractal formalism and statistical thermodynamics (Arneodo *et al.*, 1995), ζ_p plays the role of a thermodynamical potential which intrinsically contains only some degenerate information about the 'Hamiltonian' of the problem, i.e. the underlying cascading process. Therefore, it is not surprising that previous experimental determinations of the ζ_p spectrum have failed to provide a selective test to discriminate between various (deterministic or random) cascade models.

In order to go beyond the multifractal description, Castaing and co-workers (Castaing *et al.* 1990, 1993; Gagne *et al.* 1994; Naert *et al.* 1994; Chabaud *et al.* 1994; Castaing & Dubrulle 1995; Chillà *et al.* 1996) have proposed an intermittency phenomenon approach which relies on the validity of Kolmogorov's refined hypothesis (Kolmogorov, 1962) and which consists in looking for a multiplicative cascade process directly on the velocity field. This approach amounts to modelling the evolution of the shape of the velocity increment probability distribution function (PDF), from Gaussian at large scales to more intermittent profiles with stretched exponential-like tails at smaller scales (Gagne, 1987; Castaing *et al.*, 1990; Kailasnath *et al.*, 1992; Tabeling *et al.*, 1996; Belin *et al.*, 1996), by a functional equation that relates the two scales using a kernel G. This description relies upon the ansatz that the velocity increment PDF at a given scale l, $P_l(\delta v)$, can be expressed as a weighted sum of dilated PDFs at a larger scale $l' > l$:

$$P_l(\delta v) = \int G_{ll'}(\ln \sigma) \frac{1}{\sigma} P_{l'}\left(\frac{\delta v}{\sigma}\right) d \ln \sigma, \qquad (1.2)$$

where $G_{ll'}$ is a kernel that depends on l and l' only. Indeed, most of the well-known cascade models can be reformulated within this approach (Castaing & Dubrulle, 1995; Chillà *et al.*, 1996). This amounts to (i) specifying the shape of the kernel $G(u)$ which is determined by the nature of the elementary step in the cascade; and (ii) defining the way $G_{ll'}$ depends on both l and l'. In their original work, Castaing *et al.*(see Castaing *et al.* 1990, 1993; Gagne *et al.* 1994; Naert *et al.* 1994; Chabaud *et al.* 1994) mainly focused on the estimate of the variance of G and its scale behaviour. A generalization of the Castaing *et al.*ansatz to the wavelet transform (WT) of the velocity field has been proposed in previous works (Arneodo *et al.* 1997, 1999; Roux 1996) and shown to provide direct access to the entire shape of the kernel G. This wavelet-based method has been tested on synthetic turbulent signals and preliminarily applied to turbulence data. In Section 2, we use this new method to process large-velocity records in high-Reynolds-number turbulence (Arneodo *et al.*, 1998*c*). We start by briefly recalling our numerical method to estimate G. We then focus on the precise shape of G and show that, for the analysed turbulent flows, G is Gaussian within a very good approximation. Special attention is paid to statistical convergence; in particular, we show that when exploring larger samples than in previous studies (Arneodo *et al.* 1997, 1999; Roux 1996), one is able to discriminate between lognormal and log-Poisson statistics. However, in the same way the ζ_p and $D(h)$ multifractal spectra provide rather degenerate information about the nature of the underlying process; equation (1.2) is a necessary but not sufficient condition for the existence of a cascade. As emphasized in a recent work (Arneodo *et al.*, 1998*a*), one can go deeper in fractal analysis by studying correlation functions in both space and scales using the continuous wavelet transform. This 'two-point' statistical analysis has proved to be particularly well suited for studying multiplicative random cascade processes for which the correlation functions take a very simple form. In Section 3, we apply space-scale correlation functions to high-Reynolds-number turbulent velocity signals. This method confirms the existence of a cascade structure that extends over the inertial range and that this cascade is definitely not scale-invariant. In Section 4, we revisit the multifractal description of turbulent velocity fluctuations under the objective of the WT microscope. Going back to the WT coefficient PDFs and to the ζ_p spectrum, we get additional confirmation of the relevance of a lognormal cascading process. Furthermore, we discuss its robustness when varying the scale range or the Reynolds number. We conclude in Section 5 by discussing the asymptotic validity of the multifractal description of the intermittency phenomenon in fully developed turbulence.

   Throughout this study, we will compare the results obtained on experimental data with the results of similar statistical analysis of lognormal and log-Poisson numerical processes of the same length generated using an algorithm of multiplicative cascade defined on an orthonormal wavelet basis. We refer the reader to Roux 1996, Arneodo *et al.* (1997, 1998*b*, *c*, 1999) and Section 3.2, where the main points of this synthetic turbulence generator are described.

2 Experimental evidence for lognormal statistics in high-Reynolds-number turbulent flows

2.1 A method for determining the kernel G

As pointed out in Muzy *et al.* (1991, 1994) and Arneodo *et al.* (1995), the WT provides a powerful mathematical framework for analysing irregular signals in both space and scale without loss of information. The WT of the turbulent velocity spatial field v at point x and scale $a > 0$, is defined as (Meyer, 1990; Daubechies, 1992)

$$T_\psi[v](x, a) = \frac{1}{a} \int_{-\infty}^{+\infty} v(y) \psi\left(\frac{x - y}{a}\right) dy, \qquad (2.1)$$

where ψ is the analysing wavelet. Note that the velocity increment $\delta v_l(x)$ is simply $T_\psi[v](x, l)$ computed with the 'poor man's' wavelet

$$\psi_{(0)}^{(1)}(x) = \delta(x - 1) - \delta(x).$$

More generally, ψ is chosen to be well localized not only in direct space but also in Fourier space (the scale a can thus be seen as the inverse of a local frequency). Throughout this study, we will use the set of compactly supported analysing wavelets $\psi_{(m)}^{(n)}$ defined in Roux (1996) and Arneodo *et al.* (1997). The $\psi_{(m)}^{(1)}$ are smooth versions of $\psi_{(0)}^{(1)}$ obtained after m successive convolutions with the box function χ. $\psi_{(m)}^{(n)}$ are higher-order analysing wavelets with n vanishing moments. The WT associates to a function in $\mathbb{R}$, its transform defined on $\mathbb{R} \times \mathbb{R}^+$ and is thus very redundant. Following the strategy proposed in Arneodo *et al.* (1997, 1998*c*), we restrict our analysis to the *modulus maxima* of the WT (WTMM) so that the amount of data to process is more tractable (see figure 1). A straightforward generalization of equation (1.2) in terms of the WTMM PDF at scale a, $P_a(T)$, then reads

$$P_a(T) = \int G_{aa'}(u) P_{a'}(e^{-u}T) e^{-u} du, \quad \text{for } a' > a. \qquad (2.2)$$

From (2.2) one can show that, for any decreasing sequence of scales $(a_1, \ldots, a_n)$, the kernel G satisfies the composition law

$$G_{a_n a_1} = G_{a_n a_{n-1}} \otimes \cdots \otimes G_{a_2 a_1}, \qquad (2.3)$$

where $\otimes$ denotes the convolution product. According to Castaing and co-workers (Castaing *et al.*, 1990; Castaing & Dubrulle, 1995), the cascade is *self-similar* if there exists a decreasing sequence of scales $\{a_n\}$ such that $G_{a_n a_{n-1}} = G$ is independent of n. The cascade is said to be *continuously self-similar* if there exists a positive decreasing function $s(a)$ such that $G_{aa'}$ depends on a and a' only through $s(a, a') = s(a) - s(a')$: $G_{aa'}(u) = G(u, s(a, a'))$. $s(a, a')$ actually accounts for the number of elementary cascade steps from scale a' to scale a ($s(a)$ can be seen as the number of cascade steps from the integral scale L down to the considered scale a). In the Fourier space, the convolution property (equation (2.3)) turns into a multiplicative property for $\hat{G}$, the Fourier transform of G:

$$\hat{G}_{aa'}(p) = \hat{G}(p)^{s(a,a')}, \quad \text{for } a' > a. \tag{2.4}$$

From this equation, one deduces that $\hat{G}$ has to be the characteristic function of an infinitely divisible PDF. Such a cascade is referred to as a log-infinitely divisible cascade (Novikov 1990, 1995; Dubrulle 1994; She & Waymire 1995; Castaing & Dubrulle 1995). According to Novikov's definition (Novikov 1990, 1995), the cascade is *scale-similar* (or *scale-invariant*) if

$$s(a, a') = \ln(a'/a), \tag{2.5}$$

i.e. $s(a) = \ln(L/a)$. Let us note that in their original work Castaing *et al.* (1990) developed a formalism, based on an extremum principle, which is consistent with the KO62 general ideas of lognormality (Kolmogorov, 1962; Obukhov, 1962), but which predicts an anomalous power-law behaviour of the depth of the cascade $s(a) \sim (L/a)^{\beta}$. From the computation of the scaling behaviour of the variance of the kernel $G_{aa'}$, they have checked whether the above-mentioned power-law behaviour could provide a reasonable explanation for the deviation from scaling observed experimentally on the velocity fluctuation statistics (Castaing *et al.* 1990, 1993; Gagne *et al.* 1994; Naert *et al.* 1994; Chabaud *et al.* 1994; Chillà *et al.* 1996).

Our numerical estimation of G (Arneodo *et al.* 1997, 1998c) is based on the computation of the characteristic function $M(p, a)$ of the WTMM logarithms at scale a:

$$M(p, a) = \int e^{ep \ln |T|} P_a(T) \, dT. \tag{2.6}$$

From equation (2.2), it is easy to show that $\hat{G}$ satisfies

$$M(p, a) = \hat{G}_{aa'}(p) M(p, a'). \tag{2.7}$$

After the WT calculation and the WTMM detection, the real and imaginary parts of $M(p, a)$ are computed separately as $\langle \cos(p \ln |T|) \rangle$ and $\langle \sin(p \ln |T|) \rangle$, respectively. The use of the WTMM skeleton instead of the continuous WT prevents $M(p, a')$ from getting too small compared with numerical noise over a reasonable range of values of p, so that $\hat{G}_{aa'}(p)$ can be computed from the ratio

$$\hat{G}_{aa'}(p) = \frac{M(p, a)}{M(p, a')}. \tag{2.8}$$

We refer the reader to Roux (1996) and Arneodo *et al.* (1997) for test applications of this method to synthetic turbulent signals.

2.2 Experimental determination of the kernel G

The turbulence data were recorded by Gagne and co-workers in the S1 wind tunnel of
ONERA at Modane. The Taylor-scale-based Reynolds number is about $R_\lambda \simeq 2000$
and the Kolmogorov $k^{-5/3}$ law for the energy spectrum approximately holds on an
'inertial range' (in Section 5 we propose an objective definition of inertial range on
the basis of which the inertial range is significantly less than four decades; we will
implicitly use this definition in the remainder of this paper) of about four decades
(from the integral scale $L \simeq 7$ m down to the dissipative scale $\eta \simeq 0.27$ mm). The
overall statistical sample is about 25×10^7 points long, with a resolution of roughly
3η, corresponding to about 25 000 integral scales. Temporal data are identified to
spatial fluctuations of the longitudinal velocity via the Taylor hypothesis (Frisch,
1995; Tabeling & Cardoso, 1995). Figure 1 illustrates the WT and its skeleton
of a sample of the (longitudinal) velocity signal of length of about two integral
scales. The analysing wavelet $\psi_{(3)}^{(1)}$ is a first-order compactly supported wavelet.
We have checked that all the results reported below are consistent when changing
both the regularity and the order of ψ. With the specific goal of investigating the
dependence of the statistics on the Reynolds number, we will also report results of
similar analysis of wind-tunnel ($R_\lambda \simeq 3050$), jet ($R_\lambda \simeq 800$ and 600) and grid
($R_\lambda \simeq 280$) turbulences, but for statistical samples of smaller sizes.

2.2.1 *Uncovering a continuously self-similar cascade (Arneodo et al.1998c)*

In order to test the validity of equation (2.4), we first focus on the scale dependence
of $\hat{G}_{aa'}$ as calculated with equation (2.8). Figure 2a shows the logarithm of the
modulus $\ln |\hat{G}_{aa'}|$ and figure 2b shows the phase $\phi_{aa'}$ of $\hat{G}_{aa'}$ for various pairs of
scales $a < a'$ in the inertial range. In figure 2c, d, we succeed in collapsing all
these different curves onto a single kernel $\hat{G} = \hat{G}_{aa'}^{1/s(a,a')}$, in very good agreement
with equation (2.4) and the continuously self-similar cascade picture. In the inserts
of figure 2a, b, we compare our estimation of $\hat{G}_{aa'}$ for the turbulent signal and for
a lognormal numerical process of the same length (Arneodo et al., 1998b). On the
numerical lognormal cascade, deviations from the expected parabolic behaviour of
$\ln |\hat{G}_{aa'}|$, as well as from the linear behaviour of $\phi_{aa'}$ (see equation (2.12)), become
perceptible for $|p| > 5$. Very similar features are observed for the turbulence data,
showing that the slight dispersion at large values of p on the curves in figure 2c, d
can be attributed to a lack of statistics. Thus, from now on, we will restrict our
analysis of $\hat{G}(p)$ to $p \in [-4, 4]$.

   In order to test scale-similarity or more generally the pertinence of equation (2.4),
we plot in figure 3a, c,

$$m(a, a') = \partial \frac{\text{Im}(\hat{G}_{aa'})}{\partial p}\bigg|_{p=0} \quad \text{and} \quad \sigma^2(a, a') = -\frac{\partial^2 (\ln |\hat{G}_{aa'}|)}{\partial p^2}\bigg|_{p=0},$$

respectively, as functions of $s(a, a') = \ln(a'/a)$ for different couples of scales (a, a')
in the inertial range. It is striking for the jet data ($R_\lambda \simeq 800$), but also noticeable for

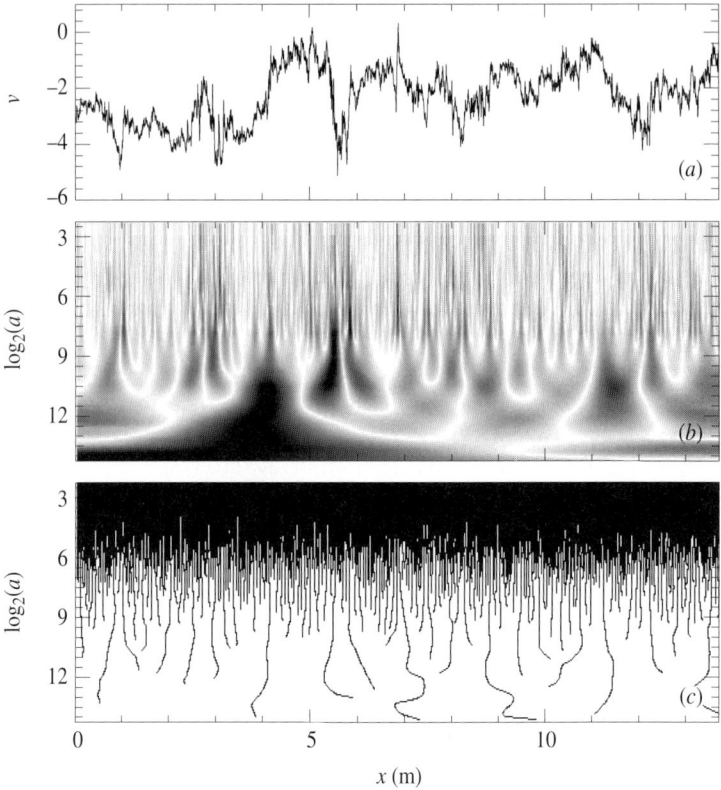

Fig. 1: Continuous WT of fully developed turbulence data from the wind-tunnel experiment ($R_\lambda \simeq 2000$). (*a*) The turbulent velocity signal over about two integral scales. (*b*) WT of the turbulent signal; the amplitude is coded, independently at each scale a, using 32 grey levels from white ($|T_\psi[v](x,a)| = 0$) to black ($\max_x |T_\psi[v](x,a)|$). (*c*) WT skeleton defined by the set of all the WTMM lines. In (*b*) and (*c*), the small scales are at the top. The analysing wavelet is $\psi_{(3)}^{(1)}$.

the wind-tunnel data ($R_\lambda \simeq 3050$), that the curves obtained when fixing the largest scale a' and varying the smallest scale a, have a clear bending and do not merge on the same straight line as expected for scale-similar cascade processes. In figure 3*b*, *d*, the same data are plotted versus $s(a,a') = (a^{-\beta} - a'^{-\beta})/\beta$ with $\beta = 0.08$ for the wind-tunnel flow and $\beta = 0.19$ for the jet flow. In this case, the data for the mean $m(a,a')$ and the variance $\sigma^2(a,a')$ fall, respectively, on a unique line. Those velocity fields are therefore not scale-similar but rather are characterized by some anomalous behaviour of the number of cascade steps between scale a' and scale a:

$$s(a,a') = (a^{-\beta} - a'^{-\beta})/\beta, \qquad (2.9)$$

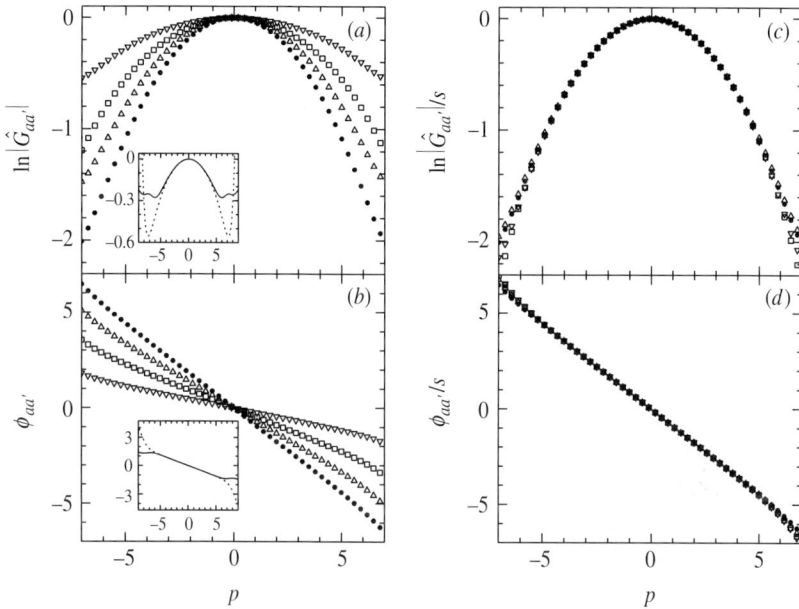

Fig. 2: Estimation of $\hat{G}_{aa'}(p)$ for the Modane turbulent velocity signal ($R_\lambda \simeq 2000$) using equation (2.8). The analysing wavelet is $\psi_{(3)}^{(1)}$. (a) $\ln|\hat{G}_{aa'}(p)|$ versus p. (b) $\phi_{aa'}(p)$ versus p for: $a = 271\eta$, $a' = 4340\eta$ (•); $a = 385\eta$, $a' = 3080\eta$ ($\triangle$); $a = 540\eta$, $a' = 2170\eta$ ($\square$); $a = 770\eta$, $a' = 1540\eta$ ($\triangledown$). Inserts: the experimental $\hat{G}_{aa'}(p)$ for $a = 770\eta$ and $a' = 1540\eta$ (dotted line) compared with the computation of $\hat{G}_{aa'}(p)$ for a lognormal numerical process of parameters $m = 0.39$ and $\sigma^2 = 0.036$ with $a = 2^8$ and $a' = 2^9$ (solid line). (c) and (d) The same curves after being rescaled by a factor of $1/s(a, a')$ with $s = 1$ (•), $s = 0.754$ ($\triangle$), $s = 0.508$ ($\square$), $s = 0.254$ ($\triangledown$).

where the exponent β somehow quantifies the departure from scale-similarity (scale-invariance being restored for $\beta \to 0$).[1] Let us point out that equation (2.9) differs from the pure power law prompted by Castaing and co-workers (Castaing et al. 1990, 1993; Gagne et al. 1994; Naert et al. 1994; Chabaud et al. 1994; Chillà et al. 1996), since when fixing the reference scale a', the number of cascade steps required to reach the scale a is not exactly $a^{-\beta}/\beta$, but some corrective term $-a'^{-\beta}/\beta$, which has to be taken into account.

---

[1] Note that in order to collapse all the curves onto a single curve in figure 2c, d, equation (2.9) was used with $\beta = 0.095$.

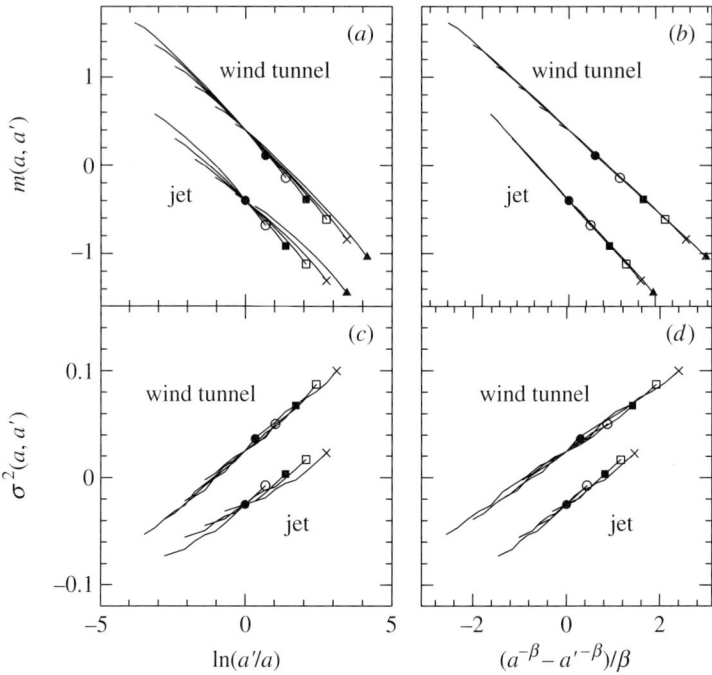

Fig. 3: $m(a, a')$ and $\sigma^2(a, a')$ as computed for the jet ($R_\lambda \simeq 800$) and wind-tunnel ($R_\lambda \simeq 3050$) velocity signals for $a' = 2^6$ ($\bullet$), 2^7 ($\circ$), 2^8 ($\blacksquare$), 2^9 ($\square$), 2^{10} ($\times$). (a) $m(a, a')$ versus $\ln(a'/a)$; (b) $m(a, a')$ versus $(a^{-\beta} - a'^{-\beta})/\beta$; (c) $\sigma^2(a, a')$ versus $\ln(a'/a)$; (d) $\sigma^2(a, a')$ versus $(a^{-\beta} - a'^{-\beta})/\beta$. In (b) and (d), $\beta = 0.19$ (jet) and $\beta = 0.08$ (wind tunnel).

2.2.2 Discriminating between lognormal and log-Poisson cascades (Arneodo et al. 1998c)

The relevance of equation (2.4) being established, let us turn to the precise analysis of the nature of G. Using the Taylor-series expansion of $\ln \hat{G}(p)$,

$$\hat{G}(p) = \exp\left(\sum_{k=1}^{\infty} c_k \frac{(ep)^k}{k!}\right), \tag{2.10}$$

equation (2.4) can be rewritten as

$$\hat{G}_{aa'}(p) = \exp\left(\sum_{k=1}^{\infty} s(a, a')c_k \frac{(ep)^k}{k!}\right), \tag{2.11}$$

where the (real-valued) coefficients c_k are the cumulants of G.

(1) Lognormal cascade process (Kolmogorov, 1962; Obukhov, 1962): a lognormal cascade is characterized by a Gaussian kernel (Roux, 1996; Arneodo *et al.*, 1997)

$$\hat{G}_{aa'}(p) = \exp[s(a, a')(-emp - \tfrac{1}{2}\sigma^2 p^2)], \qquad (2.12)$$

which corresponds to the following set of cumulants:

$$c_1 = -m, \quad c_2 = \sigma^2 \text{ and } c_k = 0 \quad \text{for } k \geq 3. \qquad (2.13)$$

(2) Log-Poisson cascade process (Dubrulle, 1994; Castaing & Dubrulle, 1995; She & Waymire, 1995): a log-Poisson cascade is characterized by the following kernel shape (Roux, 1996; Arneodo *et al.*, 1997):

$$\hat{G}_{aa'}(p) = \exp[s(a, a')(\lambda(\cos(p \ln \delta) - 1) + e(p\gamma + \lambda \sin(p \ln \delta)))], \qquad (2.14)$$

where λ, δ and γ are parameters. This log-Poisson kernel corresponds to the following set of cumulants:

$$c_1 = \gamma + \lambda \ln \delta \text{ and } c_k = \lambda \frac{(\ln \delta)^k}{k!} \quad \text{for } k \geq 2. \qquad (2.15)$$

Note that the log-Poisson process reduces to a lognormal cascade for $|p \ln \delta| \ll 1$, i.e. in the limit $\delta \to 1$, where the atomic nature of the quantized log-Poisson process vanishes.

For a given pair of inertial scales $a < a'$, we proceed to polynomial fits of $\ln |\hat{G}_{aa'}(p)|$ and $\phi_{aa'}(p)$, prior to the use of equation (2.11) to estimate the first three cumulants $C_k = s(a, a')c_k$ as a function of the statistical sample length for the wind-tunnel turbulence data at $R_\lambda \simeq 2000$ and for both a lognormal and a log-Poisson synthetic numerical process. Figure 4 shows that statistical convergence is achieved up to the third-order coefficient. However, our sample total length does not allow us to reach statistical convergence for higher-order cumulants. Note that the third cumulant computed for a synthetic lognormal process with parameters m and σ^2 chosen equal to the asymptotic values of C_1 and C_2 (figure 4a, b), namely $m = 0.39$ and $\sigma^2 = 0.036$, cannot be distinguished from the experimental C_3 (figure 4c). In the log-Poisson model, by setting $\lambda = 2$ (according to She & Leveque (1994), λ is the codimension of the most intermittent structures that are assumed to be filaments), we are able to find values of δ and γ close to those proposed in She & Leveque (1994) ($\delta = (\tfrac{2}{3})^{1/3}$ and $\gamma = -\tfrac{1}{9}$) that perfectly fit the first two cumulants. However, as seen in figure 4c, this set of parameters yields a third-order cumulant that is more than one order of magnitude larger than the experimental cumulant. Actually, when taking λ as a free parameter, good log-Poisson approximations of the first three cumulants are obtained for unrealistic values of λ of order 100 and for values of δ very close to 1, i.e. when the log-Poisson process reduces to the lognormal model. From these

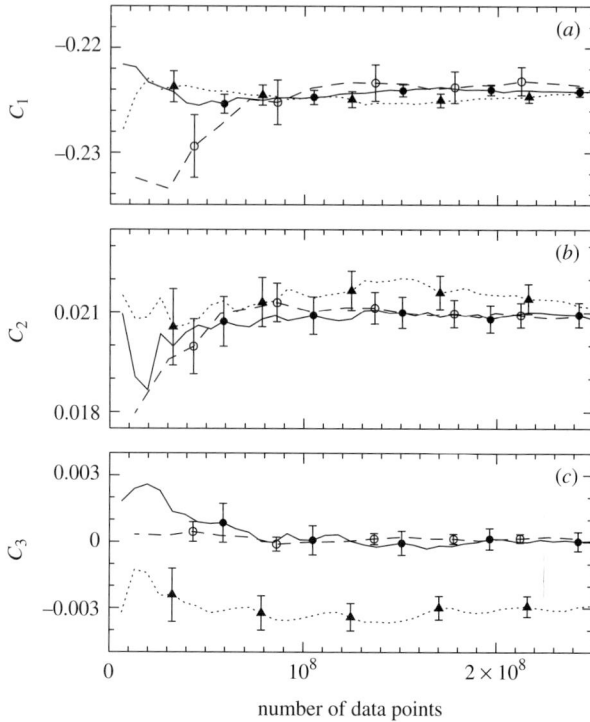

Fig. 4: The first three cumulants of $G_{aa'}$ versus the sample length. Turbulent velocity signal for $a = 770\eta$ and $a' = 1540\eta$ ($\circ$ and dashed line), lognormal numerical process of parameters $m = 0.39$ and $\sigma^2 = 0.036$ ($\bullet$ and solid line) and log-Poisson numerical process of parameters $\lambda = 2$, $\delta = 0.89$ and $\gamma = -0.082$ ($\blacktriangle$ and dotted line) for the two corresponding scales $a = 2^8$ and $a' = 2^9$. Error bars are estimates of the RMS deviations of the cumulants from their asymptotical values.

results, we conclude that, for the analysed wind-tunnel velocity signal ($R_\lambda \simeq 2000$), G is a Gaussian kernel since $C_3 = 0$ implies $C_k = 0$ for $k > 2$. Therefore, the large size of our statistical sample allows us to exclude log-Poisson statistics with the parameters proposed in She & Leveque (1994).

3   Experimental evidence for a non-scale-invariant lognormal cascading process in high-Reynolds-number turbulent flows

3.1   Space-scale correlation functions from wavelet analysis

Correlations in multifractals have already been experienced in the literature (Cates & Deutsch, 1987; Siebesma, 1988; Neil & Meneveau, 1993). However, all these studies rely upon the computation of the scaling behaviour of some partition functions involving different points; they thus mainly concentrate on spatial correlations of the local singularity exponents. The approach developed in Arneodo *et al.* (1998*a*) is different since it does not focus on (nor suppose) any scaling property but rather consists in studying the correlations of the *logarithms* of the amplitude of a space-scale decomposition of the signal. For that purpose, the wavelet transform is a natural tool to perform space-scale analysis. More specifically, if $\chi(x)$ is a bump function such that $||\chi||_1 = 1$, then by taking

$$\Sigma^2(x, a) = a^{-2} \int \chi((x-y)/a)|T_\psi[v](y, a)|^2 \, dy, \tag{3.1}$$

one has

$$||v||_2^2 = \int\int \Sigma^2(x, a) \, dx \, da, \tag{3.2}$$

and thus $\Sigma^2(x, a)$ can be interpreted as the local space-scale energy density of the considered velocity signal v (Morel-Bailly *et al.*, 1991). Since $\Sigma^2(x, a)$ is a positive quantity, we can define the *magnitude* of the field v at point x and scale a as

$$\omega(x, a) = \tfrac{1}{2} \ln \Sigma^2(x, a). \tag{3.3}$$

Our aim in this section is to show that a cascade process can be studied through the correlations of its space-scale magnitudes (Arneodo *et al.*, 1998*a*):

$$C(x_1, x_2, a_1, a_2) = \overline{\tilde{\omega}(x_1, a_1)\tilde{\omega}(x_2, a_2)}, \tag{3.4}$$

where the overline stands for ensemble average and $\tilde{\omega}$ for the centred process $\omega - \overline{\omega}$.

3.2 Analysis of random cascades using space-scale correlation functions (Arneodo *et al.* 1998*a, b*)

Cascade processes can be defined in various ways. Periodic wavelet orthogonal bases (Meyer, 1990; Daubechies, 1992) provide a general framework in which they can be constructed easily (Benzi *et al.* 1993*a*; Roux 1996; Arneodo *et al.* 1997, 1998*b, c*). Let us consider the following wavelet series:

$$f(x) = \sum_{j=0}^{+\infty} \sum_{k=0}^{2^j-1} c_{j,k}\psi_{j,k}(x), \tag{3.5}$$

where the set $\{\psi_{j,k}(x) = 2^{j/2}\psi(2^j x - k)\}$ is an orthonormal basis of $L^2([0, L])$ and the coefficients $c_{j,k}$ correspond to the WT of f at scale $a = L2^{-j}$ (L is the 'integral' scale that corresponds to the size of the support of $\psi(x)$) and position $x = ka$. The above sampling of the space-scale plane defines a dyadic tree (Meyer, 1990; Daubechies, 1992). If one indexes by a dyadic sequence $\{\epsilon_1, \ldots, \epsilon_j\}$ ($\epsilon_k = 0$

or 1), with each of the 2^j nodes at depth j of this tree, the cascade is defined by the multiplicative rule:

$$c_{j,k} = c_{\epsilon_1 \dots \epsilon_j} = c_0 \prod_{i=1}^{j} W_{\epsilon_i}.$$

The law chosen for the weights W (accounting for their possible correlations) determines the nature of the cascade and the multifractal (regularity) properties of f (Benzi *et al.*, 1993*a*; Arneodo *et al.*, 1998*b*). From the above multiplicative structure, if one assumes that there is no correlation between the weights at a given cascade step, then it is easy to show that for $a_p = L2^{-j_p}$ and $x_p = k_p a_p$ ($p = 1$ or 2), the correlation coefficient is simply the variance $V(j)$ of $\ln c_{j,k} = \sum \ln W_{\epsilon_i}$, where (j, k) is the deepest common ancestor to the nodes (j_1, k_1) and (j_2, k_2) on the dyadic tree (Arneodo *et al.* 1998*a, b*). This 'ultrametric' structure of the correlation function shows that such a process is not stationary (nor ergodic). However, we will generally consider uncorrelated consecutive realizations of length L of the same cascade process, so that, in good approximation, C depends only on the space lag $\Delta x = x_2 - x_1$ and one can replace ensemble average by space average. In that case, $C(\Delta k, j_1, j_2) = \langle C(k_1, k_1 + \Delta k, j_1, j_2) \rangle$ can be expressed as

$$C(\Delta k, j_1, j_2) = 2^{-(j-n)} \sum_{p=1}^{j-n} 2^{j-n-p} V(j - n - p), \qquad (3.6)$$

where $j = \sup(j_1, j_2)$ and $n = \log_2 \Delta k$. Let us illustrate these features on some simple cases (Arneodo *et al.* 1998*a, b*).

3.2.1　Scale-invariant random cascades

First let us choose, as in classical cascades, i.i.d. random variables $\ln W_{\epsilon_i}$ of variance λ^2 (e.g. lognormal). Then $V(j) = \lambda^2 j$ and it can be established that, for $\sup(a_1, a_2) \leq \Delta x < L$,

$$C(\Delta x, a_1, a_2) = \lambda^2 \left(\log_2 \left(\frac{L}{\Delta x} \right) - 2 + 2 \frac{\Delta x}{L} \right). \qquad (3.7)$$

Thus, the correlation function decreases very slowly, independently of a_1 and a_2, as a logarithm function of Δx. This behaviour is illustrated in figure 5*a, b*, where a lognormal cascade has been constructed using Daubechies compactly supported wavelet basis (D-5) (Arneodo *et al.*, 1998*b*). The correlation functions of the magnitudes of $f(x)$ have been computed as described above (equation (4.5)) using a simple box function for $\chi(x)$. Let us note that all the results reported in this section concern the increments of the considered signal and that we have checked that they are actually independent of the specific choice of the analysing wavelet ψ. In figure 5*a* are plotted the 'one-scale' ($a_1 = a_2 = a$) correlation functions for three different scales $a = 4$, 16 and 64. One can see that, for $\Delta x > a$, all the curves collapse to a single curve,

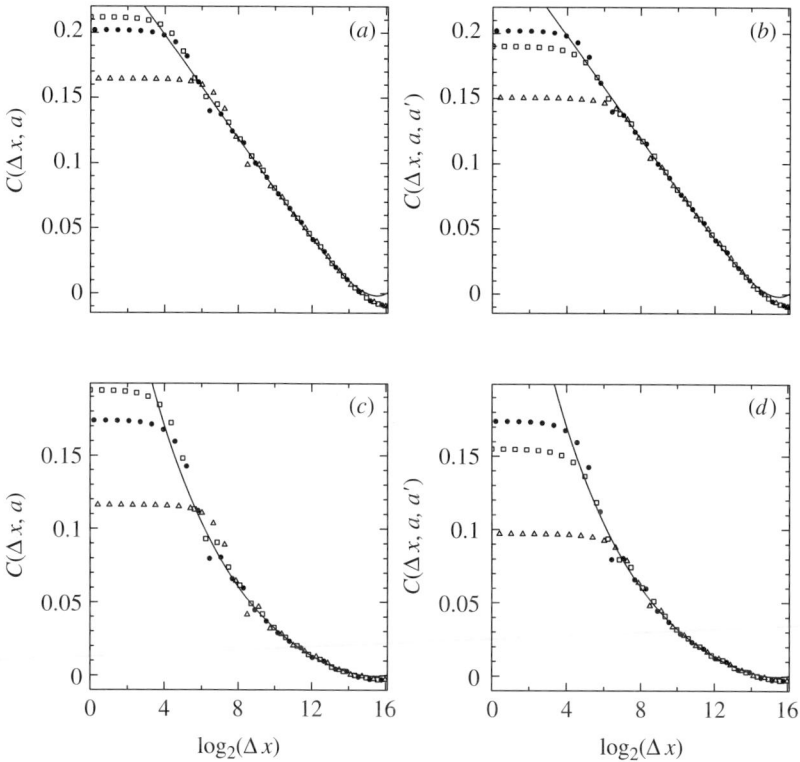

Fig. 5: Numerical computation of magnitude correlation functions for lognormal cascade processes built on an orthonormal wavelet basis. Scale-invariant cascade: (a) 'one-scale' correlation functions $C(\Delta x, a, a)$ for $a = 4$ ($\square$), 16 ($\bullet$) and 64 ($\triangle$); (b) 'two-scale' correlation functions $C(\Delta x, a, a')$ for $a = a' = 16$ ($\bullet$), $a = 4$, $a' = 16$ ($\square$) and $a = 16$, $a' = 64$ ($\triangle$). The solid lines represent fits of the data with the lognormal prediction (equation (3.7)) using the parameters $\lambda^2 = 0.03$ and $\log_2 L = 16$. Non-scale-invariant cascade: (c) 'one-scale' correlation functions; (d) 'two-scale' correlation functions. Symbols have the same meaning as in (a) and (b). The solid lines correspond to equation (3.8) with $\beta = 0.3$, $\lambda^2 = 0.2$ and $\log_2 L = 16$.

which is in perfect agreement with expression (3.7): in semi-log-coordinates, the correlation functions decrease almost linearly (with slope λ^2) up to the integral scale L, that is of order 2^{16} points. In figure 5b are displayed these correlation functions when the two scales a_1 and a_2 are different. One can check that, as expected, they still do not depend on the scales provided $\Delta x \geq \sup(a_1, a_2)$; moreover, they are again very well fitted by the above theoretical curve (except at very large Δx where finite size effects show up). The linear behaviour of $C(\Delta x, a_1, a_2)$ versus $\ln(\Delta x)$ is

characteristic for 'classical' scale-invariant cascades for which the random weights
are uncorrelated.

3.2.2 Non-scale-invariant random cascades

One can also consider non-scale-invariant cascades where the weights are not identically distributed and have an explicit scale dependence (Arneodo *et al.*, 1998*a*).
For example, we can consider a lognormal model whose coefficients $\ln c_{j,k}$ have a
variance that depends on j as $V(j) = \lambda^2(2^{j\beta} - 1)/(\beta \ln 2)$. This model is inspired
by the ideas of Castaing and co-workers (Castaing *et al.* 1990, 1993; Gagne *et al.*
1994; Naert *et al.* 1994; Chabaud *et al.* 1994; Castaing & Dubrulle 1995; Chillà *et
al.* 1996) and the experimental results reported in Section 2. Note that it reduces to
a scale-invariant model in the limit $\beta \to 0$. For finite β and $\sup(a_1, a_2) \leq \Delta x < L$,
the correlation function becomes

$$C(\Delta x, a_1, a_2) = \frac{\lambda^2}{\beta \ln 2} \left(\frac{(L/\Delta x)^\beta - (\Delta x/L)}{2^{\beta+1} - 1} - 1 + \frac{\Delta x}{L} \right). \tag{3.8}$$

As for the first example, we have tested our formalism on this model constructed
using the same Daubechies wavelet basis and considering, for the sake of simplicity,
i.i.d. lognormal weights W_{ϵ_i}. Figure 5*c, d* is the analogue of figure 5*a, b*. One can
see that, when scale-invariance is broken, our estimates of the magnitude correlation
functions are in perfect agreement with equation (3.8), which predicts a power-law
decrease of the correlation functions versus Δx.

3.2.3 Distinguishing 'multiplicative' from 'additive' processes
(Arneodo et al., 1998a)

The two previous examples illustrate the fact that magnitudes in random cascades
are correlated over very long distances. Moreover, the slow decay of the correlation
functions is independent of scales for large enough space lags ($\Delta x > a$). This is
reminiscent of the multiplicative structure along a space-scale tree. These features
are not observed in 'additive' models like fractional Brownian motions whose long-
range correlations originate from the sign of their variations rather than from the
amplitudes. In figure 6 are plotted the correlation functions of an 'uncorrelated'
lognormal model constructed using the same parameters as in the first example but
without any multiplicative structure (the coefficients $c_{j,k}$ have, at each scale j, the
same lognormal law as before but are independent) and for a fractional Brownian
motion with $H = \frac{1}{3}$. Let us note that from the point of view of both the multifractal
formalism and the increment PDF scale properties, the 'uncorrelated' and 'multiplicative' lognormal models are indistinguishable since their one-point statistics at
a given scale are identical. As far as the magnitude space-scale correlations are
concerned, the difference between the cascade and the other models is striking: for
$\Delta x > a$, the magnitudes of the fractional Brownian motion and of the lognormal
'white-noise' model are found to be uncorrelated.

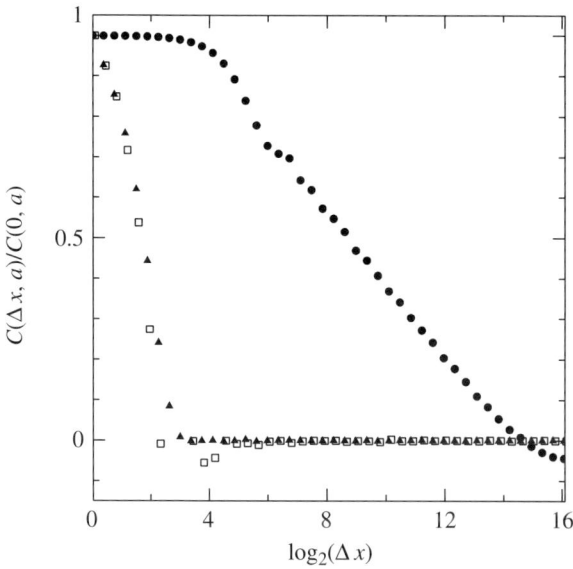

Fig. 6: 'One-scale' ($a = 4$) magnitude correlation functions: lognormal cascade process ($\bullet$); lognormal 'white noise' ($\square$); $H = \frac{1}{3}$ fractional Brownian motion ($\blacktriangle$). Magnitudes are correlated over very long distances for the cascade process while they are uncorrelated when $\Delta x > a$ for the two other processes.

3.3 Analysis of velocity data using space-scale correlation functions (Arneodo *et al.*, 1998a)

In this subsection, we report preliminary application of space-scale correlation functions to Modane wind-tunnel velocity data at $R_\lambda \simeq 2000$, which correspond to the highest statistics accessible to numerical analysis. In figure 7a, b are plotted (to be compared with figure 5) the 'one-scale' and 'two-scale' correlation functions. Both figures clearly show that space-scale magnitudes are strongly correlated. Very much like previous toy cascades, it seems that for $\Delta x > a$, all the experimental points $C(\Delta x, a_1, a_2)$ fall onto a single curve. We find that this curve is nicely fitted by equation (3.8) with $\beta = 0.3$, $\lambda^2 = 0.27$ and $L \simeq 2^{14}$ points. This latter length-scale corresponds to the integral scale of the experiment that can be estimated from the power spectrum. It thus seems that the space-scale correlations in the magnitude of the velocity field are in very good agreement with a cascade model that is not scale-invariant. This corroborates the results of Section 2 from 'one-point' statistical studies. However, we have observed several additional features that do not appear in wavelet cascades. (i) For $\Delta x > L$, the correlation coefficient is not in the noise level ($C = 0$ as expected for uncorrelated events) but remains negative up to a distance of about three integral scales. This observation can be interpreted as

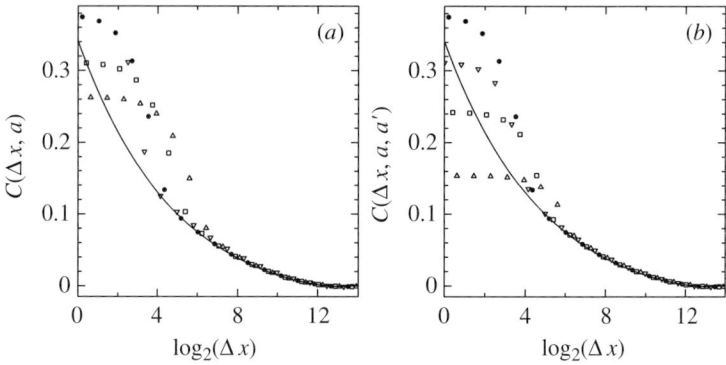

Fig. 7: Magnitude correlation functions of Modane fully developed turbulence data ($R_\lambda \simeq 2000$): (*a*) 'one-scale' correlation functions at scales $a = 24\eta$ ($\triangledown$), 48η ($\bullet$), 96η ($\square$) and 192η ($\triangle$); (*b*) 'two-scale' correlations functions at scales $a = 24\eta$, $a' = 48\eta$ ($\triangledown$), $a = 48\eta$, $a' = 48\eta$ ($\bullet$), $a = 48\eta$, $a' = 96\eta$ ($\square$) and $a = 48\eta$, $a' = 192\eta$ ($\triangle$). The solid lines correspond to a fit using equation (3.8) with $\beta = 0.3$, $\lambda^2 = 0.27$ and $\log_2 L = 13.6$.

an anticorrelation between successive eddies: very intense eddies are followed by weaker eddies and vice versa. (ii) For $\Delta x \simeq a$, there is a crossover from the value $C(\Delta x = 0, a, a)$ (which is simply the variance of ω at scale a) down to the fitted curve corresponding to the cascade model. This was not the case in previous cascade models (figure 5). This observation suggests that simple self-similar (even non-scale-invariant) cascades are not sufficient to account for the space-scale structure of the velocity field. The interpretation of this feature in terms of correlations between weights at a given cascade step or in terms of a more complex geometry of the tree underlying the energy cascade is under progress. The possible importance of spatially fluctuating viscous smoothing effects (Frisch & Vergassola, 1991) is also under consideration.

4 The multifractal description of intermittency revisited with wavelets

4.1 WTMM probability density functions (Arneodo *et al.*, 1998*c*)

A first way to check the consistency of our results is to test the convolution formula (2.2) on the WTMM PDFs using a Gaussian kernel. The results of this test application are reported in figure 8 (Arneodo *et al.*, 1998*c*). Let us mention that a naive computation of the PDFs of the (continuous) WT coefficients at different scales in the inertial range (Roux, 1996), leads to distributions that are nearly centred with

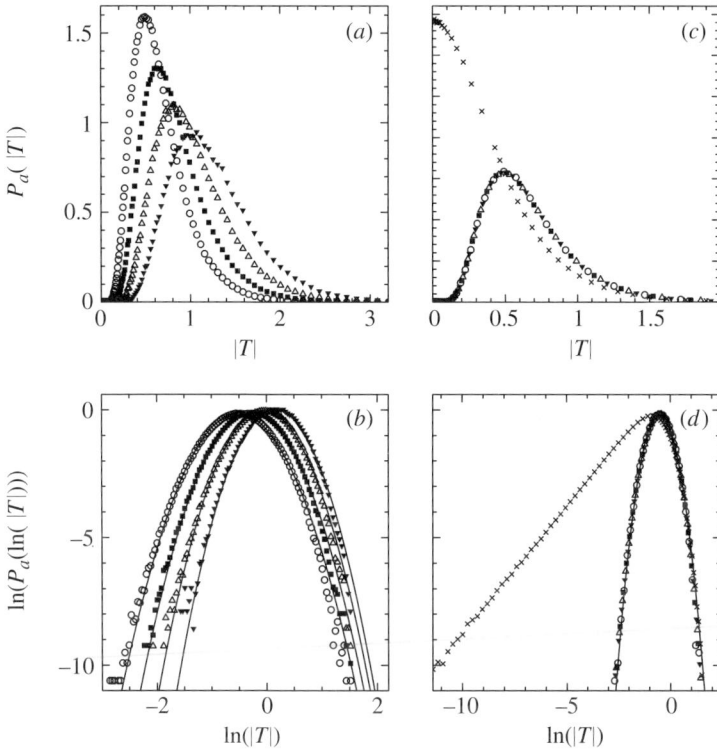

Fig. 8: Probability density functions of the WTMM for the Modane turbulent velocity signal ($R_\lambda \simeq 2000$): (a) $P_a(|T|)$ versus $|T|$ as computed at different scales $a = 385\eta$ (○), 770η (■), 1540η (△) and 3080η (▼); (b) $\ln(P_a(\ln(|T|)))$ versus $\ln|T|$ at the same scales; (c) and (d) the PDFs after being transformed according to equation (2.2) with a Gaussian kernel $G_{aa'}$ and $s(a, a') = (a^{-\beta} - a'^{-\beta})/\beta$ where $\beta = 0.095$. The (×) in (c) and (d) represent the velocity increment PDF at scale $a = 308\eta$. The solid lines in (b) and (d) correspond to the Gaussian approximations of the histograms. The analysing wavelet is $\psi_{(3)}^{(1)}$.

a shape that goes from Gaussian at large scales to stretched exponential-like tails at smaller scales, very much like the evolution observed for the velocity increment PDFs (Gagne, 1987; Castaing *et al.*, 1990; Kailasnath *et al.*, 1992; Frisch, 1995; Tabeling *et al.*, 1996; Belin *et al.*, 1996). But the wavelet theory (Meyer, 1990; Daubechies, 1992) tells us that there exists some redundancy in the continuous WT representation. Indeed, for a given analysing wavelet, there exists a reproducing kernel (Grossmann & Morlet 1984, 1985; Daubechies *et al.* 1986) from which one can express any WT coefficient at a given point x and scale a as a linear combination of the neighbouring WT coefficients in the space-scale half-plane. As emphasized in Muzy *et al.* (1991, 1993, 1994), Mallat & Hwang (1992) and Bacry *et al.* (1993), a way to break free from this redundancy is to use the WTMM representation. In figure 8*a* are reported the results of the computation of the WTMM PDFs when restricting our analysis to the WT skeleton (figure 1*c*) defined by the WT maxima lines. Since by definition the WTMM are different from zero, the so-obtained PDFs decrease very fast to zero at zero, which will make the estimate of the exponents ζ_q tractable for $q < 0$ in Section 4.2. When plotting $\ln P_a(\ln(|T|))$ versus $\ln |T|$, one gets in figure 8*b* the remarkable result that for any scale in the inertial range all the data points fall, within a good approximation, on a parabola, which is a strong indication that the WTMM have a lognormal distribution. In figure 8*c* we have succeeded in collapsing all the WTMM PDFs, computed at different scales, onto a single curve when using equation (2.2) with a Gaussian kernel $G(u, s(a, a'))$, where $s(a, a')$ is given by equation (2.9) with $\beta = 0.095$ in order to account for the scale-invariance breaking mentioned above (Section 2.2). This observation corroborates the lognormal cascade picture. Let us point out that, as illustrated in figure 8*c, d*, the velocity increment PDFs are likely to satisfy the Castaing and co-workers convolution formula (1.2) with a similar Gaussian kernel, even though their shape evolves across the scales (Roux, 1996). The fact that the WTMM PDFs turn out to have a shape which is the fixed point of the underlying kernel has been numerically revealed in previous works (Roux, 1996; Arneodo *et al.*, 1997) for various synthetic log-infinitely divisible cascade processes. So far, there exists no mathematical demonstration of this remarkable numerical observation.

4.2 ζ_q scaling exponents

A second test of the lognormality of the velocity fluctuations lies in the determination of the ζ_q spectrum. As discussed in previous studies (Muzy *et al.* 1993, 1994), the structure-function approach pioneered by Parisi & Frisch (1985) has several intrinsic insufficiencies which mainly result from the poorness of the underlying analysing wavelet $\psi_{(0)}^{(1)}$. Here we use instead the so-called WTMM method (Muzy *et al.* 1991, 1993 1994; Bacry *et al.* 1993; Arneodo *et al.* 1995) that has proved to be very efficient in achieving multifractal analysis of very irregular signals. The WTMM method consists of computing the following partition functions:

$$Z(q, a) = \sum_{l \in \mathcal{L}(a)} \left(\sup_{\substack{(x,a') \in l, \\ a' \leq a}} |T_\psi[v](x, a')| \right)^q, \quad \forall q \in \mathbb{R}, \tag{4.1}$$

where $\mathcal{L}(a)$ denotes the set of all WTMM lines of the space-scale half-plane that exist at scale a and contain maxima at any scale $a' \leq a$. A straightforward analogy with the structure functions $S_q(l)$ (equation (1.1)) yields

$$\mathcal{S}(q, a) = \frac{Z(q, a)}{Z(0, a)} \sim a^{\zeta_q}. \tag{4.2}$$

However, there exist two fundamental differences between $S_q(l)$ and $\mathcal{S}(q, a)$. (i) The summation in equation (4.1) is over the WT skeleton defined by the WTMM.

Since by definition the WTMM do not vanish, equation (4.1) allows us to extend the computation of the scaling exponents ζ_q from positive q values only when using the structure functions (as shown in Section 4.1, the velocity increment PDFs do not vanish at zero), to positive as well as negative q values without any risk of divergences (Muzy *et al.* 1993, 1994). (ii) By considering analysis of wavelets that are regular enough and have some adjustable degree of oscillation, the WTMM method allows us to capture singularities in the considered signal ($0 \leq h \leq 1$) like the structure functions can do, but also in arbitrary high-order derivatives of this signal (Muzy *et al.* 1993, 1994). In that respect, the WTMM method gives access to the entire $D(h)$ singularity spectrum and not only to the strongest singularities as the structure-function method is supposed to do from Legendre transforming ζ_q for $q > 0$ only (Muzy *et al.* 1991, 1993 1994; Bacry *et al.* 1993; Arneodo *et al.* 1995).

Since scale-invariance is likely to be broken, one rather expects the more general scale dependence of $S(q, a)$ (Roux 1996; Arneodo *et al.* 1997, 1999):

$$S(q, a) = \kappa_q \exp(-\zeta_q s(a)), \tag{4.3}$$

where κ_q is a constant that depends only on q and $s(a) = (a^{-\beta} - 1)/\beta$ consistently with the observed anomalous behaviour of $s(a, a')$ given by equation (2.9). Indeed, $S(q, a)$ can be seen as a generalized mean of $|T|^q$ so that formally, from the definition of the characteristic function $M(q, a)$ (equation (2.6)), one gets

$$S(q, a) \sim M(-eq, a). \tag{4.4}$$

From expression (2.8) of the Fourier transform of the kernel G and from equation (4.4), one deduces

$$\frac{S(q, a)}{S(q, a')} = \hat{G}_{aa'}(-eq). \tag{4.5}$$

When further using equation (2.11), this last equation becomes

$$\frac{S(q, a)}{S(q, a')} = \exp\left(\sum_{k=1}^{\infty} s(a, a') c_k \frac{q^k}{k!}\right), \tag{4.6}$$

which is consistent with equation (4.3) provided that

$$\zeta_q = -\sum_{k=1}^{\infty} \frac{c_k q^k}{k!}. \tag{4.7}$$

We have checked that fitting $S(q, a)/S(q, a')$ versus q for the two scales of figure 4 leads to the same estimates of $C_k = s(a, a') c_k$ as above to within less than 1%.

Remark 4.1

(a) Let us emphasize that for $\psi = \psi_{(0)}^{(1)}$, equation (4.3) is simply the general exponential self-similar behaviour predicted by Dubrulle (1996) (for the structure functions) by simple symmetry considerations.

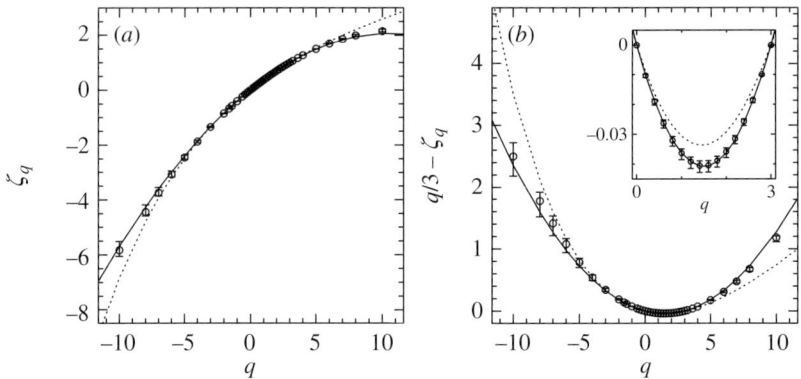

Fig. 9: WTMM estimation of the ζ_q spectrum for the Modane turbulent velocity signal ($R_\lambda \simeq 2000$). The analysing wavelet is $\psi_{(3)}^{(1)}$: (*a*) ζ_q versus q; (*b*) deviation of the experimental spectrum from the K41 $\zeta_q = \frac{1}{3}q$ prediction. The experimental ESS measurements (○) are compared with the theoretical quadratic spectrum of a lognormal process with $m = 0.39$ and $\sigma^2 = 0.036$ (solid line) and to the She & Leveque (1994) log-Poisson prediction with $\lambda = 2$, $\delta = (\frac{2}{3})^{1/3}$ and $\gamma = -\frac{1}{9}$ (dotted line).

   (b) As expressed by equation (4.3), the observed breaking of scale-invariance does not invalidate the ESS hypothesis (Benzi *et al.* 1993*b*, 1993*c*, 1995). Actually, equation (4.3) is equivalent to the ESS ansatz.

To estimate the ζ_q spectrum, we thus use the concept of ESS developed by Benzi *et al.* (1993*b*, *c*, 1995) i.e. we set $\zeta_3 = 1$ and plot $\mathcal{S}(q, a) = (\kappa_q/\kappa_3)\mathcal{S}(3, a)^{\zeta_q}$ versus $\mathcal{S}(3, a)$ in log-coordinates (for more details see Arneodo *et al.* (1999)). As shown in figure 9*a*, the experimental spectrum obtained from linear regression procedure remarkably coincides with the quadratic lognormal prediction $\zeta_q = mq - \frac{1}{2}\sigma^2 q^2$ with the same parameters as in Section 2.2 (figure 4), up to $|q| = 10$. We have checked that statistical convergence is achieved for $|q| \leq 8$; but even if the convergence becomes questionable for larger values of q, the 'error bars' obtained by varying the range of scales used for the ESS determination of ζ_q show the robustness of the spectrum. Let us point out that the log-Poisson prediction $\zeta_q = -\gamma q + \lambda(1 - \delta^q)$, with the She & Leveque (1994) parameter values: $\lambda = 2$, $\delta = (\frac{2}{3})^{1/3}$ and $\gamma = -\frac{1}{9}$, provides a rather good approximation of ζ_q for $q \in [-6, 6]$, in agreement with the structure-function estimations of ζ_q (She & Leveque, 1994; She & Waymire, 1995; Arneodo *et al.*, 1996; Belin *et al.*, 1996) and with our results on the first two cumulants of G (figure 4). However, when plotting the deviation of the ζ_q from the K41 linear $\zeta_q = \frac{1}{3}q$ spectrum (figure 9*b*), one reveals a systematic departure of the log-Poisson prediction from the experimental spectrum, and this even for $q \in [0, 3]$ as shown in the insert of figure 9*b*, whereas the

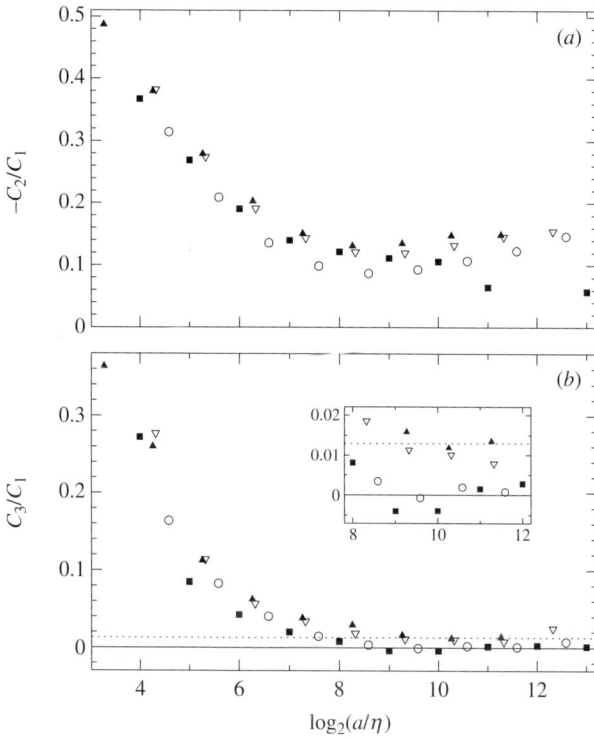

Fig. 10: Cumulant ratios $-C_2/C_1$ (a) and C_3/C_1 (b), estimated from $\hat{G}_{aa'}$ with $a' = 2a$, as a function of $\log_2(a/\eta)$ for four turbulent flows of different Reynolds numbers $R_\lambda \simeq 2000$ (o), 800 ($\triangledown$), 600 ($\blacktriangle$) and 280 ($\blacksquare$). In (b), the solid and dotted lines correspond, respectively, to the lognormal and to the She & Leveque (1994) log-Poisson predictions for C_3/C_1.

lognormal model still perfectly fits the experimental data. This nicely corroborates our findings on the third-order cumulant of G (figure 4) and shows that very long statistical samples are needed to discriminate between lognormal and log-Poisson statistics in fully developed turbulence data. Note that, according to the quadratic fit reported in figure 9, the ζ_q spectrum should decrease for $q \geq 11$, in qualitative agreement with previous discussions (Castaing *et al.*, 1990; Belin *et al.*, 1996). However, since statistical convergence is not achieved for such high values of q, one has to be careful when extrapolating the ζ_q behaviour. As reported in Belin *et al.* (1996), the number of data points needed to estimate ζ_q increases exponentially fast with q. Reaching an acceptable statistical convergence for $q \simeq 12$ would thus require velocity records about 10 times bigger than those processed in this work.

5   Conclusions and perspectives

To complete our study, we must address the issue of the robustness of our results when one varies the Reynolds number. We have reproduced our WT-based analysis on the turbulent velocity signal at Reynolds number $R_\lambda \simeq 800$ (of about the same length as the previous statistical sample at $R_\lambda \simeq 2000$ and with a resolution of 2.5η) obtained by Gagne *et al.* (1994) in a laboratory jet experiment (Arneodo *et al.*, 1998c). Because of scale invariance breaking, the notion of inertial range is not well defined. Thus, we may rather call 'inertial range' the range of scales on which equation (2.4) holds with the same kernel G. As illustrated in figure 10, for $R_\lambda \simeq 800$, $C_3/C_1 \simeq 0.01$ is significantly higher than for $R_\lambda \simeq 2000$, whereas $-C_2/C_1$ remains about equal to 0.15. An inertial range can still be defined ($128\eta \leq a \leq \frac{1}{4}L$), on which $G_{aa'}$ keeps a constant 'inertial' shape, but for $R_\lambda \simeq 800$, this shape becomes compatible with a log-Poisson distribution as proposed in She & Leveque (1994). We have checked that in that case, the She–Leveque model provides a better approximation of the ζ_q spectrum than the lognormal model (Arneodo *et al.*, 1998c). This result seems to contradict previous studies (Arneodo *et al.*, 1996; Belin *et al.*, 1996), suggesting that turbulent flows may be characterized by a universal ζ_q spectrum, independent of the Reynolds number, at least for $0 \leq q \leq 6$. However, as seen in figure 9a, for that range of q values, the various models can hardly be distinguished without plotting $q/3 - \zeta_q$. From our WT-based approach, which allows the determination of ζ_q for negative q values, when using very long statistical samples to minimize error bars, we can actually conclude that lognormal statistics no longer provide a perfect description of the turbulent velocity signals at Reynolds numbers $R_\lambda \lesssim 800$. This result, together with previous numerical (Leveque & She 1995, 1997; Benzi *et al.* 1996) and experimental (She & Leveque, 1994; Ruiz-Chavarria *et al.*, 1995) evidence for the relevance of log-Poisson statistics at low and moderate Reynolds numbers, strongly suggests that there might be some transitory regime ($R_\lambda \lesssim 1000$) towards asymptotic lognormal statistics, which could be accounted for by a quantized log-Poisson cascade or by some other cascade models that predict the correct relative order of magnitude of the higher-order cumulants (mainly c_3 and c_4) of the kernel G (equation (2.10)).

In figure 11 is reported the estimate of the scale-breaking exponent β (equation (2.9)), as a function of the Reynolds number (Arneodo *et al.*, 1999); the five points correspond to the results obtained for the two previous experiments and for three additional datasets corresponding to wind-tunnel ($R_\lambda \simeq 3050$), jet ($R_\lambda \simeq 600$) and grid ($R_\lambda \simeq 280$) turbulences. In figure 11a, β is plotted versus $1/\ln(R_\lambda)$ in order to check experimentally the validity of some theoretical arguments developed in Castaing *et al.* (1990) and Dubrulle (1996), which predict a logarithmic decay of β when increasing R_λ. Indeed the data are very well fitted by $\beta \sim 1/\ln(R_\lambda) - 1/\ln(R_\lambda^*)$, where $R_\lambda^* \simeq 12\,000$, which suggests that scale-similarity is likely to be attained at finite Reynolds numbers. However, as shown in figure 11b, for the range of Reynolds numbers accessible to today experiments the data are equally very well fitted by a power-law decay with an exponent which is close to $\frac{1}{2}$: $\beta \sim R_\lambda^{-0.556}$.

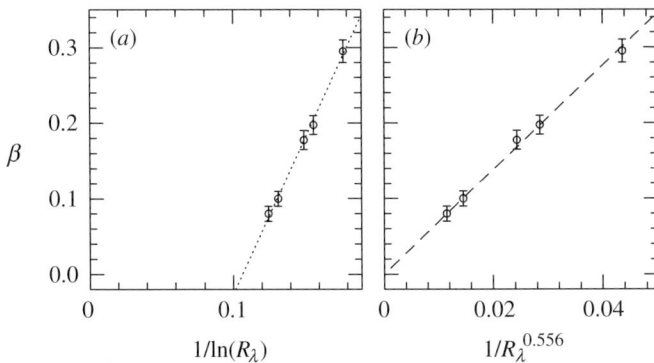

Fig. 11: β as a function of the Reynolds number. (a) β versus $1/\ln(R_\lambda)$; the dotted line corresponds to a fit of the data with $\beta = B(1/\ln(R_\lambda) - 1/\ln(R_\lambda^*))$ with $R_\lambda^* = 12\,000$. (b) β versus $R_\lambda^{-0.556}$; the dashed line corresponds to a linear regression fit of the data. Error bars account for variation of β according to the definition of the inertial range.

This second possibility brings the clue that scale-similarity might well be valid only in the limit of infinite Reynolds number. Whatever the relevant β behaviour, our findings for the kernel $G_{aa'}$ at $R_\lambda \simeq 2000$ (high statistics in the present work) and 3050 (moderate statistics in Arneodo *et al.* (1997, 1999)), strongly indicate that at very high Reynolds numbers the intermittency phenomenon can be understood in terms of a continuous self-similar multiplicative process that converges towards a scale-similar lognormal cascade.

Remark 5.1 Let us note that in figure 10*b* the estimate of C_3/C_1 for the lowest Reynolds-number velocity signal ($R_\lambda \simeq 280$) we have at our disposal cannot be distinguished from the results obtained for the wind-tunnel experiment at $R_\lambda \simeq 2000$. This observation of lognormal statistics at low Reynolds number contradicts the above conclusions. This might well be the consequence of the presence of some anisotropy at large scales in this grid turbulence where the velocity increment PDFs were found to depart significantly from a symmetric Gaussian shape (Gagne & Malecot, personal communication).

To summarize, this study has revealed the existence of a scale domain that we call 'inertial range', where a high-Reynolds-number turbulent-velocity signal ($R_\lambda \simeq 2000$) displays lognormal statistics. Our results confirm the relevance of the continuously self-similar lognormal cascade picture initiated by Castaing and co-workers (Castaing *et al.* 1990, 1993; Gagne *et al.* 1994; Naert *et al.* 1994; Chabaud *et al.* 1994; Castaing & Dubrulle 1995; Chillà *et al.* 1996). We also emphasize the fact that such an analysis requires very long statistical samples in order to get a good convergence of the cumulants of the kernel G and of the ζ_q spectrum. Our last results about the dependence of the statistics on the Reynolds number suggest that perfect lognormality may be reached only for $R_\lambda \to \infty$. A similar result is obtained concerning the breaking of scale-invariance (Roux 1996; Arneodo *et al.* 1997, 1998*c*, 1999): scale-invariance is likely to be restored only for very large Reynolds numbers. As emphasized by Frisch (1995), scale-invariance together with lognormal statistics for the velocity fluctuations imply that the Mach number of the flow increases indefinitely, which violates a basic assumption needed in deriving the incompressible Navier–Stokes equations. Let us note that this observation does not, however, violate the basic laws of hydrodynamics since it is conceivable that, at extremely high Reynolds numbers, supersonic velocity may appear. A systematic investigation of the evolution of the statistics with both the scale range and the Reynolds number is currently under progress. Further analysis of numerical and experimental data should provide new insights on the departure of $G_{aa'}$ from its 'inertial' shape outside the inertial range and on the way it converges towards a Gaussian kernel at high Reynolds numbers.

6 Acknowledgements

We are very grateful to Y. Gagne and Y. Malecot for the permission to use their experimental turbulent signals. We acknowledge very stimulating discussions with E. Bacry, B. Castaing, S. Ciliberto, Y. Couder, S. Douady, B. Dubrulle, Y. Gagne, F. Graner, J. F. Pinton, P. Tabeling and H. Willaime. This work was supported by 'Direction des Recherches, Etudes et Techniques' under contract DRET no. 95/111.

7 References

Anselmet, F., Gagne, Y., Hopfinger, E. & Antonia, R. 1984 *J. Fluid. Mech.* **140**, 63.

Arneodo, A., Bacry, E. & Muzy, J. 1995 *Physica* A **213**, 232.

Arneodo, A. (and 24 others) 1996 *Europhys. Lett.* **34**, 411.

Arneodo, A., Muzy, J. & Roux, S. 1997 *J. Physique* II **7**, 363.

Arneodo, A., Bacry, E., Manneville, S. & Muzy, J. 1998*a Phys. Rev. Lett.* **80**, 708.

Arneodo, A., Bacry, E. & Muzy, J. 1998*b J. Math. Phys.* **39**, 4142.

Arneodo, A., Manneville, S. & Muzy, J. 1998*c Eur. Phys. Jl* B **1**, 129.

Arneodo, A., Manneville, S., Muzy, J. & Roux, S. 1999 *Appl. Comput. Harmonic Analysis* **6**, 374.

Bacry, E., Muzy, J. & Arneodo, A. 1993 *J. Statist. Phys.* **70**, 635.

Belin, F., Tabeling, P. & Willaime, H. 1996 *Physica* D **93**, 52.

Benzi, R., Biferale, L., Crisanti, A., Paladin, G., Vergassola, M. & Vulpiani, A. 1993*a Physica* D **65**, 352.

Benzi, R., Ciliberto, S., Trippiccione, R., Baudet, C., Massaioli, F. & Succi, S. 1993*b Phys. Rev.* E **48**, R29.

Benzi, R., Ciliberto, S., Baudet, C., Ruiz-Chavarria, G. & Trippiccione, R. 1993*c Europhys. Lett.* **24**, 275.

Benzi, R., Ciliberto, S., Baudet, C. & Ruiz-Chavarria, G. 1995 *Physica* D **80**, 385.

Benzi, R., Biferale, L. & Trovatore, E. 1996 *Phys. Rev. Lett.* **77**, 3114.
Briscolini, M., Santangelo, P., Succi, S. & Benzi, R. 1994 *Phys. Rev.* E **50**, R1745.
Castaing, B. & Dubrulle, B. 1995 *J. Physique* II **5**, 895.
Castaing, B., Gagne, Y. & Hopfinger, E. 1990 *Physica* D **46**, 177.
Castaing, B., Gagne, Y. & Marchand, M. 1993 *Physica* D **68**, 387.
Cates, M. & Deutsch, J. 1987 *Phys. Rev.* A **35**, 4907.
Chabaud, B., Naert, A., Peinke, J., Chillà, F., Castaing, B. & Hebral, B. 1994 *Phys. Rev. Lett.* **73**, 3227.
Chillà, F., Peinke, J. & Castaing, B. 1996 *J. Physique* II **6**, 455.
Daubechies, I. 1992 *Ten lectures on wavelets.* Philadelphia, PA: SIAM.
Daubechies, I., Grossmann, A. & Meyer, Y. 1986 *J. Math. Phys.* **27**, 1271.
Dubrulle, B. 1994 *Phys. Rev. Lett.* **73**, 959.
Dubrulle, B. 1996 *J. Physique* II **6**, 1825.
Frisch, U. 1995 *Turbulence.* Cambridge University Press.
Frisch, U. & Orzag, S. 1990 *Physics Today*, January, p. 24.
Frisch, U. & Vergassola, M. 1991 *Europhys. Lett.* **14**, 439.
Gagne, Y. 1987 PhD thesis. University of Grenoble.
Gagne, Y., Marchand, M. & Castaing, B. 1994 *J. Physique* II **4**, 1.
Grossmann, A. & Morlet, J. 1984 *SIAM Jl Math. Analyt. Appl.* **15**, 723.
Grossmann, A. & Morlet, J. 1985 In *Mathematics and physics. Lecture on recent results* (ed. L. Streit), p. 135. Singapore: World Scientific.
Kailasnath, P., Sreenivasan, K., & Stolovitzky, G. 1992 *Phys. Rev. Lett.* **68**, 2766.
Kida, S. 1990 *J. Phys. Soc. Jap.* **60**, 5.
Kolmogorov, A. 1941 *C.R. Acad. Sci. USSR* **30**, 301.
Kolmogorov, A. 1962 *J. Fluid Mech.* **13**, 82.
Leveque, E. & She, Z. 1995 *Phys. Rev. Lett.* **75**, 2690.
Leveque, E. & She, Z. 1997 *Phys. Rev.* E **55**, 2789.
Mallat, S. & Hwang, W. 1992 *IEEE Trans. Inform. Theory* **38**, 617.
Meneveau, C. & Sreenivasan, K. 1991 *J. Fluid. Mech.* **224**, 429.
Meyer, Y. 1990 *Ondelettes.* Paris: Hermann.
Monin, A. & Yaglom, A. 1975 *Statistical fluid mechanics*, vol. 2. MIT Press.
Morel-Bailly, F., Chauve, M., Liandrat, J. & Tchamitchian, P. 1991 *C.R. Acad. Sci. Paris* II **313**, 591.
Muzy, J., Bacry, E. & Arneodo, A. 1991 *Phys. Rev. Lett.* **67**, 3515.
Muzy, J., Bacry, E. & Arneodo, A. 1993 *Phys. Rev.* E **47**, 875.
Muzy, J., Bacry, E. & Arneodo, A. 1994 *Int. Jl Bifur. Chaos* **4**, 245.
Naert, A., Puech, L., Chabaud, B., Peinke, J., Castaing, B. & Hebral, B. 1994 *J. Physique* II **4**, 215.
Neil, O. & Meneveau, C. 1993 *Phys. Fluids* A **5**, 158.
Novikov, E. 1990 *Phys. Fluids* A **2**, 814.
Novikov, E. 1995 *Phys. Rev.* E **50**, 3303.
Obukhov, A. 1962 *J. Fluid Mech.* **13**, 77.
Parisi, G. & Frisch, U. 1985 In *Turbulence and predictability in geophysical fluid dynamics and climate dynamics* (ed. M. Ghil, R. Benzi & G. Parisi), p. 84. Amsterdam: North-Holland.
Pedrizetti, G., Novikov, E. & Praskovsky, A. 1996 *Phys. Rev.* E **53**, 475.
Richardson, L. 1926 *Proc. R. Soc. Lond.* A **110**, 709.

Roux, S. 1996 PhD thesis, University of Aix-Marseille II.

Ruiz-Chavarria, G., Baudet, C. & Ciliberto, S. 1995 *Phys. Rev. Lett.* **74**, 1986.

Schertzer, D. & Levejoy, S. 1987 *J. Geophys. Res.* **92**, 9693.

She, Z. & Leveque, E. 1994 *Phys. Rev. Lett.* **72**, 336.

She, Z. & Waymire, E. 1995 *Phys. Rev. Lett.* **74**, 262.

Siebesma, A. 1988 In *Universality in condensed matter physics* (ed. R. Julien, L. Peliti, R. Rammal & N. Boccara), p. 188. Heidelberg: Springer.

Tabeling, P. & Cardoso, O. (eds) 1995 *Turbulence: a tentative dictionary.* New York: Plenum.

Tabeling, P., Zocchi, G., Belin, F., Maurer, J. & Willaime, H. 1996 *Phys. Rev.* E **53**, 1613.

Vincent, A. & Meneguzzi, M. 1995 *J. Fluid Mech.* **225**, 1.

3
Wavelets for the study of intermittency and its topology

F. Nicolleau[1] and J. C. Vassilicos
Department of Applied Mathematics and Theoretical Physics,
University of Cambridge, Silver Street, Cambridge CB3 9EW, UK

Abstract
We make a distinction between two topologically different types of intermittency: isolated, as in spirals, and non-isolated, as in fractals. For a broad class of isolated and a broad class of non-isolated intermittent topologies, the flatness $F(r)$ of velocity differences and the eddy capacity D_E obtained from a wavelet analysis are related by

$$F(r) \sim r^{D_E - 1}.$$

Inertial range intermittency is such that $D_E \leq 0.94$ and $F(r) \sim r^{-0.11}$ for jet turbulence with $Re_\lambda = 835$ and for grid turbulence with $Re_\lambda = 3050$.

Keywords: fractals; spirals; singularities; intermittency; wavelets; turbulence

1 Introduction

The intermittency of a statistically homogeneous signal $u(x)$ is often characterized by the flatness (Batchelor 1953; Frisch 1995),

$$F(r) = \frac{\langle \Delta u^4(r) \rangle}{\langle \Delta u^2(r) \rangle^2}, \tag{1.1}$$

where $\Delta u(r) = u(x + r) - u(x)$, the brackets $\langle \dots \rangle$ denote an average over x. A signal is often said to be intermittent when $F(r)$ increases with decreasing r. This is because an intermittent signal displays activity (in the sense that Δu is significantly non-zero) over only a fraction of space (or time) x, and this portion decreases with the scale r under consideration. However, such a property does not shed much light on the actual topology of the signal's intermittency. In this paper we use the wavelet transform to study what intermittency can actually look like in space, how we can measure its geometry, and we introduce a distinction between two different topologies of intermittency.

These different topologies can give rise to the same flatness properties. When the signal $u(x)$ is statistically scale invariant, then we may talk of scale-invariant

---

[1]Department of Mechanical Engineering, University of Sheffield, Mappin Street, Sheffield S1 3JD, UK.

intermittency and $F(r)$ must have a power-law dependence on r, i.e.

$$F(r) \sim r^{-q}. \tag{1.2}$$

This power q is a global statistical quantity, but we show that it is determined by the local geometry of the intermittency of the signal. The geometry of a signal and the degree to which this geometry is space filling are usually studied in terms of Kolmogorov capacities (fractal or box dimensions) of zero-crossings of the signal. However, we show here that the geometry of the *intermittency* of the signal is best captured by the zero-crossings of the *second derivative* of the signal, and that the Kolmogorov capacity of these zero-crossings can determine q.

2   Intermittency and eddy capacity

The zero-crossings of the second derivative $(d^2/dx^2)u(x)$ are related to both the geometry and the statistics of the intermittency of the signal $u(x)$. The signal $u(x)$ being assumed statistically homogeneous, nth-order moments $\langle |\Delta u(r)|^n \rangle$ can, therefore, be calculated as follows (see Frisch 1995):

$$\langle |\Delta u(r)|^n \rangle = \lim_{T \to \infty} \frac{1}{T} \int_0^T |u(x + r) - u(x)|^n \, dx. \tag{2.1}$$

In the limit where $r \to 0$, $\Delta u(r) = u(x + r) - u(x)$ is extremal at inflection points of the signal $u(x)$, that is, at points where $(d^2/dx^2)u(x) = 0$. In those cases where $|\Delta u(x)|$ has the same order of magnitude at all these inflection points, we can estimate that

$$\langle |\Delta u(r)|^n \rangle \sim |\Delta u(r)|^n r M_E(r), \tag{2.2}$$

where $M_E(r)$ is the minimum number of segments of size r needed to cover the zero-crossings of $(d^2/dx^2)u(x)$ per unit length. If the signal $u(x)$ is statistically scale invariant, then $M_E(r)$ has a power-law dependence on r, and this power-law dependence defines the eddy capacity D_E as follows:

$$M_E(r) \sim r^{-D_E}. \tag{2.3}$$

Hence, D_E is the Kolmogorov capacity of the zero-crossings of $(d^2/dx^2)u(x)$, and note that $0 \leq D_E \leq 1$. From (2.2) and (2.3),

$$\langle |\Delta u(r)|^n \rangle \sim |\Delta u(r)|^n r^{1 - D_E},$$

and the r dependence of non-dimensionalized structure functions is given by

$$\frac{\langle |\Delta u(r)|^n \rangle}{\langle |\Delta u(r)|^2 \rangle^{n/2}} \sim r^{(1 - D_E)(1 - (n/2))}. \tag{2.4}$$

For $n = 4$ we obtain

$$F(r) \sim r^{D_E - 1},\tag{2.5}$$

and a comparison of (2.5) with (1.2) gives

$$q = 1 - D_E.\tag{2.6}$$

These conclusions are similar to the results of the β-model,

$$\frac{\langle |\Delta u(r)|^n \rangle}{\langle \Delta u^2(r) \rangle^{n/2}} \sim r^{(1-D)(1-(n/2))},$$

where D is a fractal dimension defined in terms of the volume fraction $p(r)$ of eddies of size r (see Frisch 1995). However, in the β model, the concept of an eddy of size r remains abstract and no operative definition is given by which to identify and measure an eddy of size r.

In this respect, the situation is radically different here. An operative definition is given by which to measure D_E in terms of accessible properties of the signal, the inflection points of the signal. These properties are indirectly accessible in practice if use is made of a wavelet transform,

$$\tilde{u}(x_0, a) = a^{-3} \int u(x) \psi^* \left(\frac{x - x_0}{a} \right) dx,\tag{2.7}$$

of the signal $u(x)$, where $\psi(x)$ is the 'mother' wavelet (ψ^* its complex conjugate). The first relatively minor advantage in using a wavelet transform is of avoiding calculation of double derivatives of the signal. This is explained in the next paragraph. The second major advantage in using a wavelet transform is that it can provide an estimate of the importance of the drop in $u(x)$ at scale r across the inflection point. This is explained in conclusion (v) in Section 3.

The wavelet transform is a function of position x_0 and length-scale a. Choosing the mother wavelet in (2.7) to be a Mexican hat, that is

$$\psi(x) = \frac{d^2}{dx^2} e^{-x^2/2},$$

the zeros of $\tilde{u}(x_0, a)$ tend towards zeros of $(d^2/dx^2)u(x_0)$ as $a \to 0$. Hence, the eddy capacity D_E can be measured in practice as follows: a Mexican hat wavelet transform $\tilde{u}(x_0, a)$ is performed on the signal $u(x)$ and a box-counting algorithm is applied on the zero-crossings of $\tilde{u}(x_0, a)$ for the smallest scale a permitted by the discretization. The box counting yields $M_E(r)$ and an eddy capacity is well defined if $M_E(r) \sim r^{-D_E}$ over a significant range of scales. This wavelet-box-counting algorithm to measure D_E is applied successfully in the following section to test the validity of $F(r) \sim r^{D_E-1}$ against a variety of test signals. We have checked in all cases that the value of D_E can be obtained unchanged from the zero-crossings of $\tilde{u}(x_0, a)$ for many values of a even larger than the discretization.

In this paper, we restrict ourselves to the study of scale-invariant intermittency for which $D_E > 0$. (See the appendix for a short discussion of the differences between D_E and the Kolmogorov capacity D'_K of the zero-crossings of the signal $u(x)$ itself.)

3   Validity of $F(r) \sim r^{D_E - 1}$

Hunt & Vassilicos (1991) emphasize that signals with power spectra $E(k) \sim k^{-2p}$, where $p < 1$ (such as Kolmogorov's $k^{-5/3}$), must contain singularities (or rather near-singularities if we take into account small-scale smoothing effects such as those of viscosity or diffusivity) that are worse than mere isolated discontinuities in the signal or its derivatives. These singularities can be qualitatively classified as follows (Hunt & Vassilicos 1991): isolated cusp singularities such as $1/x$; isolated accumulating singularities such as $\sin(1/x)$ (see figures 4b and 5); and non-isolated singularities such as can be found in fractal signals (see figures 1, 4a, 7a and 8b). Simple models of small-scale turbulence structure have been proposed for each one of these singularities: the Burgers vortex (see Batchelor 1967) has a point vortex near-singularity that is a cusp near-singularity; the Lundgren vortex (Lundgren 1982) has a spiral-vortex-sheet near-singularity that is an isolated accumulating near-singularity; and fractal and multifractal models of turbulence, such as the β model and its generalizations (see Frisch 1995), assume the existence of non-isolated singularities.

We test the validity of (2.5) numerically on several model signals $u(x)$, some regular and others scale invariant, with qualitatively different types of singularity. The conclusions are outlined below.

(i) *For continuous signals with continuous derivatives, for signals with isolated discontinuities in the signal itself or its derivatives, and for isolated cusps, $D_E = 0$ and $F(r) =$ const*

This conclusion implies that signals in which $D_E \neq 0$ or $F(r) \neq$ const must necessarily have either isolated accumulating singularities or non-isolated singularities, thus leading to a topological classification of intermittency for which $D_E \neq 0$: *isolated intermittency* when the intermittent signal carries isolated accumulating singularities but no non-isolated singularities; and *non-isolated intermittency* when the intermittent signal carries non-isolated singularities.

It should be noted, however, that isolated accumulating singularities with $D_E = 0$ and $F(r) =$ const are possible, e.g. $u(x) = \sin(e^{1/x})$, and that non-isolated singularities with $D_E = 0$ are also possible.

(ii) *For fractal signals with $F(r) =$ const, $D_E = 1$ irrespective of the power spectrum's scaling.*

There exists a class of fractal signals (non-isolated singularities) for which

$$F(r) = \text{const and } D_E = 1,$$

irrespective of their power spectrum's scaling, $E(k) \sim k^{-2p}$. Examples of such fractal signals are the Weierstrass function,

$$u(x) = \sum_{j \geq 1} \lambda^{(\alpha - 2)j} \sin(\lambda^j x), \tag{3.1}$$

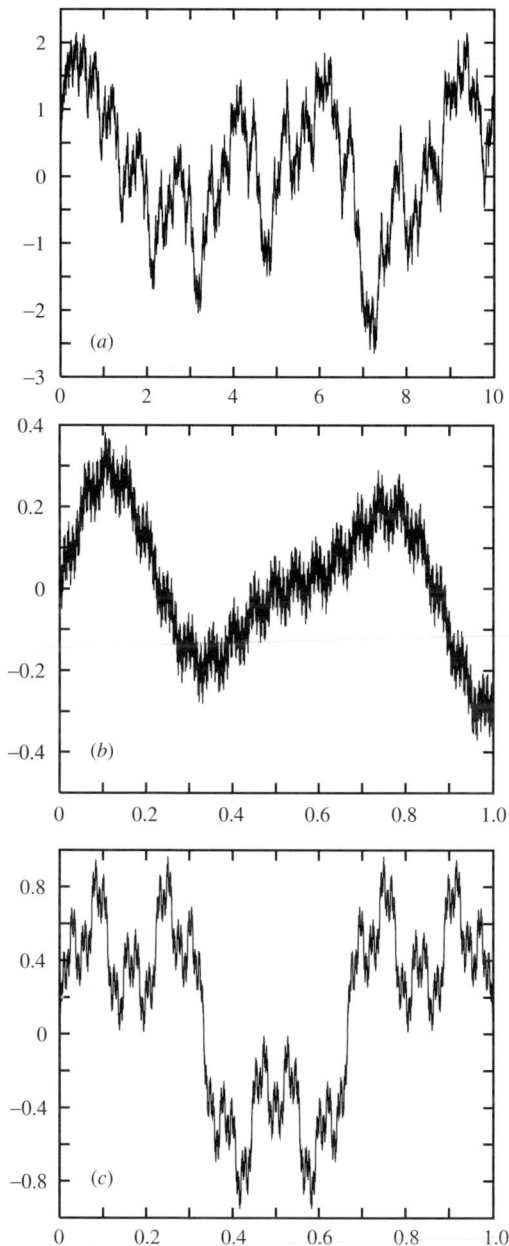

Fig. 1: Non-intermittent signals. (*a*) The Weierstrass function (3.1) (here $\lambda = \alpha = 1.5$ and $j \leq 30$). (*b*) The random-phase function (here $\lambda = \alpha = 1.5$ and $j \leq 30$). (*c*) The centred sinusoid function (3.2) (here $\lambda = 3$, $\alpha = 1.5$ and $j \leq 25$). These are signals with non-isolated singularities and are such that $D_E = 1$ (see figure 2).

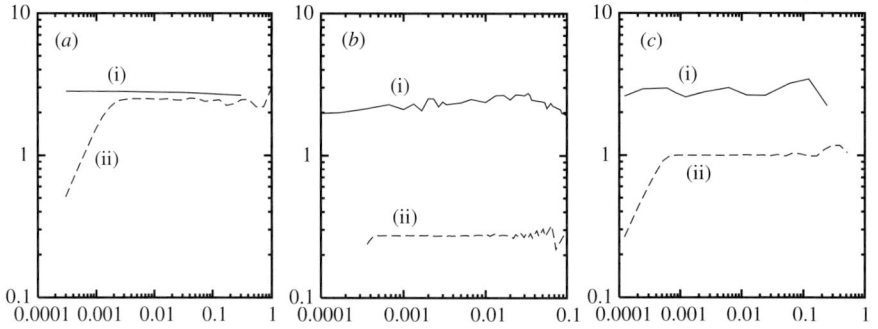

Fig. 2: Comparison of (i) $r^{D_E} M_E(r)$ and (ii) $r^{1-D_E} F(r)$ as functions of r; $M_E(r)$ is the minimum number of segments of size r needed to cover the zero-crossings of $(\mathrm{d}^2/\mathrm{d}x^2) u(x)$ per unit length. (*a*) Weierstrass function; (*b*) random-phase function; (*c*) centred sinusoid function. We have set $D_E = 1$ in all these graphs to demonstrate that $D_E = 1$ in all the cases presented in figure 1.

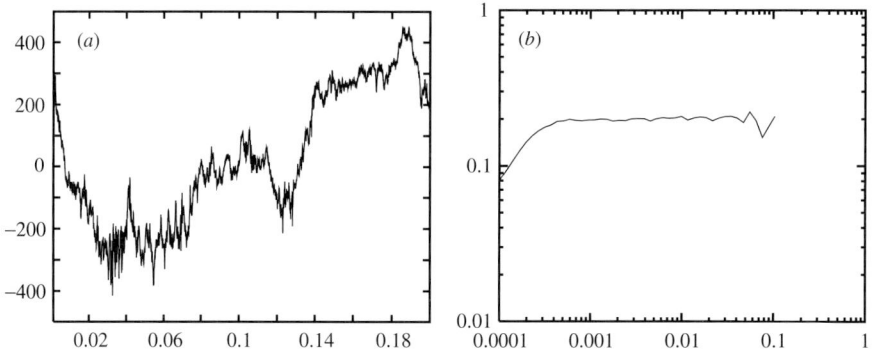

Fig. 3: (*a*) Function with non-isolated singularities defined by (3.3), for which $F(r) \sim r^{-0.27}$ but $D_E = 1$. The probability density function of η is $P(\eta) = y\eta_0 + (1 - y)\eta_1$ with $y = 0.2924$, $\eta_0 = 0.70$, $\eta_1 = 0.45$, $\sigma = 0.18$. (*b*) Plot of $r^{D_E} M_E(r)$ against r for this function, where we have set $D_E = 1$.

where $\lambda > 1$ and $1 < \alpha < 2$ (figure 1*a*); the random-phase fractal signal obtained by replacing $\sin(\lambda^j x)$ with $\sin(\lambda^j x + \phi_j)$ in (3.1), where ϕ_j is a random phase between 0 and 2π (figure 1*b*); and the sum of centred sinusoids,

$$u(x) = \sum_{j \geq 1} \lambda^{(\alpha-2)j} \sin(\lambda j \pi x), \qquad (3.2)$$

where λ is an integer and $1 < \alpha < 2$ (figure 1*c*). For all these signals, and irrespective of the values of λ and α that determine the power spectrum (see Falconer 1990), we find that $D_E = 1$ and $F(r) = $ const (figure 2).

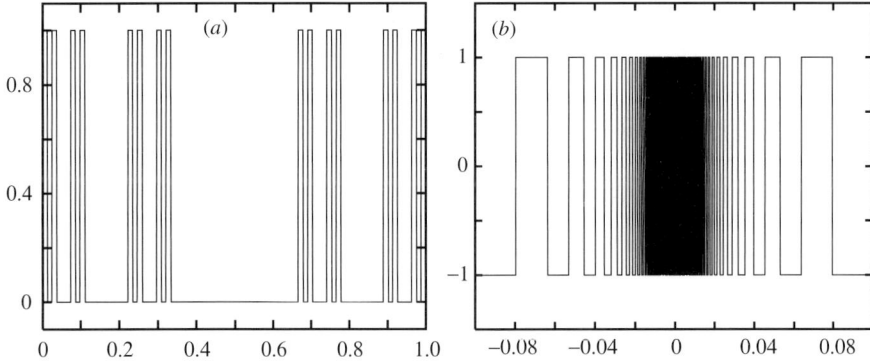

Fig. 4: Intermittent signals with non-isolated (a) and isolated (b) singularities. (a) The Cantor dust on–off function (here, $D'_K = D_E = \ln 2 / \ln 3$). ($b$) On–off signal (3.4) (here, $D'_K = D_E = \frac{1}{3}$).

(iii) *However, there exist signals with non-isolated singularities for which $F(r) \sim r^{-q}$ but $D_E = 1$.*

This is the case of the function $u(x)$ defined in Benzi *et al.*(1993) (see figure 3) as

$$u(x) = \sum_{j=0}^{15} \sum_{k=0}^{2^j-1} \alpha_{i,j} \Psi(2^j x - k), \qquad (3.3)$$

where

$$\Psi(x) = -\frac{\partial^2}{\partial x^2} e^{-(x^2)/(2\sigma^2)}, \qquad \alpha_{j,k} = \epsilon_{j,k}\eta_{j,k}\alpha_{j-1,k/2},$$

the $\eta_{j,k}$ are independent random variables and $\epsilon_{j,k} = \pm 1$ with equal probability. For this function $u(x)$, there exist sets of random variables $\eta_{i,j}$ for which $F(r) \sim r^{-q}$ with $q \neq 0$ (Benzi *et al.*1993), but we invariably find that $D_E = 1$.

(iv) *If D_E is well defined and $D_E < 1$ and if $F(r) = r^{-q}$, then $q = 1 - D_E$ for a broad class of isolated and a broad class of non-isolated intermittent topologies.*

Let us start with the restricted class of on–off signals, such as those pictured in figure 4. These signals take one of two values, say -1 and $+1$. They can have either isolated intermittency, as in the function

$$u(x) = H(\sin(x^{-t})) - H(-\sin(x^{-t})) \qquad (3.4)$$

(figure 4b), where H is the Heaviside function, or non-isolated intermittency as in the well-known Cantor set (figure 4a). When the set of points where the signal

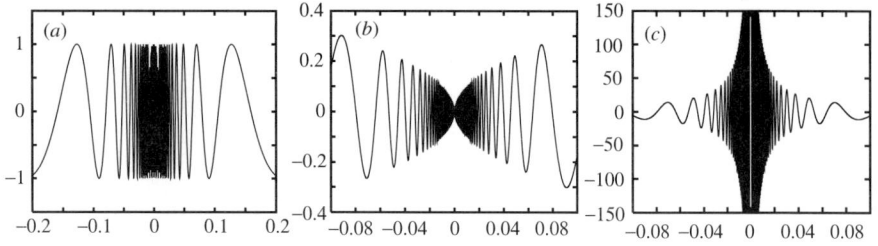

Fig. 5: Isolated intermittency for different spiral functions: (a) $\sin x^{-1/2}$ ($D'_K = D_E = \frac{1}{3}$); (b) $x^{1/2} \sin x^{-1}$ ($D'_K = D_E = \frac{1}{2}$); (c) $x^{-1} \sin x^{-1}$ ($D'_K = 0.5$, $D_E = 0.6$).

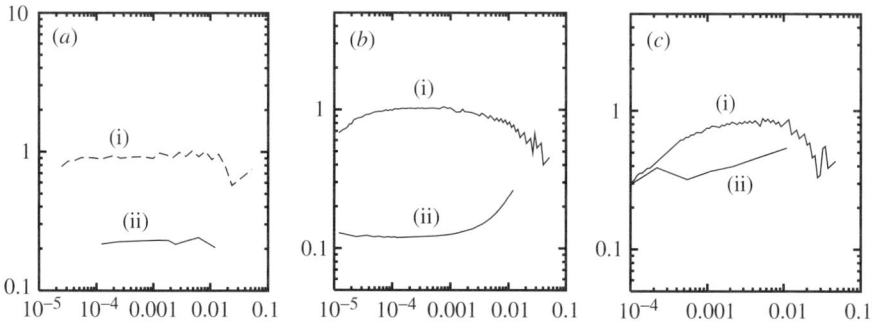

Fig. 6: Comparison of (i) $r^{D_E} M_E(r)$ and (ii) $r^{1-D_E} F(r)$ as functions of r, for the signals with isolated intermittency of figure 5. (a) The function $\sin x^{-1/2}$ (we set $D_E = \frac{1}{3}$). (b) The function $x^{1/2} \sin x^{-1}$, (we set $D_E = \frac{1}{2}$). (c) The function $x^{-1} \sin x^{-1}$ (we set $D_E = 0.6$) and the absence of plateaux indicates here that $F(r) \sim r^{D_E-1}$ is not valid. The plateaux in (a), (b) indicate that $F(r) \sim r^{D_E-1}$ is valid for these functions.

abruptly changes value is characterized by a well-defined Kolmogorov capacity D'_K, the flatness

$$F(r) \sim r^{D'_K-1}.$$

This result has been derived analytically by Vassilicos (1992) for on–off signals with either isolated accumulating singularities or non-isolated singularities. On–off signals are such that $D_E = D'_K$ because the zero-crossings of the second derivatives of such signals are the same as the zero-crossings of the signal itself. It is therefore an analytical result[2] that for on–off signals with either isolated or non-isolated intermittency:

$$F(r) \sim r^{D_E-1}.$$

---

[2]More generally, the analytical results of Vassilicos (1992) imply that (2.4) holds for all values of n when the signal is on–off.

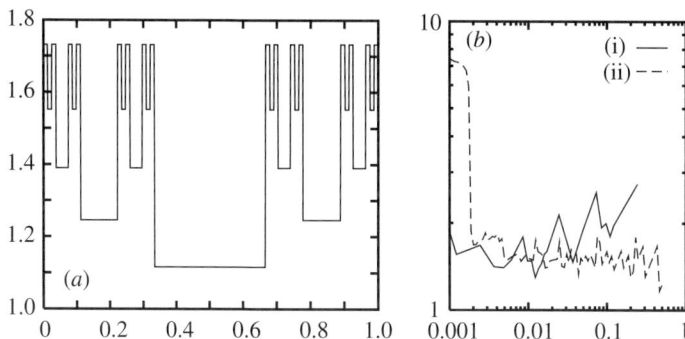

Fig. 7: (*a*) The non-isolated intermittent signal described in conclusion (iii) (here, $\sigma = 0.25$ and $D_E = 0.7$). (*b*) (i) $r^{1-D_E} F(r)$ and (ii) $r^{D_E} M_E(r)$ for this signal; the ratio of the largest to the smallest scale is 6500. $F(r) \sim r^{D_E-1}$ is verified, but, due to the properties of lacunarity of this signal, $r^{1-D_E} F(r)$ exhibits strong oscillations. $r^{D_E} M_E(r)$ keeps its plateau shape on the large scales, whereas $r^{1-D_E} F(r)$ loses its precision on these scales.

For these on–off signals, the assumption made in (2.2), that $|\Delta u(x)|$ has the same order of magnitude at all the inflection points, is indeed verified. However, we find that $F(r) \sim r^{D_E-1}$ is valid even beyond this assumption. The following numerical examples demonstrate that $F(r) \sim r^{D_E-1}$ is valid more generally when $|\Delta u(x)|$ does not have too strong a dependence on x, in the cases of both isolated and non-isolated intermittency.

An example of a family of signals with isolated intermittency for which this assumption is not verified (unless $s = 0$) is (see figure 5a–c)

$$u(x) = x^s \sin(x^{-t}). \tag{3.5}$$

The Fourier power spectrum is well defined when $-1 \leq 2s \leq t$ and $t > 0$. For these signals, $F(r) \sim r^{-q}$ and $q = 1 - D_E$ (figure 6a, b) provided that $-1 \leq 2s \leq t$ and $t > 0$. However, if the cusp singularity superimposed onto the isolated accumulation is too strong, that is if s is too negative (i.e. if $2s < -1$) as in the example of figure 5c, then $q \neq 1 - D_E$ (figure 6c).

The non-isolated intermittency of figure 7a is constructed on the basis of a Cantor set of points of Kolmogorov capacity D_K' but differently: the value of the signal between two consecutive such points is l^σ, where l is the distance between these two points and σ is a real number. This is also a signal where $|\Delta u(x)|$ depends on x. Nevertheless, $F(r) \sim r^{D_E-1}$ is valid for this signal too (figure 7b) provided that $\sigma > -\frac{1}{2}$.

However, there exist examples of isolated and non-isolated intermittency where $F(r) \sim r^{D_E-1}$ is not valid even for arbitrarily weak dependencies of $|\Delta u(x)|$ on x. This is the case for the isolated intermittencies of figures 5c and 8a, and for the

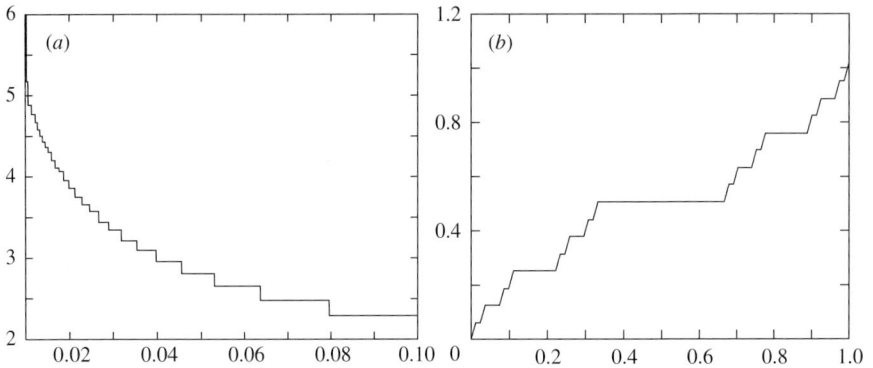

Fig. 8: Different signals with isolated (*a*) and non-isolated (*b*) intermittency for which $F(r) \sim r^{D_E^* - 1}$ is verified but $F(r) \sim r^{D_E - 1}$ is not verified. (*a*) Stair function (3.6) based on the zero-crossings of $\sin x^{-1}$ (here, $D_E^* = 0.65$, $\sigma = 0.2$). (*b*) Devil's staircase (here, $D_E^* = 0.7$ with 2000 for the ratio of the largest to the smallest scale).

non-isolated intermittency of figure 8*b*. The signal of figure 8*a* is based on the zero-crossings of signals (3.4) and (3.5), which are given by $x_n = (n\pi)^{-1/t}$, *n* being a positive integer. This signal (figure 8*a*) is defined as follows:

$$u(x) = \sum_{n \geq 1} (l_n)^\sigma H(x - x_{n+1}) H(x_n - x), \qquad (3.6)$$

where $l_n = x_n - x_{n+1}$ and σ is a real number. For such signals, $F(r) \sim r^{-q}$ but $q \neq 1 - D_E$, irrespective of the value of σ. Furthermore, figure 8*b* is a Devil's staircase constructed on a Cantor set of points of Kolmogorov capacity D_K' (see, for example, Frisch 1995), and for this case of non-isolated intermittency, we also find that $F(r) \sim r^{-q}$, but $q \neq 1 - D_E$.

The definition of D_E can be refined in such a way that $q = 1 - D_E$ can be made valid for these last two types of signals too.

  (v) *The definition of D_E can be refined using additional information from wavelet transforms to extend the domain of validity of $F(r) \sim r^{D_E - 1}$.*

The wavelet transform $\tilde{u}(x_0, a)$ of the signal $u(x)$ is

$$\tilde{u}(x_0, a) = a^{-3} \int u(x) \psi^* \left(\frac{x - x_0}{a} \right) dx,$$

and if the 'mother' wavelet is a Mexican hat, that is if

$$\psi(x) = \frac{d^2}{dx^2} e^{-x^2/2},$$

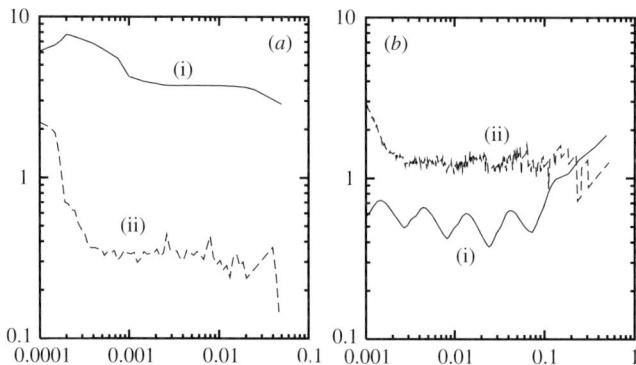

Fig. 9: Plots of (i) $r^{1-D_E^*} F(r)$ and (ii) $r^{D_E^*} M_E(r)$ as functions of r. (a) Case 8a (we set $D_E^* = 0.65$). (b) Devil's staircase with a scale ratio of 2000 (we set $D_E^* = 0.7$). $F(r) \sim r^{D_E^*-1}$ is verified for both cases, but, for the signal in figure 8b, $r^{1-D_E^*} F(r)$ exhibits strong oscillations due to the properties of lacunarity of this signal; note that in this case $r^{D_E^*} M_E(r)$ keeps its plateau shape on the large scales, whereas $r^{1-D_E^*} F(r)$ loses its precision on these scales.

then the zeros of $\tilde{u}(x_0, a)$ tend towards zeros of $(d^2/dx^2)u(x_0)$ as $a \to 0$. Zero-crossing curves of $\tilde{u}(x_0, a)$ in the x_0–a plane start at a point (x_0', a_{max}), and, as a is decreased towards $a = 0$, end at a point $(x_I, 0)$, where $(d^2/dx^2)u(x_I) = 0$. Hence, a characteristic length-scale a_{max} is assigned to every inflection point x_I, and it is therefore possible to define turbulent eddies that are positioned at x_I and that have a size and intensity related to a_{max} (see Kevlahan & Vassilicos 1994). The minimum number $M_E^*(r)$ of segments of size r needed to cover the inflection points x_I with $a_{max} \geq r$ (this sole condition differentiates $M_E^*(r)$ from $M_E(r)$) gives a measure of the number of eddies of size larger than r, and an eddy capacity D_E^* can be defined if the signal is self-similar, in which case

$$M_E^*(r) \sim r^{-D_E^*}. \tag{3.7}$$

Clearly, $M_E^*(r) \leq M_E(r)$. Furthermore, there can exist no inflection point x_I for which $a_{max} = 0$. This is because zero-crossings of $\tilde{u}(x_0, a)$ tend towards *all* zero-crossings of $(d^2/dx^2)u(x_0)$ as $a \to 0$, and, therefore, if an inflection point x_I existed for which $a_{max} = 0$, this inflection point could not be approached continuously by zeros of $\tilde{u}(x_0, a)$ as $a \to 0$. Hence, there exists a minimum value of a_{max}, which we call a_{min}, and which is different from zero. Noting that

$$M_E^*(a_{min}) = M_E(a_{min}),$$
$$M_E^*(r) = M_E^*(a_{min})(r/a_{min})^{-D_E^*},$$
$$M_E(r) = M_E(a_{min})(r/a_{min})^{-D_E},$$

we obtain

$$(r/a_{\min})^{-D_E^*} \le (r/a_{\min})^{-D_E}, \qquad (3.8)$$

for $r/a_{\min} \ge 1$, and we therefore conclude that

$$D_E^* \ge D_E. \qquad (3.9)$$

D_E^* is a refined definition of an eddy capacity, which, unlike D_E, takes into account the intensity of $|\Delta u|$ at inflection points x_I in such a way that

$$F(r) \sim r^{D_E^*-1}$$

is valid over a wider class of signals than

$$F(r) \sim r^{D_E-1}.$$

In particular, we find numerically that $D_E^* = D_E$ in all the cases where we find that $q = 1 - D_E$ (see table 1). For signals of type (3.6) (figure 8a) and for the Devil's staircase (figure 8b), $D_E^* > D_E$ and $q = 1 - D_E^*$, whereas $q \ne 1 - D_E$ (see figure 9).

In the remainder of this paper (including figures), the eddy capacity is invariably defined in terms of the more refined wavelet algorithm giving D_E^*, which is in fact a very practical tool for computing the eddy capacity, and we replace the notation D_E^* by D_E for the sake of simplicity.

(vi) *In the absence of noise, D_E is a better measure of intermittency than $F(r)$.*

Our conclusions (iv) and (v) can be summarized with the statement that $F(r) \sim r^{D_E-1}$ for a broad class of isolated and non-isolated intermittent topologies.

We now find that D_E is a better *measure* of intermittency than $F(r)$ in the absence of noise. We deal with the problem of noise in Section 4.

In all the model signals in which we find that $F(r) \sim r^{D_E-1}$, D_E appears to be a more sensitive measure of intermittency than $F(r)$ in two respects. Firstly, the measurement of D_E only requires a well-resolved, but relatively small, part of the signal, whereas $F(r)$ needs large amounts of data to converge. Secondly, $F(r)$ is more sensitive than D_E to the outer cut-off scale of the self-similar range and can exhibit oscillations (Smith *et al.* 1986). In all our model signals, D_E remains well defined and accurate even close to the outer cut-off scale, whereas $F(r)$ does not. This means that $F(r)$ needs a larger self-similar range of scales than D_E to ensure a good accuracy on the measurement of the power law. Typically, the ratio of the outer cut-off (sampling size) to the inner cut-off (resolution) needs to be of the order of 10–100 for an accurate measurement of D_E. Whereas, a ratio of 100 is not enough for the dependence of $F(r)$ on r to reach its asymptotic form. This is particularly clear in the comparison of figures 7b, 9b and 10, which display the measurement of D_E and $F(r)$ for the signals of figures 7a and 8b. These two signals are fractal constructions, and a fractal resolution is defined by the ratio of the outer cut-off to the

Table 1 Different eddy capacities and flatness factors of the signals in figures 1, 4, 5, 7a, 8 and 12(D_E is the Kolmogorov capacity of the zero-crossings of $(\mathrm{d}^2/\mathrm{d}x^2)u(x)$; D_E^* is computed using the refined definition (3.7) based on the wavelet transform (2.7).)

figure	type	D_E^*	D_E	$F(r)$
1a	non-isolated	1	1	$r^{D_E-1} \sim 3$
1b	non-isolated	1	1	$r^{D_E-1} \sim 2$
1c	non-isolated	1	1	$r^{D_E-1} \sim 3$
4a	non-isolated	0.63	0.63	$6 \neq r^{D_E-1}$
5c	non-isolated	0.6	0.5	$r^{1/6} \neq r^{D_E-1}$
3a	non-isolated	1	1	$r^{-0.27}$
5a	isolated	0.33	0.33	r^{D_E-1}
5b	isolated	0.50	0.50	r^{D_E-1}
4b	isolated	0.33	0.33	r^{D_E-1}
8a	isolated	0.65	0.5	$r^{D_E^*-1}$
7a	non-isolated	0.7	0.63	$r^{D_E^*-1}$
8b	non-isolated	0.7	0.7	r^{D_E-1}
12a	jet	0.92	—	$r^{-0.11}$
12b	wind-tunnel	0.95	—	$r^{-0.11}$

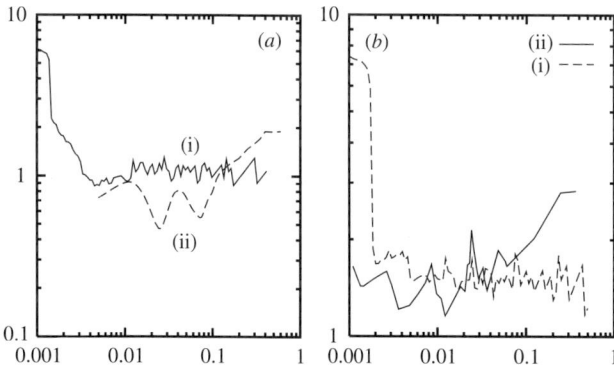

Fig. 10: Comparison of the accuracy of D_E and $F(r)$. (a) Devil's staircase with a ratio of scales of only 81 ($D_E = 0.7$). (b) Case 7a with a ratio of scales of only 2000 ($D_E = 0.7$). (i) $r^{D_E}M(r)$ keeps its plateau shape in both (a) and (b), whereas (ii) $r^{1-D_E}F(r)$ exhibits no plateau slope in (a) and does but on a very limited range of scales in (b).

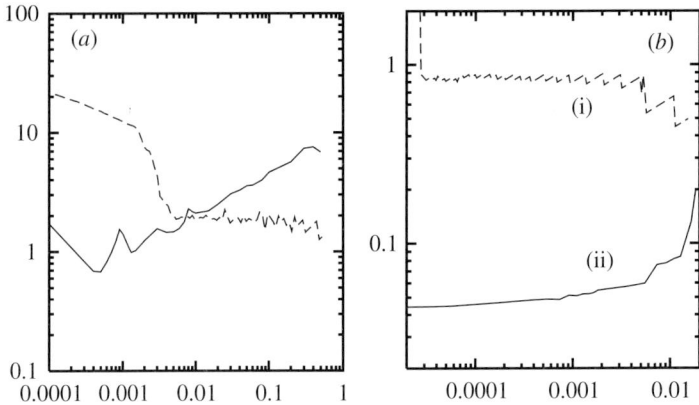

Fig. 11: Plots of $r^{D_E} M_E(r)$ (dashed line) and $r^{1-D_E} F(r)$ (solid line) as functions of r for the two signals in figure 4. (*a*) Cantor dust on–off function, $D_E = 0.63$. (*b*) On–off spiral signal, $D_E = \frac{1}{3}$; the plateau is observed, indicating that $F(r) \sim r^{D_E-1}$.

inner cut-off scales of the fractal process. The highest resolution is that of figures 7*b* and 9*b*, and the smallest that of figure 10 (see figure captions for details), and it turns out that D_E is accurately measured and well defined for both resolutions, whereas the r dependence of $F(r)$ is closer to its asymptotic form only in the cases of figures 7*b* and 9*b*, where the resolution is best.

Particularly striking are the cases of some fractal on–off signals where we know that $F(r) \sim r^{D_E-1}$ analytically (see figure 4*a*). The numerical results obtained for various resolutions indicate that $F(r)$ does not exhibit, in practice, its asymptotic form r^{D_E-1} unless the resolution is really extraordinary. Nevertheless, D_E is always well defined even for low resolutions of the fractal on–off structure, and is found, as expected, to be equal to D'_K, the Kolmogorov capacity of the set of points where the signal changes values. Of course, this class of intermittent signals is extreme in that these signals are equal to, say, zero nearly everywhere except on a fractal set (a Cantor set in the case of figure 4*a*). Nevertheless, D_E can capture the intermittency even in such extreme situations, whereas $F(r)$ cannot do so except if the resolution is enormous (see figure 11*a*).

Hence, in the absence of noise, as is the case of all our model signals, D_E requires smaller datasets and smaller self-similar ranges than $F(r)$ to be determined accurately. However, in the presence of even a small amount of noise, a direct measurement of D_E gives $D_E = 1$ irrespective of the underlying intermittent structure onto which the noise is superimposed (Kevlahan & Vassilicos 1994). In the following section, we show how to make D_E robust to the presence of noise, and we measure the eddy capacity D_E and flatness $F(r)$ of the two one-point turbulent velocity signals of figure 12.

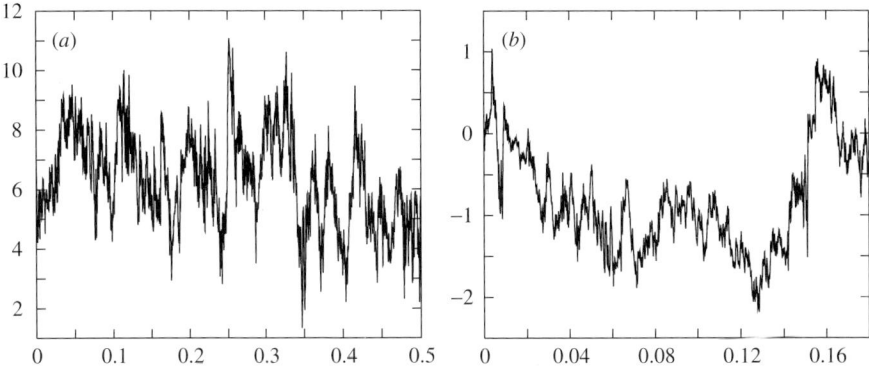

Fig. 12: Experimental one-point turbulent velocity signals from Modane. u against time. (*a*) Jet turbulent velocity. (*b*) Wind-tunnel turbulent velocity (see table 2.)

4 Application to high resolution turbulent velocity signals

Signals measured in practice always contain some noise strongly affecting their eddy-capacity measurement. To deal with this problem, we propose to replace the direct measure of the eddy capacity on the measured signal $u_{\mathrm{meas}}(x)$ by a measure on a 'distilled' signal $u_J(x)$ extracted from the measured signal. Our aim is to discard in this intermediary signal $u_J(x)$ the effect of noise and retrieve part of the structure of the underlying signal $u(x)$ (that is the ideal physical signal without noise). To construct this intermediary signal, we start by defining a series of increasing thresholds $U_0, U_1, U_2, \ldots, U_J$ evenly spaced between the minimum and maximum values (u_{min} and u_{max}) of the measured signal $u_{\mathrm{meas}}(x)$ ($U_j = u_{\mathrm{min}} + j[(u_{\mathrm{max}} - u_{\mathrm{min}})/J]$ for $j = 0, \ldots, J$). The points x_i, where $u_{\mathrm{meas}}(x_i)$ equals one of these thresholds, say $u_{\mathrm{meas}}(x_i) = U_j$, are given in increasing order ($x_i < x_{i+1}$) and so that $u_{\mathrm{meas}}(x_{i+1})$ equals either U_{j+1} or U_{j-1}. The intermediary signal is defined as follows:

$$u_J(x) = \sum_{i \leq J} u_{\mathrm{meas}}(x_i)(H(x - x_i) - H(x - x_{i+1})). \qquad (4.1)$$

Figure 13*a* shows this construction for the experimental signal of figure 12*b* with $J = 7$.

The measured signal $u_{\mathrm{meas}}(x)$ results from the superimposition of noise on $u(x)$. We may express this noise in the form $\epsilon(x)u(x)$, where $\epsilon(x)$ is a dimensionless random function of x, in which case

$$u_{\mathrm{meas}}(x) = u(x)(1 + \epsilon(x)). \qquad (4.2)$$

We can then compute the quantity $M_E(r)$ both for the measured signal $u_{\mathrm{meas}}(x)$ and for each intermediary signal $u_J(x)$, and from now on we restrict the notation $M_E(r)$ to the signal without noise $u(x)$ and use $M_E^{\mathrm{meas}}(r)$ and $M_E^J(r)$ for, respectively, the

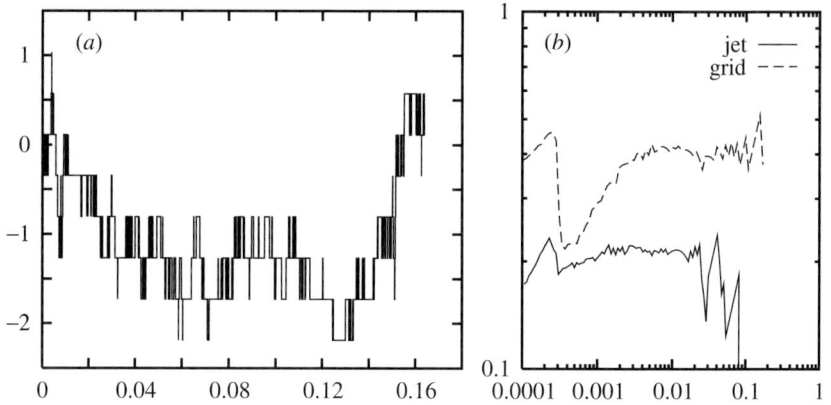

Fig. 13: (*a*) $u_J(x)$ in case 12*b* with $J = 8$, based on the 8192 first points that correspond to $0 < x < 0.114$. (*b*) $r^{D_E} M_E(r)$ measured on figure 12*a, b* with $J = 8$, as functions of r, $D_E = 0.95$.

measured signal $u_{\text{meas}}(x)$ and the intermediary signal $u_J(x)$. The construction of the intermediary signals introduces a new scale $r_{\text{min}}^J = \min(|x_{i+1} - x_1|)$, below which $u_J(x)$ contains no information about $u_{\text{meas}}(x)$. Hence, any measurement's noise below r_{min}^J is not in $u_J(x)$. The computation of $M_E^J(r)$ is based on the algorithm introduced in conclusion (v) of Section 3, for which only zero-crossing curves (in the x_0–a plane) of $\tilde{u}_J(x_0, a)$ with $a_{\text{max}} > r$ are counted in $M_E^J(r)$. Any zero-crossing curves of $\tilde{u}_{\text{meas}}(x_0, a)$ caused by noise and such that $a_{\text{max}} < r_{\text{min}}^J$, are, therefore, eliminated from the calculation of $M_E^J(r)$ for $r > r_{\text{min}}^J$. One may expect that the remaining zero-crossing curves that are taken into account in the calculation of $M_E^J(r)$ are those of $\tilde{u}_{\text{meas}}(x_0, a)$ with $a_{\text{max}} \geq r > r_{\text{min}}^J$, along with a few extra zero-crossing curves introduced by the discontinuous nature of the construction of $u_J(x)$ itself. But we conjecture that for sufficiently large r_{min}^J, that is sufficiently small J, the procedure by which we construct $u_J(x)$ effectively removes noise even at scales larger than r_{min}^J, so that the zero-crossing curves that are counted in the calculation of $M_E^J(r)$ are in fact those of $\tilde{u}(x_0, a)$ with $a_{\text{max}} \geq r > r_{\text{min}}^J$, along with the extra zero-crossing curves introduced by the construction of $u_J(x)$. On this basis we can write

$$M_E(r) \leq M_E^J(r) \leq M_{\text{meas}}(r),$$

for $r > r_{\text{min}}^J$, and we may conclude that where eddy capacities are well defined,

$$D_E \leq D_E^J \leq D_E^{\text{meas}},$$

with obvious notation. To validate the above conjecture, we verify the inequality

$$D_E \leq D_E^J \leq D_E^{\text{meas}},$$

on three different types of signals, each with a different singularity structure: fractal (figure 14*a*), spiral (figure 14*c*) and spirals on a fractal set (figure 14*e*). We also verify on these signals that where D_E^J is constant over a range of J, then $D_E = D_E^J$ for these values of J, and where D_E^J is not constant for a range of J, $D_E < D_E^J$. Finally, we also verify that the maximum value of J for which $D_E^J \leq D_E^{\text{meas}}$ may be expected to hold is determined by the requirement that $U_j \epsilon_{\max} < U_{j+1} - U_j$ for all j, which implies that

$$J < [(u_{\max} - u_{\min})/u_{\max}]_{\epsilon_{\max}}$$

(assuming that a maximum value $\epsilon_{\max}$ exists such that $\epsilon_{\max} \geq |\epsilon(x)|$). Of course, it is also necessary that $1 \leq J$ for $u_J(x)$ not to be trivial.

The left-hand plots of figure 14 show the signals with some added noise corresponding to $\epsilon_{\max} = 0.1$ (current experimental hot-wire probes have a typical accuracy of the order of 1%). Figure 14*a* corresponds to the case of figure 7*a*, figure 14*c* to a spiral accumulation with $D'_K = 0.5$, and figure 14*e* to a compounded signal obtained by placing spiral accumulations on a Cantor set. In the right-hand plots of figure 14, we report measurements of D_E^J, D_E^{meas} against D_E. Specifically, in figure 14*b, d, f*, we plot $r^{D_E} M_E(r)$, $r^{D_E} M_E^J(r)$ and $r^{D_E} M_E^{\text{meas}}(r)$. The lowest curves in these right-hand plots correspond to $r^{D_E} M_E(r)$ (no noise), the uppermost curves to $r^{D_E} M_E^{\text{meas}}(r)$, and intermediate curves with intermediate slopes to $r^{D_E} M_E^J(r)$. Note that in all the cases of figure 14, the value of $r^{D_E} M_E^J(r)$ is, at the largest scales, the same for all

$$J < [(u_{\max} - u_{\min})/(u_{\max}\epsilon_{\max})].$$

This observation supports the claim that the extra zero-crossings in the wavelet plane introduced by the very construction of $u_J(x)$ do not affect the r dependence of $M_E^J(r)$ for $r > r_{\min}^J$. In figure 14*b*, we plot $r^{D_E} M_E^J(r)$ for the signal in figure 14*a* for different values of J. In this case,

$$[(u_{\max} - u_{\min})/(u_{\max}\epsilon)] = 3.7,$$

and, indeed, a loss in precision is observed for $J > 3$. The signal without noise has an eddy capacity $D_E = 0.7$, whereas, due to the noise, the signal of figure 14*a* has an eddy capacity equal to 1. In figure 14*f*, we plot $r^{D_E} M_E^J(r)$ corresponding to figure 14*e* for different values of J. In this example,

$$[(u_{\max} - u_{\min})/(u_{\max}\epsilon_{\max})] = 20,$$

and we can draw the same conclusion as with the signal of figure 14*a*, though here, $D_E = 0.83$. For the signal in figure 14*c*, it turns out that D_E^J is not constant over an interval of J but we do observe, nevertheless, that $M_E(r) \leq M_E^J(r) \leq M_E^{\text{meas}}(r)$. In this case,

$$\frac{u_{\max} - u_{\min}}{u_{\max}\epsilon_{\max}} = 20,$$

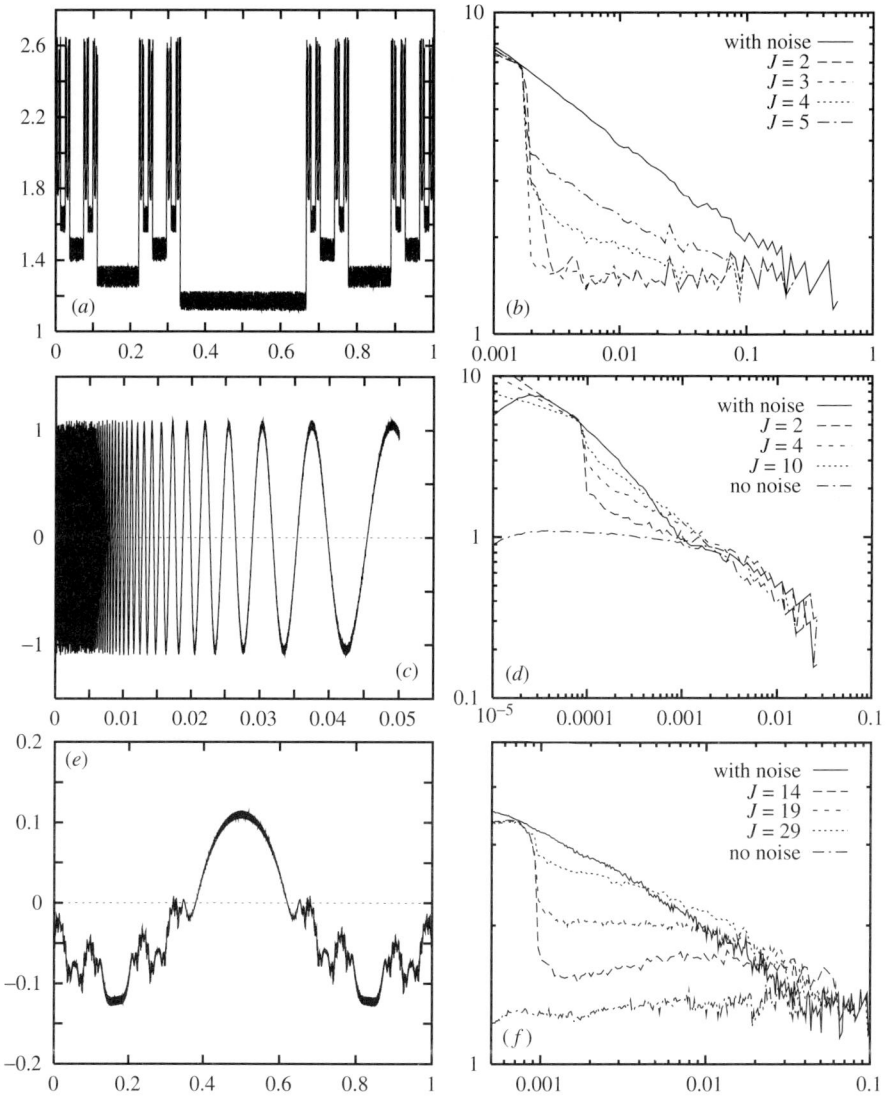

Fig. 14: Control of the effect of noise on D_E.

and $D_E = 0.3$. This analysis of these three topologically different intermittent signals with noise confirms our expectation that $D_E \leq D_E^J \leq D_E^{\text{meas}}$, and that our measurement of D_E^J on the basis of the intermediary signal $u^J(x)$ leads to significant improvement of the measurement of D_E by yielding an upper limit D_E^J less affected by the noise and smaller than D_E^{meas}.

Table 2 Main characteristics of the two experimental signals from Modane

figure	Re_λ	resolution	η (m)	η (s)	λ (m)	λ (s)	u' (ms^{-1})
12a	835	2η	1.40×10^{-4}	2.01×10^{-5}	7.8×10^{-3}	1.20×10^{-3}	1.65
12b	3050	1.2η	3.5×10^{-4}	1.75×10^{-5}	3.85×10^{-2}	1.93×10^{-3}	0.8

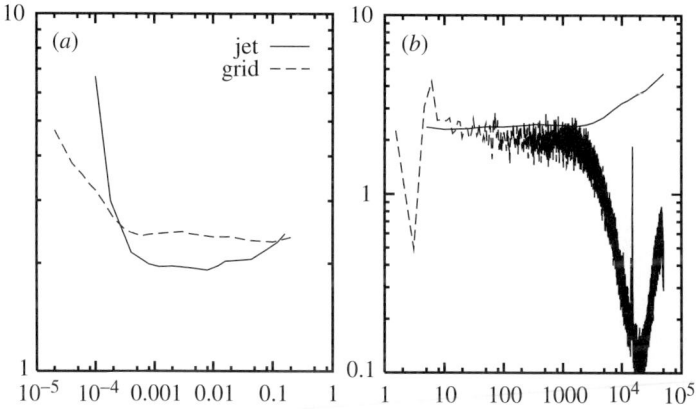

Fig. 15: (a) $r^{0.11}F(r)$ as a function of r for jet and grid turbulence. (b) $k^{-5/3}E(k)$ (dashed line) compared with $r^{0.11}F(r)$ as a function of $1/r$ for case 12b.

We now apply this method to two high-resolution one-point turbulent velocity signals obtained by Y. Gagne and his team in Grenoble (see Arneodo et al.1996). The first experimental signal was measured in a jet, its Reynolds number based on the Taylor microscale is $Re_\lambda = 835$ and its resolution is 2η, where η is the Kolmogorov length-scale. The second experimental signal was measured in a grid turbulence and has a higher Reynolds number ($Re_\lambda = 3050$) and a higher resolution (1.2η). These signals are referred to as 12a and 12b throughout this paper. Their main characteristics are given in table 2.

Figure 15a shows the curve $F(r)$ as a function of r for both signals. One can observe a range of scales where $F(r) \sim r^{-0.11}$. Figure 15b shows that $E(k) \sim k^{-5/3}$ ($E(k)$ is the energy spectrum and k the wavenumber) and $F(r) \sim r^{-0.11}$ are valid over the same range of scales. If an inertial range of scales exists, this range should be it, and we observe that it lies one or two decades above η for $Re_\lambda = 3050$.

We measure D_E^J for the two experimental signals in figure 12a, b, and assume that we can extend to these cases the result $D_E^J < D_E^{\text{meas}}$. Supporting this assumption, we find that the eddy capacity D_E^J is independent of J in the range $3 < J < 9$. The value of D_E^J is the same for all values of J between 4 and 8, but this value is defined over a range of scales that grows as J increases from 4 to 8. When $J \geq 9$, the noise is abruptly retrieved, and $D_E^J = 1 = D_E^{meas}$ over the entire range.

Figure 13*b* shows $r^{D_E^J} M_E^J(r)$ as a function of r in the cases of jet and grid turbulence; eddy capacities are clearly well defined over nearly two decades. The eddy capacity measured as a least-squares value over this range of scales is 0.94 (± 0.01) for both jet and grid turbulence. $J = 8$ gives the smallest scale down to which we can measure D_E^J, and figure 13*b* shows that this inner scale is 1.6λ in the present datasets. This method gives no information about the dissipation range ($r \ll \lambda$).

It is interesting to note that D_E^J and q are well defined over approximately the same range of length-scales, and that, because $q \approx 0.11$ and $D_E^J \approx 0.94$ for $3 < J < 9$, $q + D_E^J \approx 1.05$ and

$$q + D_E \leq 1.05$$

(to compare with (2.6)). Note also that D_E is clearly different from 1, and, therefore, neither the jet nor the grid turbulence can be modelled by signals such as that in figure 3, described in Benzi *et al.*(1993).

5   Conclusion

In this paper we have introduced a distinction between two topologically different types of intermittency: isolated and non-isolated.

Whereas the scalings of power spectra and second-order structure functions are influenced and sometimes even determined by the Kolmogorov capacity D_K' of the zero-crossings of the signal (Vassilicos & Hunt 1991; Hunt & Vassilicos 1991), the scalings of higher-order structure functions, and, in particular, that of the flatness factor, are often determined by the eddy capacity D_E of the signal, that is the Kolmogorov capacity of the zero-crossings of the second derivative of the signal. The capacities D_E and D_K' are, in general, different. *The eddy capacity is a direct measure of the geometry of intermittency.* Non-intermittent signals are such that $F(r) = const$ and $D_E = 1$, but the Kolmogorov capacity D_K' of their zero-crossings can take any value between 0 and 1; and if their power spectrum $E(k) \sim k^{-2p}$, then the power $2p$ is also insensitive to $D_E = 1$, but is influenced and, in fact, often determined by D_K' (the relationships between D_K' and $2p$ are discussed in many of the references cited at the end of this paper). The eddy capacity of intermittent signals can, however, determine the scaling of the flatness factor, and we find that

$$F(r) \sim r^{-q}, \quad \text{with } q + D_E = 1,$$

for a broad class of intermittent signals of both the isolated and the non-isolated topological types. The Kolmogorov capacity D_K' does not affect the scaling of the flatness factor except, of course, somehow indirectly, in the cases where $D_K' = D_E$. Examples where $D_K' = D_E$ are pictured in figure 4*a*, *b*, and examples where $D_K' \neq D_E$ are pictured in figures 1 and 7.

The results of our one-point turbulence data analysis indicate that inertial-range intermittency at high Reynolds numbers is such that $D_E = 0.94$ and $F(r) \sim r^{-0.11}$ both for grid and jet turbulence. Inertial range turbulence is, therefore, intermittent, albeit weakly so, and signals such as those of figure 1 are, therefore, not good models of turbulence fluctuations. Furthermore,

$$q + D_E \leq 1.05,$$

in the inertial range of high-Reynolds-number turbulence.

Acknowledgements

We are grateful to Yves Gagne for providing the turbulence datasets, and gratefully acknowledge financial support from the Royal Society and EPSRC grant GR/K50320.

Appendix A. D_E and D'_K

The eddy capacity D_E is a number between 0 and 1 and is a measure of how space filling eddies are in real space, whereas the Kolmogorov capacity D'_K of the zero-crossings of the signal $u(x)$ itself characterizes the distribution of energy among wavenumbers in Fourier space (Kevlahan & Vassilicos 1994). There is no general direct relation between D'_K and D_E. As can be seen from (2.4), D_E influences the scaling of moments of order $n > 2$. Instead, D'_K affects the scaling of *second-order* structure functions and power spectra. Indeed, the power spectrum $E(k)$ of self-similar signals $u(x)$ with a well-defined D'_K is given by (Vassilicos 1992),

$$E(k) \sim k^{-2+D'_K-2\sigma}, \tag{A1}$$

where σ is a scaling exponent characterizing local magnitudes of $u(x)$. For example, in the case of an on–off scalar field, such as may occur when the molecular diffusivity is very much smaller than the kinematic viscosity, $\sigma = 0$ (Vassilicos & Hunt 1991), and for signals $u(x) = x^s \sin x^{-t}$, $D'_K = (t/(t + 1))$ and (A1) is valid with $\sigma = s(1 - D'_K)$, provided that $-1 \leq 2s \leq t$ (see Kevlahan & Vassilicos 1994).

To summarize, the relevant geometrical scaling parameter for the description of intermittency is D_E and not D'_K; firstly, because $F(r) \sim r^{D_E-1}$, whereas D'_K affects only second- and not higher-order structure functions; and, secondly, because, as shown by Kevlahan & Vassilicos (1994), D_E is a measure of how space filling eddies are in real space, whereas D'_K is a measure of how evenly or unevenly the energy is distributed among wavenumbers in Fourier space.

6   References

Arneodo, A. (and 23 others) 1996 Structure functions in turbulence, in various flow configurations, at Reynolds number between 30 and 5000, using extended self-similarity. *Europhys. Lett.* **34**, 411–416.

Batchelor, G. K. 1953 *The theory of homogeneous turbulence*. Cambridge University Press.

Batchelor, G. K. 1967 *An introduction to fluid dynamics*. Cambridge University Press.

Benzi, R., Biferale, L., Crisanti, A., Paladin, G., Vergassola, M. & Vulpiani, A. 1993 A random process for the construction of multiaffine fields. *Physica* D **65**, 352–358.

Falconer, K. 1990 *Fractal geometry mathematical foundations and applications*. Wiley.

Frisch, U. 1995 *Turbulence*. Cambridge University Press.

Hunt, J. C. R. & Vassilicos, J. C. 1991 Kolmogorov's contributions to the physical and geometrical understanding of small-scale turbulence and recent developments. *Proc. R. Soc. Lond.* A **434**, 183–210.

Kevlahan, N. K.-R. & Vassilicos, J. C. 1994 The space and scale dependencies of the self-similar structure of turbulence. *Proc. R. Soc. Lond.* A **447**, 341–363.

Lundgren 1982 Strained spiral vortex model for turbulent fine structures. *Phys. Fluids* **25**, 2193–2203.

Smith, L. A., Fournier, J.-D. & Spiegel, E. A. 1986 Lacunarity and intermittency in fluid turbulence. *Phys. Lett.* A **114**, 465–468.

Vassilicos, J. C. 1992 The multi-spiral model of turbulence and intermittency. In *Topological aspects of the dynamics of fluids and plasmas*, pp. 427–442. Kluwer.

Vassilicos, J. C. & Hunt, J. C. R. 1991 Fractal dimensions and spectra of interfaces with application to turbulence. *Proc. R. Soc. Lond.* A **435**, 505–534.

4
Wavelets in statistics: beyond the standard assumptions

Bernard W. Silverman

Department of Mathematics, University of Bristol,
University Walk, Bristol BS8 1TW, UK

Abstract

The original application of wavelets in statistics was to the estimation of a curve given observations of the curve plus white noise at 2^J regularly spaced points. The rationale for the use of wavelet methods in this context is reviewed briefly. Various extensions of the standard statistical methodology are discussed. These include curve estimation in the presence of correlated and non-stationary noise, the estimation of (0–1) functions, the handling of irregularly spaced data and data with heavy-tailed noise, and deformable templates in image and shape analysis. Important tools are a Bayesian approach, where a suitable prior is placed on the wavelet expansion, encapsulating the notion that most of the wavelet coefficients are zero; the use of the non-decimated, or translation-invariant, wavelet transform; and a fast algorithm for finding all the within-level covariances within the table of wavelet coefficients of a sequence with arbitrary band-limited covariance structure. Practical applications drawn from neurophysiology, meteorology and palaeopathology are presented. Finally, some directions for possible future research are outlined.

Keywords: Bayesian non-parametric modelling; deformable templates; meteorology; neurophysiology; non-parametric regression; non-stationarity

1 Introduction

1.1 The standard assumptions

The early papers on the use of wavelets in statistics concentrated on the standard non-parametric regression problem of estimating a function g from observations

$$Y_i = g(t_i) + \epsilon_i, \quad i = 1, \ldots, n, \qquad (1.1)$$

where $n = 2^J$ for some J, $t_i = i/n$ and ϵ_i are independent identically distributed normal variables with zero mean and variance σ^2. The basic method used was the discrete wavelet transform (DWT); for convenience of reference, the notation we shall use is set out in the appendix. The DWT of a sequence x will be written $\mathcal{W}x$.

In the problem (1.1), let (d_{jk}) be the DWT of the sequence $g(t_i)$. Consider the DWT $(\eta_{jk}) = \mathcal{W}Y$ of the observed data. Since $\mathcal{W}$ is an orthogonal transform, we will have

$$\eta_{jk} = d_{jk} + \epsilon_{jk}, \tag{1.2}$$

where the ϵ_{jk} are, again, independent identically distributed normal random variables with zero mean.

On the face of it, the structure of (1.2) is the same as that of (1.1), and the DWT has not done anything to help us. However, the wavelet transform has the property that large classes of functions g likely to crop up in practice have *economical* wavelet expansions, in the sense that g is well approximated by a function, most of whose wavelet coefficients are zero. These do not just include functions that are smooth in a conventional sense, but also those that have discontinuities of value or of gradient, and those with varying frequency behaviour. This is another manifestation of the ability of wavelets to handle intermittent behaviour, and is demonstrated by mathematical results such as those discussed by Donoho *et al.*(1995).

The notion that most of the d_{jk} are zero, or may be taken to be zero, gives intuitive justification for a *thresholding rule*; if $|\eta_{jk}| \leq \tau$ for some threshold τ, then we set $\hat{d}_{jk} = 0$, on the understanding that η_{jk} is pure noise. For larger $|\eta_{jk}|$, the estimate is either η_{jk} itself, or a value shrunk towards zero in some way that depends only on the value of $|\eta_{jk}|$. Several papers (Donoho & Johnstone 1994, 1995, 1998, and references therein) show that estimators of this kind have excellent asymptotic adaptivity properties, and an example of the kind of result that can be derived is given in Section 2 below. However, for finite sample sizes, the basic recipe of thresholding the DWT can be improved and extended.

1.2   Prior information and modelling uncertainty

Before moving away from standard assumptions, we discuss a way in which the ideas behind thresholding can be developed further. The properties of wavelet expansions make it natural to model the unknown function g by placing a prior distribution on its wavelet coefficients. We focus on one particular prior model and posterior estimate (for alternative approaches, see, for example, Chipman *et al.*(1997) and Clyde *et al.*(1998)).

The approach we consider in detail is that of Abramovich *et al.*(1998), who consider a Bayesian formulation within which the wavelet coefficients are independent with

$$d_{jk} \sim (1 - \pi_j)\delta_0 + \pi_j N(0, \tau_j^2), \tag{1.3}$$

a mixture of an atom of probability at zero and a normal distribution. The mixing probability π_j and the variance of the non-zero part of the distribution are allowed to depend on the level j of the coefficient in the transform. Different choices of these hyperparameters correspond to different behaviours of the functions drawn from the prior, and, in principle, these properties can be used to choose the properties of the functions.

In practice, it is often more straightforward to have an automatic choice of the hyperparameters, and this is provided by Johnstone & Silverman (1998) who use a marginal maximum likelihood formulation. Under the prior (1.3), the marginal distribution of the wavelet coefficients η_{jk} is a mixture of a $N(0, \sigma^2)$ and a $N(0, \sigma^2 + \tau_j^2)$. The likelihood of the hyperparameters can then be maximized, most conveniently using an EM algorithm.

The posterior distribution of the individual wavelet coefficients is a mixture of an atom at zero and a general normal distribution. The traditional summary of the posterior distribution is the posterior mean, but, in this case, the posterior *median* has attractive properties. Abramovich *et al.*(1998) show that it yields a thresholding rule, in that the posterior median of d_{jk} is only non-zero if the absolute value of the corresponding coefficient η_{jk} of the data exceeds some threshold. Generally, the posterior median will have a sparse wavelet expansion, and this is in accordance with the construction of the prior. Also, the posterior median corresponds to the minimum of an L^1 loss, which is more appropriate for discontinuous and irregular functions than the squared error loss that leads to the posterior mean.

2   Correlated and non-stationary noise

One important move away from the standard assumptions is to consider data that have correlated noise. This issue was discussed in detail by Johnstone & Silverman (1997). For many smoothing methods correlated noise can present difficulties, but, in the wavelet case, the extension is straightforward.

2.1   Level-dependent thresholding

Provided the noise process is stationary, one effect of correlated noise is to yield an array of wavelet coefficients with variances that depend on the level j of the transform. This leads naturally to *level-dependent* thresholding, using for each coefficient a threshold that is proportional to its standard deviation. The variances are constant within levels. Therefore, at least at the higher levels, it is possible to estimate the standard deviation separately at each level, implicitly estimating both the standard deviation of the noise and the relevant aspects of its correlation structure. The usual estimator, in the wavelet context, of the noise standard deviation is

$$\text{(median of } |\eta_{jk}| \text{ on level } j)/0.6745, \tag{2.1}$$

where the constant 0.6745 is the upper quartile of the standard normal distribution. The motivation for the estimate (2.1) is again based on the properties of wavelets: in the wavelet domain, we can assume that the signal is sparse and so only the upper few $|\eta_{jk}|$ will contain signal as well as noise.

2.2   Adaptivity results

Johnstone & Silverman (1997) derive theoretical justification for the idea of using a method that uses thresholds proportional to standard deviations. Suppose that X has an n-dimensional multivariate normal distribution with mean vector θ and general variance matrix V, with $V_{ii} = \sigma_i$. Let $\hat{\theta}$ be a suitable estimator obtained by thresholding the X_i one by one, using thresholds proportional to standard deviations. Under very mild conditions, the mean square error of $\hat{\theta}$ is within a factor of $(1 + 2 \log n)$ of an ideal but unattainable estimator, where the optimal choice for each θ is made of 'keeping' or 'killing' each X_i in constructing the estimate; furthermore, no other estimator based on the data can improve, in order of magnitude, on this behaviour.

In our setting, the vector X consists of the wavelet coefficients of the data. Risk calculations for the 'ideal' estimator show that, for both short- and long-range-dependent noise, the level-dependent thresholding method applied in wavelet regression gives optimally adaptive behaviour. For a wide range of smoothness classes of functions g, the estimate's behaviour is close to that of the best possible estimator for each particular smoothness class. The smoothness classes include those that allow for intermittently irregular behaviour of the function; thus, the level-dependent thresholded wavelet estimator automatically adjusts to the regularity (or intermittent irregularity) of the unknown function being estimated.

3   The non-decimated wavelet transform

3.1   The transform and the average basis reconstruction

An important development in the statistical context has been the routine use of the non-decimated wavelet transform (NDWT), also called the stationary or translation-invariant wavelet transform (see, for example, Shensa 1992; Nason & Silverman 1995; Lang *et al.* 1996; Walden & Contreras Cristan 1998). Conceptually, the NDWT is obtained by modifying the Mallat DWT algorithm as described in the appendix: at each level, as illustrated in figure 7, no decimation takes place, but instead the filters $\mathcal{H}$ and $\mathcal{G}$ are repeatedly padded out with alternate zeros to double their length. The effect (roughly speaking, depending on boundary conditions) is to yield an overdetermined transform with n coefficients at each of $\log_2 n$ levels. The transform contains the standard DWT for every possible choice of time origin.

Since the NDWT is no longer (1–1), it does not have a unique inverse, but the DWT algorithm is easily modified to yield the *average basis* inverse (Coifman & Donoho 1995), which gives the average of the DWT reconstructions over all choices of time origin. Both the NDWT and the average basis reconstruction are $O(n \log n)$ algorithms.

3.2 Using the NDWT for curve estimation

In virtually every case that has been investigated in detail, the performance of wavelet regression methods is improved by the use of the NDWT. The standard paradigm is to perform the NDWT, and then to process each level of the resulting coefficients using the particular method of choice, regarding the coefficients as independent. Then the average basis reconstruction method is used to give the curve estimate. This procedure corresponds to carrying out a separate smoothing for each choice of time origin and then averaging the results, but the time and storage used only increase by an $O(\log n)$ factor.

In the Bayesian case, it should be noted that the prior model for the wavelet expansion will not be exactly consistent between different positions of the origin. If a random function has independent wavelet coefficients with a mixture distribution like (1.3) with respect to one position of the origin, then it will not, in general, have such a representation if the origin is changed. The prior distributions cannot be generated from a single underlying prior model for the curve, and so, strictly speaking, the method corresponds to a separate modelling of the prior information at each position of the origin.

Johnstone & Silverman (1998) investigate the use of the NDWT in conjunction with the marginal maximum likelihood approach discussed above. Even in the case where the original noise is independent and identically distributed, the n coefficients of the NDWT of the data at each level are not actually independent. In the case of correlated noise, even the standard wavelet coefficients are not, in general, actually independent. However, good results are obtained by proceeding *as if* the NDWT coefficients are independent, maximizing an as-if-independent likelihood function to find the mixture hyperparameters at each level, and then using an average basis reconstruction of the individual posterior medians of the wavelet coefficients. In a simulation study the improvement over the fixed basis marginal maximum likelihood method is substantial, typically around 40% in mean square error terms.

3.3 A neurophysiological example

An important problem in neurophysiology is the measurement of the very small currents that pass through the single membrane channels that control movement in and out of cells. A key statistical issue is the reconstruction of a (0–1) signal from very noisy, and correlated, data. The two levels correspond to periods within which the membrane channel is closed or open.

In connection with their encouragement (Eisenberg 1994) of the application of modern signal processing techniques to this problem, Eisenberg and Levis have supplied a generated dataset intended to represent most of the relevant challenges in such single-channel data. This generated example differs in kind from the usual kind of simulated data, in that its underlying model is carefully selected by practitioners directly involved in routine collection and analysis of real data. The reason for using a generated dataset rather than an actual dataset obtained in practice is that in the case of a 'real' dataset, the 'truth' is not known, and so it is impossible to quantify

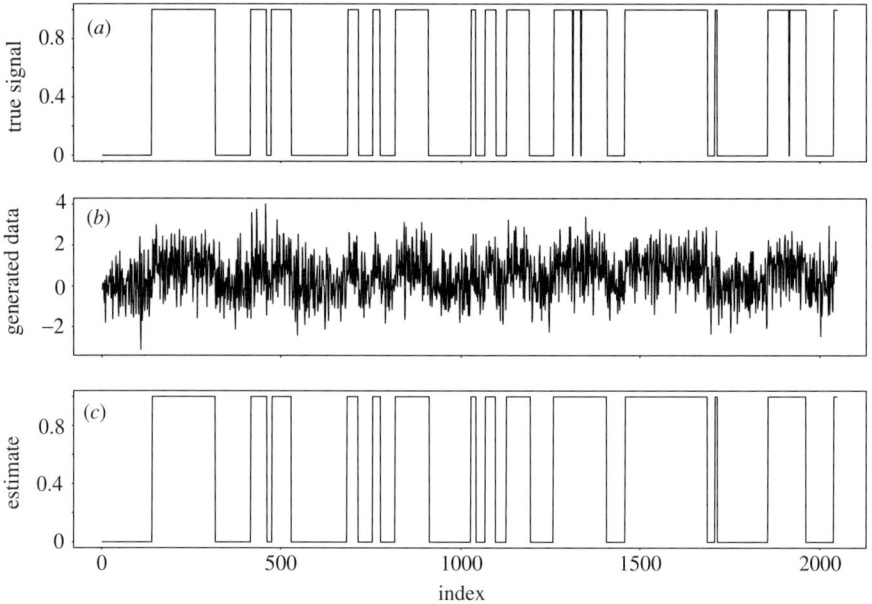

Fig. 1: (*a*) The 'true' ion channel signal for time points 1–2048. (*b*) The corresponding section of generated data (on a smaller vertical scale). (*c*) The estimate obtained by the translation-invariant marginal maximum likelihood method. Reproduced from Silverman (1998) with the permission of the author.

the performance of any particular method.

The top two panels of figure 1 illustrate the data we have to deal with. The top graph is the generated 'true' (0–1) signal, which is used to judge the quality of the reconstruction. Of course, the processing is done without any reference to this information. The middle panel shows the actual data. The vertical scale is much smaller than the top panel; the range of the displayed data is −3.1 to 4, and the signal to noise ratio is about $\frac{1}{3}$.

The bottom panel is obtained by using the translation-invariant marginal maximum likelihood Bayesian procedure, as described in Section 3 *b* above, and then rounding off the result to the nearest of 0 and 1. Only the three highest levels of the wavelet transform are processed, and the rest are passed straight through to the rounding stage. The way in which this works out is illustrated in figure 2. The output from the wavelet smoothing step is still fairly noisy, but its variance is sufficiently small that the rounding step gives very good results.

It should be pointed out that rounding the original data gives very poor results; about 28.6% of the points are misclassified, almost as bad an error rate as setting every value to zero, which would misclassify 34.3% of points. The method illustrated in the figures, on the other hand, achieves an error rate of only 2%; for fuller details of

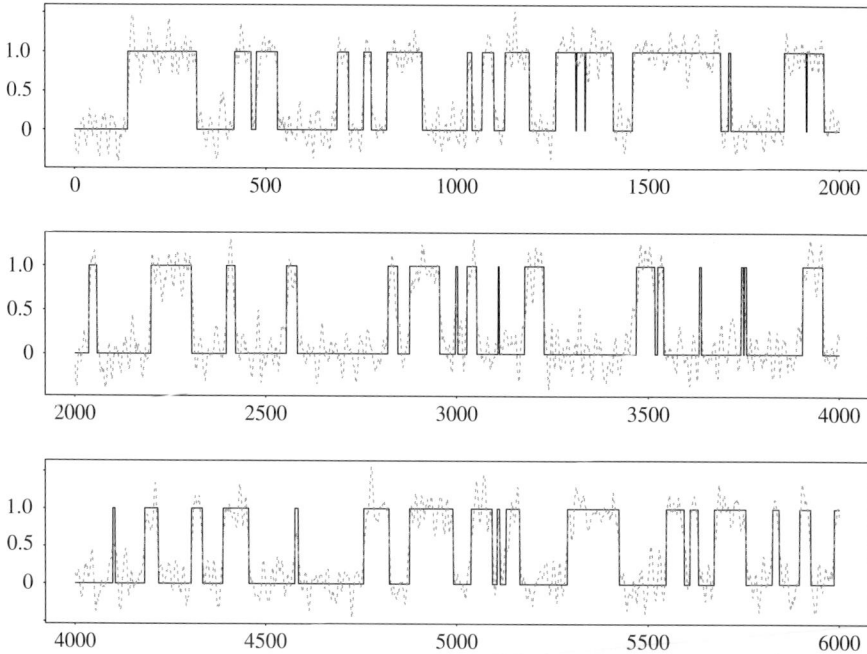

Fig. 2: The true ion channel signal, and the curve obtained by applying the translation-invariant marginal maximum likelihood approach to the top three levels only. Rounding the curve off to the nearest integer gives an excellent estimate of the original signal. Reproduced from Johnstone & Silverman (1998) with the permission of the authors.

numerical comparisons see Silverman (1998). This performance is as good as that of the special purpose method designed by Eisenberg and Levis. Close examination of figure 1 shows that the wavelet method faithfully recovers the true pattern of openings and closings, except for three very short closings, each of which is only of length two time points. The special-purpose method also misses these closings. The way in which a general purpose wavelet method can match a special purpose method that cannot be expected to work well in other contexts is particularly encouraging.

4   Dealing with irregular data

4.1   Coefficient-dependent thresholding

So far we have considered the case of stationary correlated data, but the results of Johnstone & Silverman (1997) give motivation for the treatment of much more general correlation structures. Their theory gives support to the thresholding of every coefficient in the wavelet array using a threshold proportional to its standard deviation. If the original data are heteroscedastic, or have non-stationary covariance structure, then these standard deviations will not necessarily even be constant at each level of the transform.

A heteroscedastic error structure can arise in the data in several ways. For example, it may be known that the covariance structure is of a given non-stationary form, but this will not often apply in practice. Kovac & Silverman (1998) consider other possibilities. Often, the original observations are on an irregular grid of points. One can then interpolate or average locally to obtain regular gridded data in order to use a wavelet approach. Even if the noise in the original data is independent and identically distributed, the irregularity of the data grid will lead to data that are, in general, heteroscedastic and correlated in a non-stationary way.

Another context in which heteroscedastic error structures occur is in robust versions of standard regression. Even if we start with a regular grid, downweighting or eliminating outlying observations will lead to a heteroscedastic error structure or to data on an irregular grid, and this will be discussed in Section 4 c. A final possibility considered by Kovac & Silverman (1998) is that of a number of data points that is not a power of 2. Though other methods are available, a possible approach is to interpolate to a grid of length 2^m for some m and then, as in the case of irregular data, the resulting error structure will be heteroscedastic.

4.2   Finding the variances of the wavelet coefficients

In all these cases, it is important to find the variances of the wavelet coefficients of the data given the covariance matrix Σ of the original data. Kovac & Silverman (1998) provide an algorithm that yields all the variances and within-level covariances of the wavelet coefficients. If the original Σ is band limited, which it will be in the cases described above, then the algorithm will take only $O(n)$ operations, for $n = 2^J$ data points.

Using the notation for the DWT set out in the appendix, let Σ^j denote the variance matrix of c^j and $\tilde{\Sigma}^j$ that of d^j. Then $\Sigma^J = \Sigma$ by definition. From the recursive definition of the c^j and d^j it follows that, for each $j = J - 1, J - 2, \ldots, 0$,

$$\Sigma^j = H^{j+1} \Sigma^{j+1} (H^{j+1})^{\mathrm{T}} \tag{4.1}$$

and

$$\tilde{\Sigma}^j = G^{j+1} \Sigma^{j+1} (G^{j+1})^{\mathrm{T}}. \tag{4.2}$$

Note that this gives us not only the variances

$$\sigma_{jk} = \tilde{\Sigma}^j_{k,k}$$

of the individual wavelet coefficients d^j_k, but also the covariance structure of the wavelet coefficients at each level. The sparsity structure of the matrices H_{j+1} and G_{j+1} allows the calculations (4.1) and (4.2) to be carried out in $O(n_j)$ operations.

Hence, the complexity of the entire algorithm, deriving the variance matrices for all j, is $O(2^J)$, and is also linear in the length of the wavelet filters and in the bandwidth of the original variance matrix (see Kovac & Silverman (1998) for details).

4.3 Robust estimation

Standard wavelet regression methods do not work well in the presence of noise with outliers or with a heavy-tailed distribution. This is almost inevitable, because the methods allow for abrupt changes in the signal, and so outlying observations are likely to be interpreted as indicating such abrupt changes. Bruce *et al.*(1994) and Donoho & Yu (1998) have suggested approaches based on nonlinear multi-resolution analyses, but the algorithm of Section 4*b* allows for a simple method based on classical robustness methods. The method works whether or not the original data are equally spaced. As a first step, a wavelet estimator is obtained using the method of Section 4*a*, using a fairly high threshold. Outliers are then detected by a running median method. These are removed and the method of Section 4*a* applied again to produce the final estimate (see Kovac & Silverman (1998) for a detailed discussion).

An example is given in figure 3. The data are taken from a weather balloon and describe the radiation of the sun. The high-frequency phenomena in the estimated signal are due to outlier patches in the data; these may be caused by a rope that sometimes obscured the measuring device from the direct sunlight. It is interesting to note that the robust estimator removes the spurious high-frequency effects, but still models the abrupt change in slope in the curve near time 0.8.

5 Deformable templates

5.1 Images collected in palaeopathology

We now turn to a rather different area of application, that of deformable templates in image and shape analysis. There are many problems nowadays where an observed image can be modelled as a deformed version of a standard image, or template. The assumption is that the image is a realization of the template, perhaps with additional variability that is also of interest. My own interest in this issue stems from a study of skeletons temporarily excavated from a cemetery in Humberside. Of particular interest to the palaeopathology group in the Rheumatology Department at Bristol University is the information that can be gained about patterns of osteoarthritis of the knee. Shepstone *et al.*(1999) discuss the collection and analysis of a considerable number of images of the kind shown in figure 4, using the experimental set-up shown in figure 5. Further details of the work described in this section are given by Downie *et al.*(1998).

The important features of these bones as far as the osteoarthritis study is concerned are the shape of the bone and the occurrence and position of various changes, notably eburnation (polished bone, caused by loss of cartilage) and osteophytes (bony outgrowths). The images are digitized as pixel images and are marked up

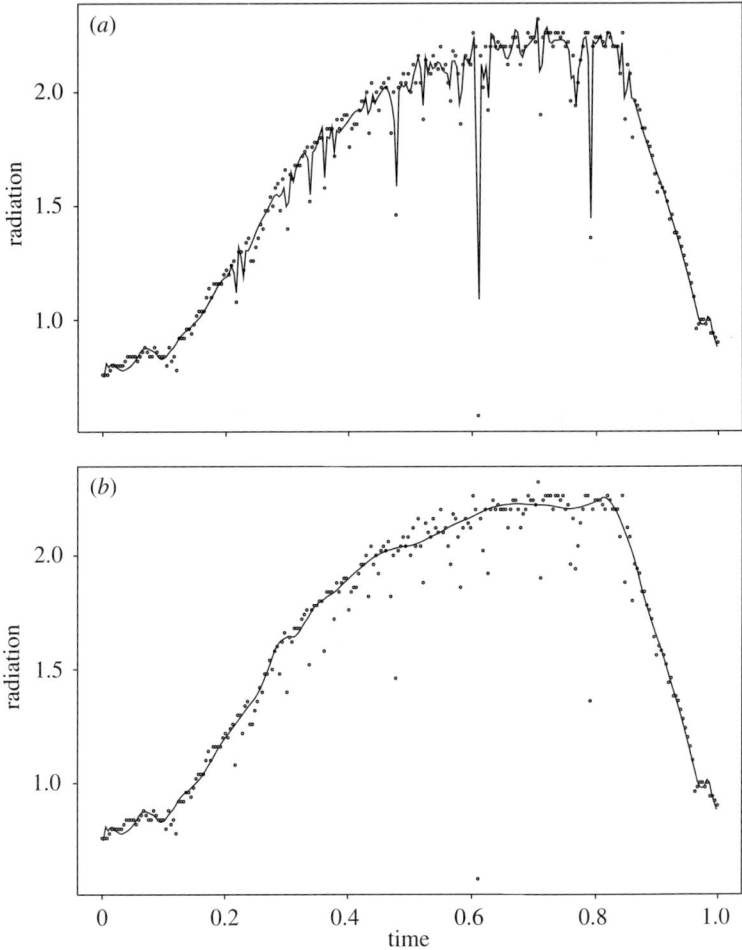

Fig. 3: (*a*) Thresholding without outlier removing: balloon data with a standard wavelet regression estimator. (*b*) Thresholding after outlier removing: balloon data with a robust wavelet estimator. From Kovac & Silverman (1998) with the permission of the authors.

by comparison with the original bone to label the pixels corresponding to the areas of these changes. The aim of any study of the deformations is twofold: firstly, to give a standard mapping to relate positions on various bones; and secondly, to gain information about the shape of individual bones. For the first purpose, we are interested only in the *effect* of the deformation, but, for the second, the details of the deformation itself are important.

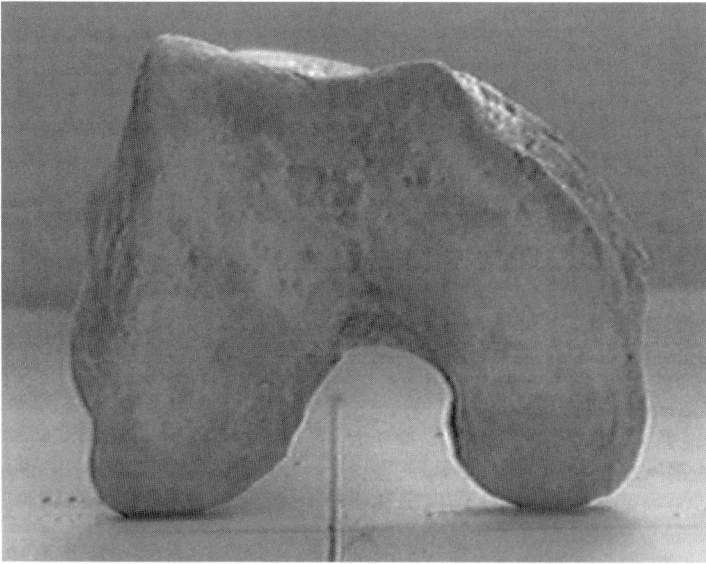

Fig. 4: A typical image of the femoral condyles from the distal end of the femur. Reproduced from Downie *et al.*(1998) with the permission of the authors.

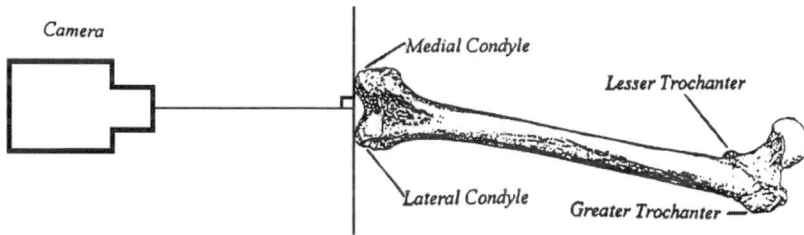

Fig. 5: The experimental set-up for collecting the femur image data. Reproduced from Downie *et al.*(1998) with the permission of the authors.

5.2   Models for deformations

Deformations can be modelled as follows. Let I and T be functions on the unit square $\mathcal{U}$, representing the image and the template, respectively. In our particular application, they will be 0–1 functions. The deformation is defined by a two-dimensional *deformation function* f such that, for u in $\mathcal{U}$, $u + f(u)$ is also in $\mathcal{U}$. The aim is then to get a good fit of the image $I(u)$ to the deformed template $T(u + f(u))$ measuring discrepancy by summed squared difference over the pixels in the image.

The deformation f is a vector of two functions (f_x, f_y) on the unit square, giving the coordinates of the deformation in the x and y directions. In our work, we expand each of them as a two-dimensional wavelet series. Because it is reasonable to assume that deformations will have localized features, this may be more appropriate than the Fourier expansion used, for example, by Amit *et al.*(1991). In two dimensions the wavelet multi-resolution analysis of an array of values (Mallat 1989*b*) yields coefficients w_κ, where the index $\kappa = (j, k_1, k_2, \ell)$; this coefficient gives information about the array near position (k_1, k_2) on scale j. Three orthogonal aspects of local behaviour are modelled, indexed by ℓ in $\{1, 2, 3\}$, corresponding to horizontal, vertical and diagonal orientation.

To model the notion that the deformation has an economical wavelet expansion, a mixture prior of the kind described in Section 1 *b* was used. Because the assumption of normal identically distributed errors is not realistic, we prefer to consider our method as being a penalized least-squares method with a Bayesian motivation, rather than a formal Bayesian approach. A particular bone unaffected by any pathology was arbitrarily designated as the template. An iterated coordinatewise maximization is used to maximize the posterior likelihood.

5.3   Gaining information from the wavelet model

Figure 6 demonstrates the information available in the wavelet expansion of a particular deformation. Only 27 of the possible 2048 coefficients are non-zero, indicating the extreme economy of representation of the deformation. For each of these coefficients, a number equal to the level j of the coefficient is plotted at the position (k_1, k_2). The size at which the number is plotted gives the absolute size of the coefficient; the orientation ℓ is indicated by colours invisible on this copy, but available on the World Wide Web version of Downie *et al.*(1998) at www.statistics.bristol.ac.uk/~bernard.

The figure shows that most of the non-zero coefficients are near the outline of the image, because of the localization properties of the wavelets. At the top of the image in the y component, coefficients at all resolution levels are present, indicating the presence of both broad-scale and detailed warping effects. The deformation is dominated by two coefficients, showing that the main effects are a fairly fine-scale effect at the middle of the top of the image, and a larger-scale deformation centred in the interior of the image. The full implications of this type of plot remain a subject for future research; in some contexts, the coefficients and their values will be candidates for subsequent statistical analysis, while elsewhere they will be valuable for the insight they give into the position and scale of important aspects of the deformation.

6   Discussion

Although there has been a recent explosion of interest in the use of wavelets in statistics, there are very many open questions and issues, both theoretical and prac-

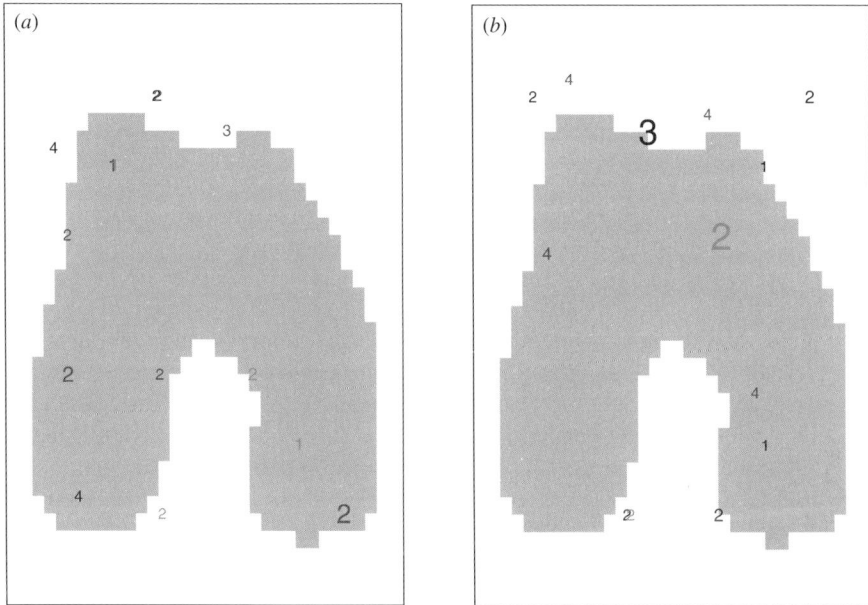

Fig. 6: The wavelet coefficient positions and sizes of a typical deformation. (*a*) *x* wavelet coefficients (i.e. showing the *x*-coordinate of the deformation); (*b*) *y* wavelet coefficients (i.e. showing the *y*-coordinate of the deformation). The numbers denote the scale of the particular coefficient, and are plotted at the centre of the support of the corresponding wavelet. The printed size of each number indicates the absolute size of the wavelet coefficient. Adapted from Downie *et al.*(1998) with the permission of the authors.

tical, to be addressed before the full value of wavelets in statistics is understood and appreciated. Some other issues, not discussed here, will of course be raised in the other papers in this issue.

Even for the standard regression problem, there is still considerable progress to be made in determining the appropriate way to process the wavelet coefficients of the data to obtain the estimate. In this paper, attention has been focused on methods that treat coefficients at least as if they were independent. However, it is intuitively clear that if one coefficient in the wavelet array is non-zero, then it is more likely (in some appropriate sense) that neighbouring coefficients will be also. One way of incorporating this notion is by some form of *block thresholding*, where coefficients are considered in neighbouring blocks (sec, for example, Hall *et al.*1998; Cai & Silverman 1998). An obvious question for future consideration is how to integrate the ideas of block thresholding and related methods within the range of models and methods considered in this paper.

It is clear that a much wider class of Bayesian approaches will be useful. There are two related directions to proceed in. As in most statistical contexts where Bayesian methods are used, careful thought about priors within the DWT and NDWT contexts is needed in order to genuinely model prior knowledge of unknown functions. More particularly, the NDWT frees the modelling from the choice of origin, but one might wish to go further and move away from the powers of two in the scale domain. The *atomic decomposition* models discussed, for example, by Abramovich *et al.*(2000), may be a good starting point here.

The present paper has discussed the NDWT, but there are many extensions and generalizations of the basic DWT idea using other bases and function dictionaries to express the functions of interest. In a statistical context, some consideration has been given to the use of multiple wavelets (Downie & Silverman 1998) and ridgelets and other functions (Candès & Donoho, this issue). There is much scope for gaining a clear understanding of the contexts in which more general function dictionaries, and developments such as wavelets for irregular grids (see, for example, Daubechies *et al.*, this issue) will be of statistical use. In the more traditional regression context (see, for example, Green & Silverman 1994), semi-parametric methods, which use a combination of classical linear methods and non-parametric regression, are often useful. Similarly, in the wavelet context, there may well be scope for the use of a combination of ideas from wavelets and from other regression methods, to give hybrid approaches that may combine the advantages of both.

Until now, most of the work on wavelets in statistics has concentrated on the standard regression problem. There has been some work on statistical inverse problems (see, for example, Abramovich & Silverman 1998) and on time-series analysis (see, for example, Walden & Contreras Cristan 1998; Nason & von Sachs, this issue). However, it is important to extend the use of wavelets to a much wider range of statistical problems. One of the major advances in statistics in recent decades has been the development and routine use of generalized linear models (see, for example, McCullagh & Nelder 1989). There has been a considerable amount of work in the application of penalized likelihood regression methods to deal non-parametrically or semi-parametrically with generalized linear model dependence (see, for example, Green & Silverman 1994, ch. 5 and 6), and it is natural to ask whether wavelet methods can make a useful contribution. One common ingredient of generalized linear model methods is iterated reweighted least squares (see, for example, Green 1984), and, in the wavelet context, any implementation of an iterated least-squares method will require algorithms like that discussed in Section 4 *b* above.

Above all, the greatest need is for advances in theoretical understanding to go hand-in-hand with widespread practical application. The wide interest in wavelets demonstrated by the papers in this issue indicates that wavelets are not just an esoteric mathematical notion, but have very widespread importance in and beyond science and engineering. Of course, they are not a panacea, but as yet we have only made a small step in the process of understanding their true statistical potential.

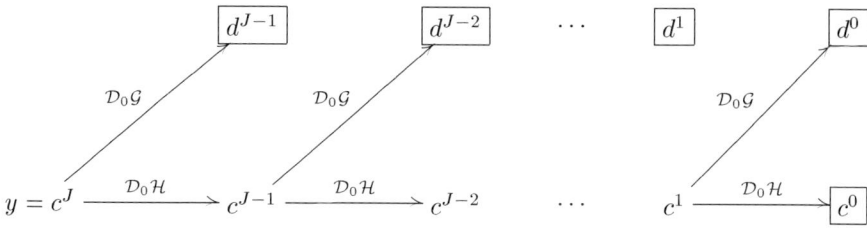

Fig. 7: The discrete wavelet transformation of a sequence y.

7 Acknowledgements

The author gratefully acknowledges the financial support of the Engineering and Physical Sciences Research Council (UK) and the National Science Foundation (USA). The comments of Graeme Ambler, Guy Nason and Theofanis Sapatinas on previous versions were greatly appreciated.

Appendix A. The discrete wavelet transform

In order to define notation, it is useful to review the standard discrete wavelet transform algorithm of Mallat (1989a, b). For further details see any standard text on wavelets, such as Daubechies (1992) or Chui (1992). The transform is defined by linear high- and low-pass filters $\mathcal{G}$ and $\mathcal{H}$, specified by coefficient sequences (g_k) and (h_k), respectively. For any sequence x, we have, for example:

$$(\mathcal{G}x)_k = \sum_i g_{i-k} x_i.$$

The coefficient sequences satisfy $g_k = (-1)^k h_{1-k}$, and have finite (and usually short) support. Denote by $\mathcal{D}_0$ the 'binary decimation' operator that chooses every even member of a sequence, so that $(\mathcal{D}_0 x)_j = x_{2j}$.

The discrete wavelet transform (DWT) of a sequence y of length 2^J will then be carried out as in the schema shown in figure 7. The vectors c^j and d^j are the *smooth* and the *detail* at level j, and are of length n_j, where $n_j \approx 2^j$, depending on the treatment of boundary conditions. We denote by H_j and G_j the $n_{j-1} \times n_j$ matrices, such that $c^{j-1} = H_j c^j$ and $d^{j-1} = G_j c^j$. The data vector y is decomposed into vectors of wavelet coefficients $d^{J-1}, d^{J-2}, \ldots, d^0, c^0$, also written as an array d_{jk} or $\mathcal{W}y$.

8 References

Abramovich, F. & Silverman, B. W. 1998 Wavelet decomposition approaches to statistical inverse problems. *Biometrika* **85**, 115–129.

Abramovich, F., Sapatinas, T. & Silverman, B. W. 1998 Wavelet thresholding via a Bayesian approach. *J. R. Statist. Soc.* B **60**, 725–749.

Abramovich, F., Sapatinas, T. & Silverman, B. W. 2000 Stochastic expansion in an overcomplete wavelet dictionary. *Prob. Theory Related Fields*. (In the press.)

Amit, Y., Grenander, U. & Piccioni, M. 1991 Structural image restoration through deformable templates. *J. Am. Statist. Ass.* **86**, 376–387.

Bruce, A., Donoho, D. L., Gao, H.-Y. & Martin, R. 1994 Smoothing and robust wavelet analysis. In *Proc. CompStat*, Vienna, Austria.

Cai, T. T. & Silverman, B. W. 1998 Incorporating information on neighboring coefficients into wavelet estimation. Technical Report, Department of Mathematics, University of Bristol.

Chipman, H. A., Kolaczyk, E. D. & McCulloch, R. E. 1997 Adaptive Bayesian wavelet shrinkage. *J. Am. Statist. Ass.* **92**, 1413–1421.

Chui, C. K. 1992 *An introduction to wavelets*. Academic.

Clyde, M., Parmigiani, G. & Vidakovic, B. 1998 Multiple shrinkage and subset selection in wavelets. *Biometrika* **85**, 391–402.

Coifman, R. R. & Donoho, D. L. 1995 Translation-invariant denoising. In *Wavelets and statistics* (ed. A. Antoniadis & G. Oppenheim). Springer Lecture Notes in Statistics, no. 103, pp. 125–150.

Daubechies, I. 1992 *Ten lectures on wavelets*. Philadelphia, PA: SIAM.

Donoho, D. L. & Johnstone, I. M. 1994 Ideal spatial adaptation via wavelet shrinkage. *Biometrika* **81**, 425–455.

Donoho, D. L. & Johnstone, I. M. 1995 Adapting to unknown smoothness by wavelet shrinkage. *J. Am. Statist. Ass.* **90**, 1200–1224.

Donoho, D. L. & Johnstone, I. M. 1998 Minimax estimation via wavelet shrinkage. *Ann. Statist.* **26**, 879–921.

Donoho, D. L. & Yu, T. P. Y. 1998 Nonlinear 'wavelet transforms' based on median-thresholding. Technical Report, Department of Statistics, Stanford University.

Donoho, D. L., Johnstone, I. M., Kerkyacharian, G. & Picard, D. 1995 Wavelet shrinkage: asymptopia? (with discussion). *J. R. Statist. Soc.* B **57**, 301–369.

Downie, T. R. & Silverman, B. W. 1998 The discrete multiple wavelet transform and thresholding methods. *IEEE Trans. Sig. Proc.* **46**, 2558–2561.

Downie, T. R., Shepstone, L. & Silverman, B. W. 1998 Economical representation of image deformation functions using a wavelet mixture model. Technical Report, Department of Mathematics, University of Bristol.

Eisenberg, R. 1994 Biological signals that need detection: currents through single membrane channels. In *Proc. 16th Ann. Int. Conf. IEEE Engineering in Medicine and Biology Society* (ed. J. Norman & F. Sheppard), pp. 32a–33a.

Green, P. J. 1984 Iterated reweighted least squares for maximum likelihood estimation, and some robust and resistant alternatives (with discussion). *J. R. Statist. Soc.* B **46**, 149–192.

Green, P. J. & Silverman, B. W. 1994 *Nonparametric regression and generalized linear models: a roughness penalty approach*. London: Chapman & Hall.

Hall, P., Kerkyacharian, G. & Picard, D. 1998 Block threshold rules for curve estimation using kernel and wavelet methods. *Ann. Statist.* **26**, 922–942.

Johnstone, I. M. & Silverman, B. W. 1997 Wavelet threshold estimators for data with correlated noise. *J. R. Statist. Soc.* B **59**, 319–351.

Johnstone, I. M. & Silverman, B. W. 1998 Empirical Bayes approaches to wavelet regression. Technical Report, Department of Mathematics, University of Bristol.

Kovac, A. & Silverman, B. W. 1998 Extending the scope of wavelet regression methods by coefficient-dependent thresholding. Technical Report, Department of Mathematics, University of Bristol.

Lang, M., Guo, H., Odegard, J. E., Burrus, C. S. & Wells, R. O. 1996 Noise reduction using an undecimated discrete wavelet transform. *IEEE Sig. Proc. Lett.* **3**, 10–12.

Mallat, S. G. 1989*a* Multiresolution approximations and wavelet orthonormal bases of $L^2(R)$. *Trans. Am. Math. Soc.* **315**, 69–89.

Mallat, S. G. 1989*b* A theory for multiresolution signal decomposition: the wavelet representation. *IEEE Trans. Patt. Analysis Mach. Intell.* **11**, 674–693.

McCullagh, P. & Nelder, J. A. 1989 *Generalized linear models*, 2nd edn. London: Chapman & Hall.

Nason, G. P. & Silverman, B. W. 1995 The stationary wavelet transform and some statistical applications. In *Wavelets and statistics* (ed. A. Antoniadis & G. Oppenheim). Springer Lecture Notes in Statistics, no. 103, pp. 281–300.

Shensa, M. J. 1992 The discrete wavelet transform: wedding the à trous and Mallat algorithms. *IEEE Trans. Sig. Proc.* **40**, 2462–2482.

Shepstone, L., Rogers, J., Kirwan, J. & Silverman, B. 1999 The shape of the distal femur: a palaeopathological comparison of eburnated and non-eburnated femora. *Ann. Rheum. Dis.* **58**, 72–78.

Silverman, B. W. 1998 Wavelets in statistics: some recent developments. In *CompStat: Proc. in Computational Statistics 1998* (ed. R. Payne & P. Green), pp. 15–26. Heidelberg: Physica.

Walden, A. T. & Contreras Cristan, A. 1998 Matching pursuit by undecimated discrete wavelet transform for non-stationary time series of arbitrary length. *Statistics Computing* **8**, 205–219.

5

Wavelets and the theory of non-parametric function estimation

Iain M. Johnstone

Department of Statistics, Sequoia Hall, Stanford University, Stanford, CA 94305, USA

Abstract

Non-parametric function estimation aims to estimate or recover or denoise a function of interest, perhaps a signal, spectrum or image, that is observed in noise and possibly indirectly after some transformation, as in deconvolution. 'Non-parametric' signifies that no *a priori* limit is placed on the number of unknown parameters used to model the signal. Such theories of estimation are necessarily quite different from traditional statistical models with a small number of parameters specified in advance.

Before wavelets, the theory was dominated by linear estimators, and the exploitation of assumed smoothness in the unknown function to describe optimal methods. Wavelets provide a set of tools that make it natural to assert, in plausible theoretical models, that the *sparsity* of representation is a more basic notion than smoothness, and that nonlinear thresholding can be a powerful competitor to traditional linear methods. We survey some of this story, showing how sparsity emerges from an optimality analysis via the game-theoretic notion of a least-favourable distribution.

Keywords: minimax; Pinsker's theorem; sparsity; statistical decision problem; thresholding; unconditional basis

1 Introduction

Within statistics, the first applications of wavelets were to theory. While the potential for statistical application to a variety of problems was also apparent, and is now being realized (as surveyed in other articles in this issue), it was in the theory of non-parametric function estimation that progress was initially fastest. The goal of this article is to survey some of this work retrospectively.

Why should developments in the theory of statistics interest a wider scientific community? Primarily, perhaps, because theory attempts to isolate concepts of broad generality that clarify in what circumstances and under what assumptions particular data analytic methods can be expected to perform well, or not. As a classical example, the most widely used statistical tools—regression, hypothesis tests, confidence intervals—are typically associated with *parametric* models, that is, probability models for observed data that depend on, at most, a (small) finite number of unknown parameters. The true scope, versatility and applicability of these tools was clarified by the development of underlying theoretical notions such as likelihood, sufficiency, unbiasedness, Cramér–Rao bounds, power, and so forth. Many of these concepts have passed into the general toolkit of scientific data analysis.

A further key point is that theory promotes *portability* of methods between scientific domains: thus, Fisher's analysis of variance was initially developed for agricultural field trials, but he also created theoretical support with such effect that most uses of analysis of variance now have nothing to do with agriculture.

What is meant by the term non-parametric function estimation? The advent of larger, and often instrumentally acquired, datasets and a desire for more flexible models has stimulated the study of *non-parametric* models, in which there is no *a priori* bound on the number of parameters used to describe the observed data. For example, when fitting a curve to time-varying data, instead of an *a priori* restriction to, say, a cubic polynomial, one might allow polynomials of arbitrarily high degree or, more stably, a linear combination of splines of local support.

Despite prolific theoretical study of such infinite-dimensional models in recent decades, the conclusions have not dispersed as widely as those for the parametric theory. While this is partly because non-parametric theory is more recent, it is certainly also partly due to the greater nuance and complexity of its results, and a relative paucity of unifying principles.

The arrival of wavelet bases has improved the situation. Wavelets and related notions have highlighted *sparsity* of representation as an important principle in estimation and testing. Through the *dyadic Gaussian sequence model*, they have bridged parametric and non-parametric statistics and re-invigorated the study of estimation for multivariate Gaussian distributions of finite (but large) dimension. Wavelets have also been the vehicle for an influx of ideas—unconditional bases, fast algorithms, new function spaces—from computational harmonic analysis into statistics, a trend that seems likely to continue to grow in future.

2 A simple model for sparsity

We begin with an apparently naive discussion of sparsity in a 'monoresolution' model. Suppose that we observe an n-dimensional data vector y consisting of an unknown signal θ, which we wish to estimate, contaminated by additive Gaussian white noise of scale σ_n. If the model is represented in terms of its coefficients in a particular orthonormal basis B, we obtain (y_k^B), (θ_k^B), etc., though the dependence on B will usually be suppressed. Thus, in terms of basis coefficients,

$$y_k = \theta_k + \sigma_n z_k, \quad k = 1, \ldots, n, \tag{2.1}$$

and $\{z_k\}$ are independently and identically distributed $N(0, 1)$ random variables. Here, we emphasize that $\theta = (\theta_k)$ is, in general, regarded as fixed and unknown. This model might be reasonable, for example, if we were viewing data as Fourier coefficients, and looking in a particular frequency band where the signal and noise spectrum are each about constant.

If, in addition, it is assumed that $\{\theta_k\}$ are random, being drawn from a Gaussian distribution with $\mathrm{Var}(\theta_k) = \tau_n^2$, then the optimal (Wiener) filter, or estimator, would involve *linear shrinkage* by a constant *linear* factor:

$$\hat{\theta}_k = \frac{\rho}{\rho+1} y_k, \qquad \rho = \frac{\tau_n^2}{\sigma_n^2}. \tag{2.2}$$

The ratio τ_n^2/σ_n^2 (or some function of it) is usually called the *signal-to-noise* ratio.

The two key features of this traditional analysis are

(a) the Gaussian prior distribution leads to linear estimates as optimal; and

(b) the linear shrinkage is invariant to orthogonal changes of coordinates: thus, the same Wiener filter is optimal, regardless of the basis chosen.

Sparsity. In contrast, sparsity has everything to do with the choice of bases. Informally, 'sparsity' conveys the idea that most of the signal strength is concentrated in a few of the coefficients. Thus, a 'spike' signal $\gamma(1, 0, \ldots, 0)$ is much sparser than a 'comb' vector $\gamma(n^{-1/2}, \ldots, n^{-1/2})$, even though both have the same energy, or ℓ_2 norm: indeed these could be representations of the same vector in two different bases. In contrast, noise, almost by definition, is not sparse in any basis, and among representations of signals in various bases, it is the ones that are sparse that will be most easily 'denoised'.

Remark 2.1 Of course, in general terms, sparsity is a familiar notion in statistics and beyond: think of parsimonious model choice, 'Occam's razor', and so forth. It is the motivation for the principal components analysis of Hotelling (1933), suitable for high-dimensional, approximately Gaussian data. However, in the specific domain of non-parametric function estimation, prior to the advent of wavelets, the role of sparsity was perhaps somewhat obscured by the focus on the related, although somewhat more special, notion of smoothness.

Figure 1 shows part of a real signal represented in two different bases: figure 1a is a subset of 2^7 wavelet coefficients θ^W, while figure 1b shows a subset of 2^7 Fourier coefficients θ^F. Evidently, θ^W has a much sparser representation than does θ^F.

The sparsity of the coefficients in a given basis may be quantified using ℓ_p norms,

$$\|\theta\|_p = \left(\sum_1^n |\theta_k|^p \right)^{1/p},$$

for $p < 2$, with smaller p giving more stringent measures. Thus, while the ℓ_2 norms of our two representations are roughly equal,

$$\|\theta^F\|_2 = 25.3 \approx 23.1 = \|\theta^W\|_2,$$

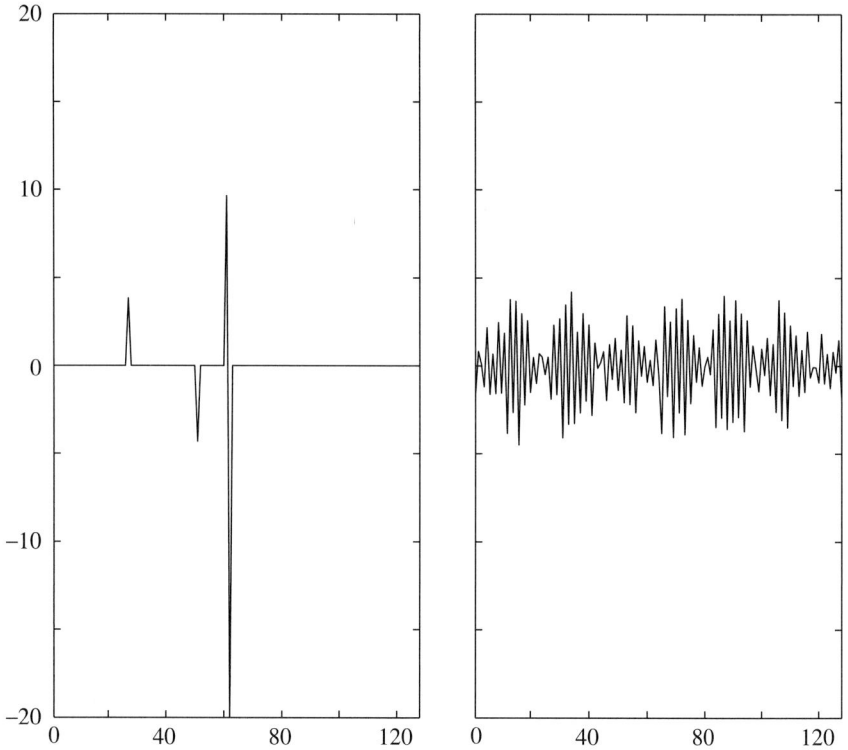

Fig. 1: (*a*) θ_k^W = level 7 of estimated NMR reconstruction g of figure 4, while in (*b*) θ_k^F = Fourier coefficients of g at frequencies $65, \ldots, 128$, both real and imaginary parts shown. While these do not represent exactly the same projections of f, the two overlap and $\|\theta^F\|_2 = 25.3 \approx 23.1 = \|\theta^W\|_2$.

the ℓ_1 norms differ by a factor of 6.5:

$$\|\theta^F\|_1 = 246.5 \gg 37.9 = \|\theta^W\|_1.$$

Figure 2 shows that the sets,

$$\left\{ \theta : \sum_1^n |\theta_k|^p \leq C^p \right\},$$

become progressively smaller and more clustered around the coordinate axes as p decreases. Thus, the only way for a signal in an ℓ_p ball to have large energy (i.e. ℓ_2 norm) is for it to consist of a few large components, as opposed to many small components of roughly equal magnitude. Put another way, among all signals with a given energy, the sparse ones are precisely those with small ℓ_p norm.

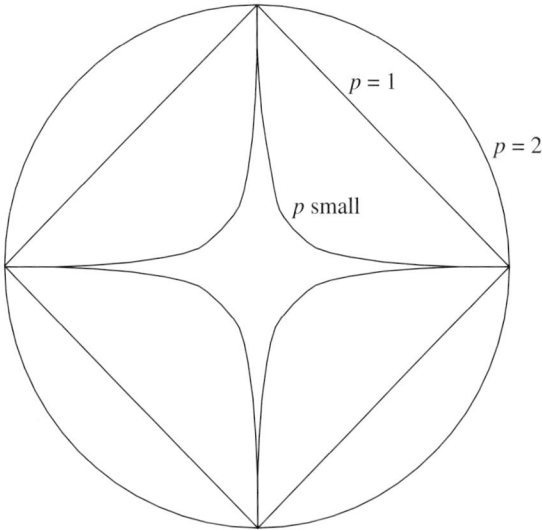

Fig. 2: Contours of ℓ_p balls.

Thus, we will use sets $\{\|\theta\|_p \leq C\}$ as models for *a priori* constraints that the signal θ has a sparse representation in the given basis. Assume, for simplicity here, that $\sigma_n = 1/\sqrt{n}$ and that $p = C = 1$: it is thus supposed that

$$\sum_1^n |\theta_k| \leq 1.$$

Other situations can be handled by developing the theory for general (p, C_n, σ_n) (see Donoho & Johnstone 1994*b*). How to exploit this sparsity information in order to better estimate θ: in other words, can we estimate θ^W better than θ^F?

Figure 3 shows an idealized case in which all θ_k are zero except for two spikes, each of size $1/2$. Two extreme examples of linear estimators are $\hat{\theta}_1(y) \equiv y$, which leaves the data unadjusted, and $\hat{\theta}_0(y) \equiv 0$, which sets every coordinate to zero. The first, a pure 'variance' estimator, has MSE $= \sigma_n^2 = 1/n$ in each of the n coordinates, for a total MSE $= 1$. The second, $\hat{\theta}_0$, a pure 'bias' estimator, is exactly correct on all but the two spikes, where it suffers a total MSE $= 2 \cdot (1/2)^2 = 1/2$. Given the symmetry of the prior knowledge and the statistical independence of the observations, the only other plausible choices for a linear estimator have the form cy, for a constant c, $0 \leq c \leq 1$. It can be shown that such estimators are effectively a combination of the two extremes, and, in particular, do not have noticeably better MSE performance.

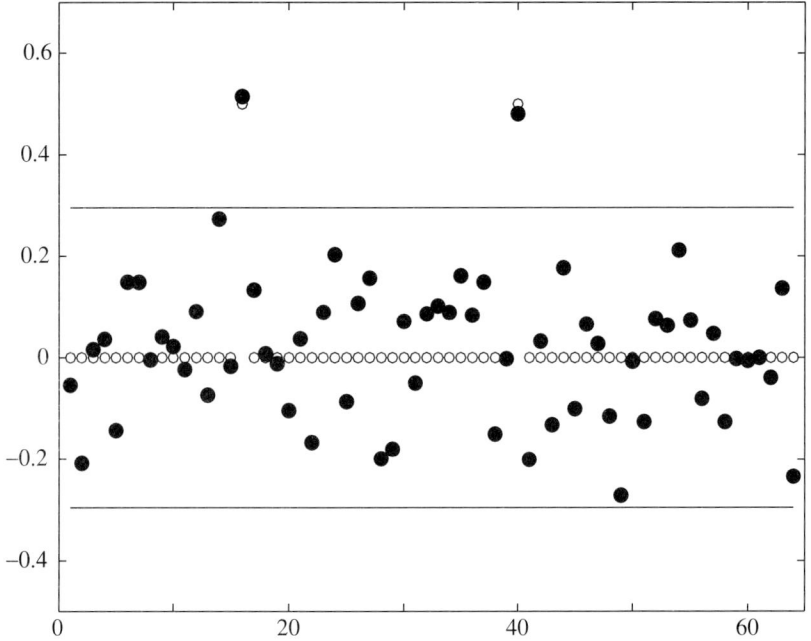

Fig. 3: Visualization of model (2.1): open circles are unknown values θ_k, solid circles are observed data y_k, $k = 1, \ldots, n = 64$. Horizontal lines are thresholds at $\lambda_n = \sqrt{\log n / n}$.

In the situation of figure 3, thresholding is natural. Define the *hard threshold* estimator by its action on coordinates,

$$\hat{\theta}_{\lambda,k}(y) = \begin{cases} y_k, & \text{if } |y_k| \geq \lambda \sigma_n, \\ 0, & \text{otherwise,} \end{cases} \tag{2.3}$$

and figure 3 shows a threshold of

$$\lambda_n = \sigma_n \sqrt{\log n} = \sqrt{\log n / n}.$$

For the particular configuration of true means θ_k shown there, the data from the two spikes pass the threshold unchanged, and as such are essentially unbiased estimators. Meanwhile, in all other coordinates, the threshold correctly sets all data to zero except for the small fraction of noise that exceeds the threshold. Thus, it can be directly verified that

$$\text{MSE}(\hat{\theta}_\lambda, \theta) \approx 2\sigma_n^2 + n\sigma_n^2 E\{Z^2, Z^2 > \log n\} \approx 2n^{-1} + 2\sqrt{\log n / n},$$

where Z is a standard Gaussian variate. This mean squared error is of course much better than for any of the linear estimators.

Statistical games and the minimax theorem. The skeptic will object that the configuration of figure 3 was chosen to highlight the advantages of thresholding, and indeed it was! It is precisely to avoid such reasoning from constructed cases that the tools of game theory have been adapted for use in statistics. A sterner and fairer test of an estimator is obtained by creating a statistical two-person zero-sum game, or *statistical decision problem.*

(i) Player I ('the scientist') is allowed to choose any estimator $\hat{\theta}(y)$, linear, threshold or of more complicated type.

(ii) Player II ('nature') may choose $\theta \in \mathbb{R}^n$ *at random*, and may choose a probability distribution π for θ subject only to the sparsity constraint that $E_\pi \|\theta\|_1 \leq 1$.

(iii) The pay-off is calculated as the expected mean squared error of $\hat{\theta}(Y)$ when θ is chosen according to π, and then Y satisfies model (2.1): $Y = \theta + \sigma_n z$ for $z \sim N_n(0, I)$. Thus, the pay-off now averages over *both* θ and Y:

$$r(\hat{\theta}, \pi) = E_\pi E_{Y|\theta} \|\hat{\theta}(Y) - \theta\|_2^2.$$

Of course, the scientist tries to minimize the pay-off and nature tries to maximize it.

Classical work in statistical decision theory (Wald 1950; Le Cam 1986; see also Johnstone 1998) shows that the minimax theorem of von Neumann can be adapted to apply here, and that the game has a well-defined value, the *minimax risk*:

$$R_n = \inf_{\hat{\theta}} \sup_{\pi} r(\hat{\theta}, \pi) = \sup_{\pi} \inf_{\hat{\theta}} r(\hat{\theta}, \pi). \tag{2.4}$$

An estimator attaining the left-hand infimum in (2.4) is called a *minimax* strategy or *estimator* for player I, while a prior distribution π attaining the right-hand supremum is called *least favourable* and is an optimal strategy for player II. It is the *structure* of these optimal strategies, and their effect on the minimax risk R_n, that is of chief statistical interest.

While these optimal strategies cannot be exactly evaluated for finite n, informative asymptotic approximations are available (Donoho & Johnstone 1994*b*), with the consequence that under our unit norm sparsity constraint,

$$R_n \sim \sqrt{\log n / n},$$

as $n \to \infty$. Indeed, an approximately least-favourable distribution is given by drawing the individual coordinates θ_k independently from a *two-point* distribution with

$$\theta_k = \begin{cases} \sigma_n \sqrt{\log n}, & \text{with probability } \varepsilon_n \doteq 1/\sqrt{n \log n}, \\ 0, & \text{otherwise.} \end{cases} \tag{2.5}$$

This amounts to repeated tossing of a coin highly biased towards zero. Thus, in n draws, we expect to see a relatively small number, namely $\sqrt{n/\log n}$ of non-zero components. The size of these non-zero values is such that they are hard to distinguish from the larger values among the more numerous remaining $n - \sqrt{n/\log n}$ noise observations. Of course, what makes this distribution difficult for player II is that the *locations* of the non-zero components are random as well.

An approximately minimax estimator for this setting is given by the hard thresholding rule described earlier, but with threshold λ_n slightly larger than $\sqrt{\log n}$: for example, $\lambda_n = \sqrt{\log(n \log n)}$ will do. It can also be verified that no linear estimator can achieve a pay-off of better than $1/2$ if nature chooses a suitably uncooperative probability distribution for θ.

It is perhaps the qualitative features of this solution that most deserve comment. Had we worked with simply a signal-to-noise constraint—$E_\pi \|\theta\|_2^2 \leq 1$, say—we would have obtained a Gaussian prior distribution as being approximately least favourable, and the linear Wiener filter (2.2) with $\sigma_n^2 = \tau_n^2 = 1/n$ as an approximately minimax estimator. The imposition of a sparsity constraint takes us far away from Gaussian priors and linear estimators.

Sparsity and improved MSE.   There is an alternative way to show how sparsity of representation affects the mean squared error of estimation using thresholding. Return to model (2.1), and observe that the MSE of $\hat{\theta}_1(y_i) = y_i$ for estimating θ_i is σ_n^2, while the MSE of $\hat{\theta}_0(y_i) = 0$ is θ_i^2. Given a choice, an omniscient 'oracle' would choose the estimator that yields the smaller of the two MSEs. Repeating this for each coordinate leads to a notion of *ideal risk*:

$$\mathcal{R}(\theta, \sigma) = \sum_i \min(\theta_i^2, \sigma^2). \qquad (2.6)$$

Suppose that the coefficients are rearranged in decreasing order:

$$\theta_1^2 \geq \theta_2^2 \geq \cdots \geq \theta_n^2.$$

The notion of 'compressibility' captures the idea that the number of large coefficients, $N_\sigma(\theta) = \#\{\theta_i : |\theta_i| \geq \sigma\}$ is small, and also that there is little energy in the tail sums

$$c_k^2 = \sum_{i>k} \theta_i^2.$$

Then, good compressibility is actually equivalent to small ideal risk:

$$\mathcal{R}(\theta, \sigma) = N_\sigma(\theta)\sigma^2 + c_{N_\sigma}^2(\theta).$$

Ideal risk cannot be attained by any estimator, which, as a function of y alone, lacks access to the oracle. However, thresholding comes relatively close to *mimicking* ideal risk: for *soft* thresholding at $\lambda_n = \sqrt{2 \log n}$, Donoho & Johnstone (1994*a*), show for all n, and for all $\theta \in \mathbb{R}^n$, that

$$E\|\hat{\theta}_{\mathrm{ST}} - \theta\|^2 \leq (2 \log n + 1)[\epsilon^2 + \mathcal{R}(\theta, \epsilon)].$$

Thus, sparsity implies good compressibility, which in turn implies the possibility of good estimation, in the sense of relatively small MSE.

Remark 2.2 The scope of model (2.1) is broader than it may at first appear. Suppose that the observed data satisfy

$$Y = A\theta + \epsilon, \tag{2.7}$$

where Y is an $N_n \times 1$ data vector, A is an $N_n \times n$ orthogonal design matrix ($A^t A = m I_n$), and ϵ has independent Gaussian components. Model (2.1) is recovered by premultiplying (2.7) by $m^{-1} A^t$. Thus, A might be a change of basis, and so our analysis covers situations where there is *some* known basis in which the signal is thought to be sparse. Indeed, this is how sparsity in wavelet bases is employed, with A being (part of the inverse of) a wavelet transform.

Remark 2.3 The sparsity analysis, motivated by wavelet methods below, considers a *sequence* of models of increasing dimensionality; indeed, the index n is precisely the number of variables. This is in sharp contrast with traditional parametric statistical theory, in which the number of unknown parameters is held fixed as the sample size increases. In practice, however, larger quantities of data N_n typically permit or even require the estimation of richer models with more parameters. Model (2.7) allows N_n to grow with n. Thus, wavelet considerations promote a style of asymptotics whose importance has long been recognized (Huber 1981, Section 7.4).

Remark 2.4 A common criticism of the use of minimax analyses in statistics holds that it is unreasonable to cast 'nature' as a malicious opponent, and that to do so risks throwing up as 'worst cases' parameters or prior configurations that are irrelevant to normal use. This challenge would be most pertinent if one were to propose an estimator on the basis of a single decision problem. Our perspective is different: we analyse *families* of statistical games, hoping to discover the common structure of optimal strategies; both estimators and least-favourable distributions. If an estimator class, such as thresholding, emerges from many such analyses, then it has a certain robustness of validity that a single minimax analysis lacks.

3   The 'signal in Gaussian white-noise' model

The multivariate Gaussian distribution $N_p(\theta, \sigma^2 I)$ with mean θ and p independent coordinates of standard deviation σ is the central model of parametric statistical inference, arising as the large sample limit of other p-parameter models, as well as in its own right.

In non-parametric statistics, the 'signal in Gaussian white-noise' model plays a similar role. The observation process $\{Y(s), \ 0 \leq s \leq 1\}$ is assumed to satisfy

$$Y(t) = \int_0^t f(s) \, ds + \sigma W(t), \tag{3.1}$$

where f is a square integrable function on $[0, 1]$ and W is a standard Brownian, or Wiener, process starting at $W(0) = 0$. In infinitesimal form, this becomes

$$dY(t) = f(t) \, dt + \sigma \, dW(t),$$

suggesting that the observations are built up from data on $f(t)$ corrupted by independent white-noise increments $dW(t)$. The unknown parameter is now f, and it is desired to estimate or test f or various functionals of f, such as point values or integrals, on the basis of Y.

As leaders of the Soviet school of non-parametric function estimation, Ibragimov & Khas'minskii (1981) gave a central place to model (3.1), arguing that its challenges are all conceptual and not merely technical. As in the finite-dimensional case, (3.1) arises as an appropriate large-sample or low-noise limit of certain other non-parametric models, such as probability density estimation, regression and spectrum estimation (see, for example, Brown & Low 1996; Nussbaum 1996). It extends to images or other objects if one replaces $t \in [0, 1]$ by a multi-parameter index $t \in D \subset \mathbb{R}^d$, and W by a Brownian sheet.

Model (3.1) has an equivalent form in the sequence space of coefficients in an orthonormal basis $\{\psi_I, I \in \mathcal{I}\}$ for $L_2([0, 1])$, the square integrable functions on the unit interval. Thus, let

$$y_I = \int \psi_I \, dY,$$

and, similarly,

$$\theta_I = \int \psi_I f \text{ and } z_I = \int \psi_I \, dW,$$

the latter being a Wiener–Ito stochastic integral. Then, (3.1) becomes

$$y_I = \theta_I + \sigma z_I, \quad I \in \mathcal{I}, \tag{3.2}$$

where $\theta = (\theta_I) \in \ell_2$, and, by the elementary properties of stochastic integrals, z_I are independent and identically distributed (i.i.d.) standard Gaussian variates.

While (3.2) looks like a straightforward infinite-dimensional extension of the Euclidean $N_p(\theta, \sigma^2 I)$ model, there are significant difficulties. For example, the sequence (y_I) is, with probability one, *not* square summable, because the noise (z_I) is i.i.d. Similarly, there is no probability distribution supported on square summable sequences that is invariant under all orthogonal transformations, or even simply under permutation of the coordinates. If the index set $\mathcal{I}$ is linearly ordered, as for the Fourier basis, one must typically work with sequences of weights, often polynomially decreasing, which lack simplifying invariance properties.

The multi-resolution character of wavelet bases is helpful here. If $\{\psi_I\}$ is now an orthonormal wavelet basis for $L^2[0, 1]$, such as those of Cohen et al.(1993), then the index $I = (j, k)$ becomes bivariate, corresponding to level j and location $k2^{-j}$ within each level. The index set $\mathcal{I}$ becomes

$$\bigcup_{j \geq 0} \mathcal{I}_j \bigcup \mathcal{I}_{-1},$$

with $|\mathcal{I}_j| = 2^j$ counting the possible values of k, and $\mathcal{I}_{-1}$ an exceptional set for the scaling function ϕ_0.

Collect the data coefficients y_{jk} in (3.2) observed at level j into a vector y_j that has a *finite*-dimensional $N_{2j}(\theta_j, \sigma^2 I)$ distribution. For many theoretical and practical purposes, it is effective to work with each of these level-wise distributions separately. Since they are of finite (but growing!) dimension, it is possible, and often a scientifically reasonable simplification, to give θ_j an orthogonally or permutation-invariant probability distribution. Indeed, the sparsity results of Section 2, derived for Gaussian distributions of large *finite* dimension, had precisely this permutation invariance character, and can be applied to each level in the dyadic sequence model.

The full non-parametric estimation conclusions are obtained by combining results across resolution levels. However, it often turns out, especially for minimax analyses, that for a given noise level σ, the 'least-favourable' behaviour occurs at a single resolution level $j = j(\sigma)$, so that conclusions from the $j(\sigma)$th permutation-invariant Gaussian model provide the key to the non-parametric situation. As the noise level σ decreases, the critical level $j(\sigma)$ increases, but in a controllable fashion. Thus, wavelet-inspired dyadic sequence models allow comparatively simple finite-dimensional Gaussian calculations to reveal the essence of non-parametric estimation theory.

In this sense, wavelet bases have rehabilitated the finite-dimensional multivariate Gaussian distribution as a tool for non-parametric theory, establishing in the process a bridge between parametric and non-parametric models.

4 Optimality in the white-noise model

Our discussion of optimality in the white-noise model illustrates the truism that the available tools, conceptual and mathematical, influence the theory that can be created at a given time. Prior to the advent of wavelet bases, formulations emphasizing good properties of linear estimators were the norm; subsequently, theoretical conclusions became possible that were more in accord with recent practical experience with algorithms and data.

As a framework for comparing estimators, we continue to use statistical games and the minimax principle. In the sequence model (3.2), a strategy for player I, the scientist, is a sequence of estimator coefficients $\hat{\theta}(y) = (\hat{\theta}_I)$, which in terms of functions becomes

$$\hat{f}(t) = \sum_I \hat{\theta}_I \psi_I(t).$$

A strategy for player II, nature, is a prior distribution π on θ, subject to a constraint that $\pi \in \mathcal{P}$. In function terms, this corresponds to choosing a random *process* model for $\{f(t), \ 0 \le t \le 1\}$. The pay-off function from the scientist to nature is

$$r(\hat{\theta}, \pi) = E_\pi \|\hat{\theta}(Y) - \theta\|^2 = E_\pi \int (\hat{f} - f)^2.$$

The constraint set $\mathcal{P} = \mathcal{P}(\Theta)$ usually requires, in some average sense usually defined by moments, that π concentrates on the set $\Theta = \Theta(\mathcal{F})$.

It is necessary to take $\mathcal{F}$ to be a compact subset of $L_2[0, 1]$, because otherwise the minimax risk does not even decrease to zero in the low-noise limit ($\epsilon \to 0$): in other words, even consistency cannot be guaranteed without restricting $\mathcal{F}$. The restrictions usually imposed have been on smoothness, requiring that f have α derivatives with bounded size in some norm. In the 1970s and 1980s, the norms chosen were typically either Hölder, requiring uniform smoothness, or Hilbert–Sobolev, requiring smoothness in a mean square sense.

4.1   Linear estimators

To describe the historical background, we start with linear methods. Estimators that are linear functions of observed data arise in a number of guises in application: they are natural because they are simple to compute and study, and already offer considerable flexibility. In the single time parameter model (3.1)–(3.2), time-shift invariance is also natural, in the absence of specific prior information to the contrary. Thus, in what follows, we switch freely between time domain $\hat{f}$ and Fourier-coefficient domain $\hat{\theta} = (\hat{\theta}_k)$. It then turns out that all shift-invariant estimators have similar structure as follows.

(i) Weighted Fourier series. Using the Fourier series form (3.2) for the data model,

$$\hat{\theta}_k = \hat{\kappa}(hk)y_k, \qquad (4.1)$$

where the shrinkage function $\hat{\kappa}$ is decreasing, corresponding to a downweighting of signals at higher frequencies. The 'bandwidth' parameter h controls the actual location of the 'cut-off' frequency band.

(ii) Kernel estimators. In the time domain, the estimator involves convolution with a window function K, scaled to have 'window width' h:

$$\hat{f}(t) = \int \frac{1}{h} K\left(\frac{t-s}{h}\right) dY(s). \qquad (4.2)$$

The representation (4.1) follows after taking Fourier coefficients.

(iii) Smoothing splines. The estimator $\hat{\theta}$ minimizes

$$\sum (y_k - \theta_k)^2 + \lambda^{2r} \sum k^{2r} \theta_k^2,$$

where the *roughness penalty* term takes the mean square form,

$$c \int_0^1 (D^r f)^2,$$

in the time domain for some positive integer r. In this case, calculus shows that $\hat{\theta}_k$ again has the form (4.1) with

$$\hat{\kappa}(\lambda k) = [1 + (\lambda k)^{2r}]^{-1}.$$

Each of these forms was studied by numerous authors, either in the white-noise model, or in asymptotically similar models—regression, density estimation—usually over Hölder or Hilbert–Sobolev $\mathcal{F}$. A crowning result of Pinsker (1980) showed that linear estimators of the form (4.1) were asymptotically minimax among *all* estimators over ellipsoidal function classes. More specifically, suppose that $\mathcal{F}$ may be represented in the sequence space model (3.2) in terms of an *ellipsoid* with semiaxes determined by a sequence $\{a_k\}$: thus,

$$\mathcal{F} = \left\{ \theta : \sum_k a_k^2 \theta_k^2 \le C^2 \right\}.$$

For example, if $\mathcal{F}$ corresponds to functions with r mean squared derivatives,

$$\int (D^r f)^2 \le L^2,$$

then

$$a_{2k-1} = a_{2k} = (2k)^r \text{ and } C^2 = L^2/\pi^{2r}.$$

We denote the resulting space $\mathcal{F}_{r,C}$, and concentrate on these special cases below. Pinsker (1980) constructed a family of *linear* shrinkage estimators $\hat{f}_\epsilon \leftrightarrow \hat{\theta}_\epsilon$ of the form (4.1) with $\hat{\kappa} = \hat{\kappa}_\epsilon$ depending also on (r, C), so that the worse case MSE of $\hat{\theta}_\epsilon(r, C)$ over $\mathcal{F}_{r,C}$ was the best possible in the small-noise limit:

$$\sup_{f \in \mathcal{F}_{r,C}} r(f, \hat{f}_\epsilon) \sim r_\epsilon(\mathcal{F}_{r,C}), \quad \epsilon \to 0.$$

Furthermore, Pinsker (1980) showed that an asymptotically least-favourable sequence of prior distributions could be described by assigning each θ_k an independent Gaussian distribution with mean zero and appropriate scale $\sigma_k^2(\epsilon, r, C)$.

This result would seem to give definitive justification for the use of linear methods: the least-favourable distributions for ellipsoids are approximately Gaussian, and for Gaussian processes, the optimal (Bayes) estimators are linear.

At about this time, however, some cracks began to appear in this pleasant linear/Gaussian picture. In the theoretical domain, Nemirovskii (1985) and Nemirovskii *et al.*(1985) showed, for certain function classes $\mathcal{F}$ in which smoothness was measured in a mean *absolute* error (L_1) sense, that linear estimators were no longer minimax, and indeed had suboptimal *rates* of convergence of error to zero as $\epsilon \to 0$.

Meanwhile, methodological and applied statistical research harnessed computing power to develop smoothing algorithms that used different, and *data-determined*, window widths at differing time points. This is in clear contrast with the fixed-width h implied by the kernel representation (4.2). For example, Cleveland (1979) investigates local smoothing, and Friedman & Stuetzle (1981), in describing the univariate smoother they constructed for projection pursuit regression, say, explicitly, the actual bandwidth used for local averaging at a particular value of [the predictor] can be larger or smaller than the average bandwidth. Larger bandwidths are used in regions of high local variability of the response.

This amounts to an implicit rejection of the ellipsoid model. The algorithms of Friedman & Stuetzle and others were iterative, involving multiple passes over the data, and were, thus, beyond theoretical analysis.

4.2   Wavelet bases and thresholding

The appearance of wavelet bases enabled a reconciliation of the Gaussian-linear theory with these divergent trends. Informally, this might be explained by 'Mallat's heuristic', quoted by Donoho (1993):

> Bases of smooth wavelets are the best bases for representing objects composed of singularities, when there may be an arbitrary number of singularities, which may be located in all possible spatial positions.

This captures the notion that a function with spatially varying smoothness (transients at some points, very smooth elsewhere) might be sparsely represented in a smooth wavelet basis and hence well estimated.

Indeed, figure 4*c* illustrates the improvement yielded by wavelet thresholding on a noisy NMR signal in comparison with figure 4*e*, which shows an arguably best near-linear estimator in the spirit of Pinsker's theorem (for further details, see Donoho & Johnstone (1995) and Johnstone (1998)). Clearly, the Pinsker-type estimator fails to adjust the (implied) window width in (4.2) to both capture the sharp peaks *and* to average out noise elsewhere.

We turn now to describe some of the theory that underlies these reconstructions and Mallat's heuristic. More technically, wavelets form an *unconditional basis* simultaneously for a vast menagerie of function spaces, allowing more flexible measures of smoothness than the Hölder and Hilbert–Sobolev spaces hitherto used in statistics. An unconditional basis for a Banach space B with norm $\| \cdot \|$ is defined by a countable family $\{\psi_I\} \subset B$ with two key properties as follows.

(i) Any element $f \in B$ has a *unique representation*:

$$f = \sum_1^\infty \theta_I \psi_I,$$

in terms of coefficients $\theta_I \in \mathbb{C}$.

(ii) *Shrinkage*: there is an absolute constant C such that if $|\theta_I'| \le |\theta_I|$ for all I, then

$$\left\| \sum \theta_I' \psi_I \right\| \le C \left\| \sum \theta_I \psi_I \right\|.$$

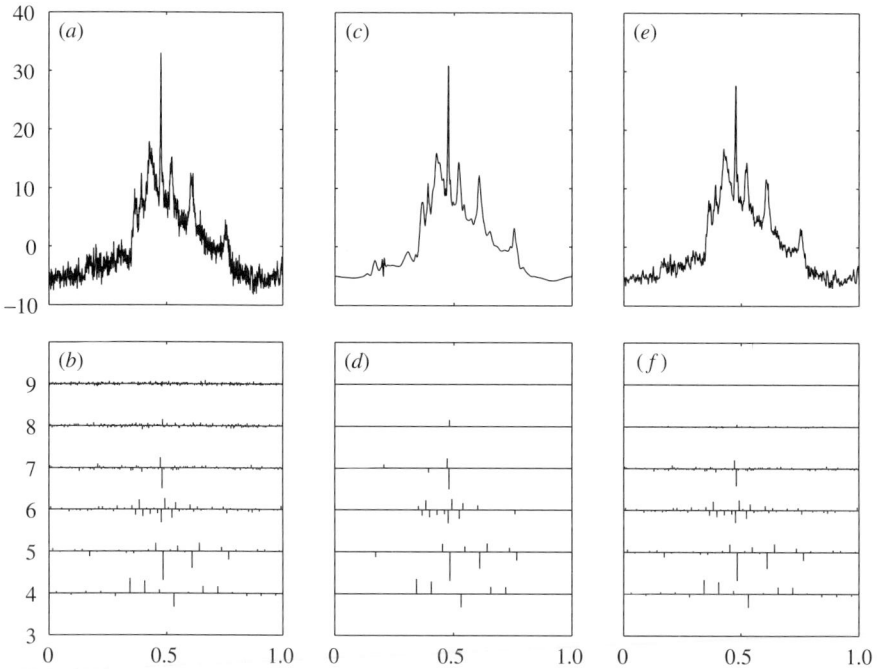

Fig. 4: (*a*) Sample NMR spectrum provided by A. Maudsley and C. Raphael; $n = 1024$. (*b*) Empirical wavelet coefficients w_{jk} displayed by nominal location and scale j, computed using a discrete orthogonal wavelet transform: Daubechies near-symmetric filter of order $N = 6$ (Daubechies 1992, ch. 6). (*c*) Reconstruction using inverse discrete wavelet transform of coefficients in (*d*). (*d*) Wavelet coefficients after hard thresholding at $\hat{\sigma}\sqrt{2\log n}$. $\hat{\sigma} = \text{med.abs.dev.}(w_{9k})/0.6745$, a resistant estimate of scale at level 9 (for details, see Donoho *et al.*1995). (*e*), (*f*) Adaptive (quasi-) linear shrinkage of wavelets coefficients in (*b*) using the James–Stein estimator on the ensemble of coefficients at each level (cf. Donoho & Johnstone 1995), and reconstruction by the discrete wavelet transform.

The statistical significance of these two properties is firstly that functions $f \in B$ may be described in terms of coefficient *sequences* $\{\theta_I\}$, and secondly that the basic statistical operation of *shrinkage* on these sequences, whether linear or via thresholding, is stable in B, in that the norms cannot be badly inflated. Notably, this property is *not* shared by the Fourier basis (Kahane *et al.*1977).[1]

Figure 5 represents the class of Besov spaces schematically in terms of the smoothness index α (equal to the number of derivatives) and the homogeneity index p, plotted as $1/p$ as is customary in harmonic analysis. Each point $(\alpha, 1/p)$

---

[1] See Donoho (1993) for a formalization of Mallat's heuristic using the unconditional basis property.

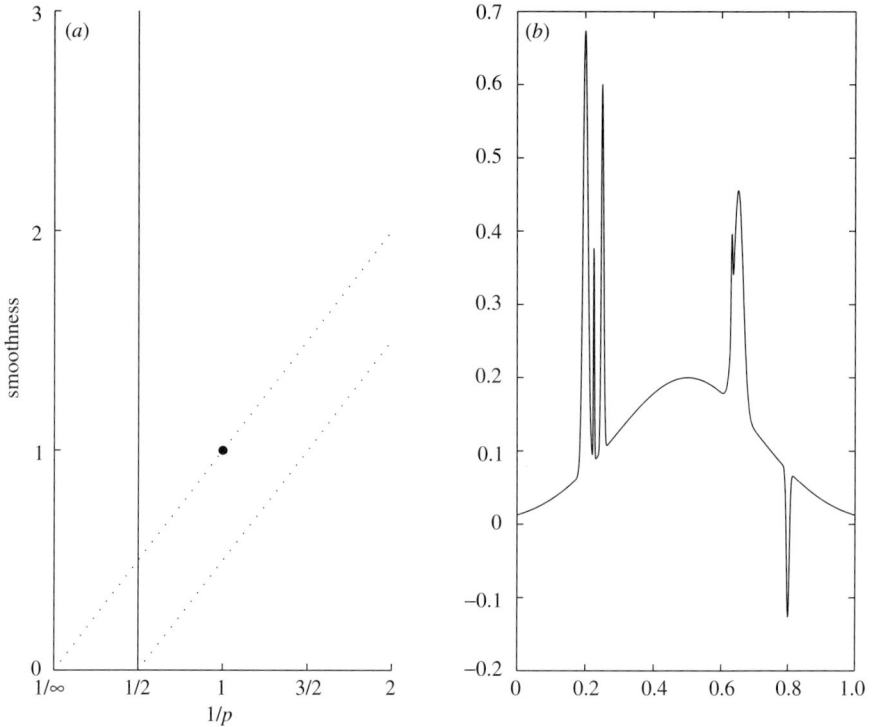

Fig. 5: (*a*) Schematic diagram of Besov spaces of varying homogeneity p and smoothness. Spaces above the diagonal line $\alpha = 1/p$ consist of functions that are at least continuous. The point at $(1, 1)$ corresponds to the bump algebra. (*b*) Caricature evoking a function in the bump algebra: a superposition of Gaussians at widely different spatial scales.

corresponds to a class of Besov spaces. The vertical line $p = 2$ represents the Hilbert–Sobolev smoothness spaces traditionally used in statistics, while points to the right are spaces with $p < 2$, hence having some degree of sparsity of representation in the wavelet domain.

To be more concrete, we consider a single example from the family of Besov spaces, the bump algebra (Meyer 1990, Section 6.6).[2] Let

$$g_{\mu,\sigma}(t) = \exp\{-(x - \mu)^2/2\sigma^2\}$$

denote a normalized Gaussian bump with location μ and scale σ. The bump algebra on $\mathbb{R}$ is the collection of all functions f representable as a convergent superposition

---

[2]For a discussion in terms of the space of bounded total variation, see Mallat (1998).

of signed, scaled and located bumps:

$$f = \sum_{1}^{\infty} \alpha_i g_{\mu_i,\sigma_i}, \qquad \sum_{1}^{\infty} |\alpha_i| < \infty.$$

This might seem a plausible model for (say) signed spectra with peaks of varying location, width and height (compare with figure 5b). As Meyer notes, while the simplicity of this description is perhaps deceptive due to lack of uniqueness in the representation, an equivalent and stable description can be given via a smooth wavelet orthobasis $\{\psi_I\}$. When restricted to $L_2[0, 1]$, we may use the index system $\mathcal{I} = \cup_j \mathcal{I}_j$ of Section 3. A subset $\mathcal{B}_C$ with norm at most C is defined by those

$$f = \sum \theta_I \psi_I$$

for which

$$\sum_{j \geq 0} 2^{j/2} \sum_{k \in \mathcal{I}_j} |\theta_{jk}| \leq C. \qquad (4.3)$$

The condition (4.3) is a scale-weighted combination of the ℓ_1 norms of the coefficients at each level j. The full bump algebra is simply the union of all $\mathcal{B}_C$. This wavelet representation is the key to statistical results.

Now consider a statistical game in the wavelet domain, with the prior constraint family $\mathcal{P}$ consisting of those priors π such that the π expectation of the left-hand side of (4.3) is bounded by C. In view of the analysis in Section 2, it is perhaps now not surprising that no linear estimator can achieve optimal rates of convergence. In fact, the minimax risk $R_\epsilon(\mathcal{B}_C)$ decreases like $C^{2/3}\epsilon^{4/3}$, whereas the best rate possible for linear estimators is much slower, namely $O(\epsilon)$.

The sequence space structure provided by the unconditional basis property also implies that optimal estimators in the statistical game are *diagonal*: $\hat{\theta}_I(y) = \delta_I(y_I)$ depends on y_I alone. While these optimal estimators cannot be described explicitly, this separation of variables is an important simplification. For example, the least-favourable distributions are, at least asymptotically, obtained by making the wavelet coefficients *independent* and, indeed, identically distributed within each level. Furthermore, it turns out that thresholding estimators, as described in Section 2 but now with thresholds depending on level, have MSE always within a constant multiple (less than or equal to 2.2) of the optimal value.

Thresholding in a wavelet basis automatically has the spatial adaptivity that previous algorithmic work sought: the effective window width at a given time point is proportional to $2^{-j(t_0)}$, where $j(t_0)$ is the finest level wavelet coefficient that survives thresholding among those wavelets whose support contains t_0.

Sample paths from approximately least-favourable distributions for $\mathcal{B}_C$ are informative. A sample realization of f can be plotted by substituting a sample draw of coefficients, θ_{jk}, into

$$f = \sum \theta_{jk} \psi_{jk}.$$

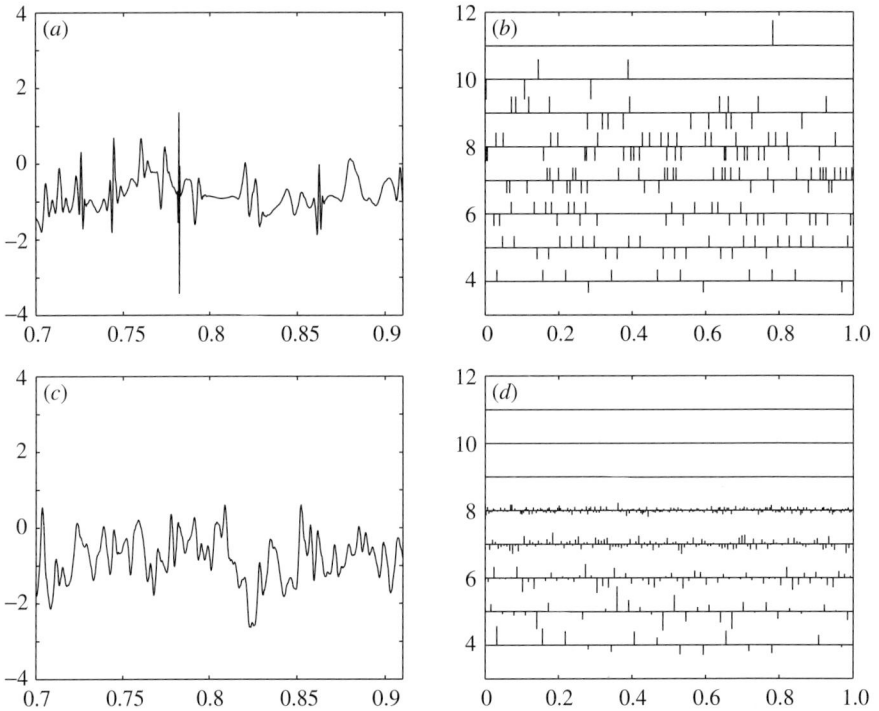

Fig. 6: (*a*) A segment of a sample path on [0, 1] from a prior distribution on wavelet coefficients that is approximately least favourable for a bump algebra ball of form (4.3). (*b*) The corresponding wavelet coefficients—those at level *j*—are i.i.d. draws from a three-point distribution $(1 - \epsilon_j)\delta_0 + \epsilon_j(\delta_{\mu_j} + \delta_{-\mu_j})$, as described in the text. The wavelets ψ_{jk} are derived from the $N = 8$ instance of the Daubechies (1992, ch. 6) 'closest to linear phase' filter. (*c*) Sample path from the Gaussian process that is the least-favourable distribution for an ellipsoid $\mathcal{F}_{m,C}$ with $m = 1$ square integrable derivatives. (*d*) Corresponding wavelet coefficients with variance σ_j^2 decreasing with *j*.

By considering only threshold rules, which are nearly minimax as just mentioned, it can be shown that a near-least-favourable distribution can be constructed from three point distributions, $(1 - \epsilon_j)\delta_0 + \epsilon_j(\delta_{\mu_j} + \delta_{-\mu_j})$, that are quite similar to that given at (2.5). Now, however, the location μ_j and size $\epsilon_j/2$ of the non-zero atom and its reflection depend on the level *j*, but, within level, the 2^j draws are independent, as they are in (2.5). A numerical optimization (Johnstone 1994) allows evaluation of μ_j and ϵ_j for a given ϵ and $\mathcal{B}_C$. Figure 6a shows a representative sample path and figure 6b shows the corresponding individual wavelet coefficients for this distribution.

Figure 6c shows a corresponding sample path drawn from the Gaussian least-favourable distribution on wavelet coefficients, figure 6d corresponding to the ellipsoid $\mathcal{F}_{m,C}$ with $m = 1$ derivatives assumed to be square integrable. Again, the wavelet coefficients are i.i.d. within level, but are now drawn from a Gaussian distribution with variance $\sigma_j^2(m, C, \epsilon)$ determined by Pinsker's solution. The two plots are calibrated to the same indices of smoothness $\alpha = 1$, scale $C = 1$ and noise level $\epsilon = 1/64$.

The qualitative differences between these plots are striking: the Gaussian sample path has a spatially homogeneous irregularity, with the sample wavelet coefficients being 'dense', though decreasing in magnitude with increasing scale or 'frequency octave'. In contrast, the bump algebra sample path has a greater spikiness: the sample wavelet coefficients have increasing sparsity and magnitude with each increasing scale. These differences become even more pronounced if one increases the smoothness α and decreases the homogeneity index p; see, for example, the plots for $\alpha = 2$, $p = 1/2$ in Johnstone (1994).

To summarize: with only the sparsity in mean constraint, and no other restriction on estimators or prior distributions, coordinatewise thresholding and sparse priors emerge as the near optimal strategies for the bump algebra statistical game. Thresholding has better MSE, indeed faster rates of convergence, than any linear estimate over $\mathcal{B}$. This may also be seen visually in the relatively much more noise-free reconstruction using wavelet thresholding, shown in figure 4.

We mention briefly the problem of *adaptation*. The near optimality of thresholding and sparse priors holds in similar fashion for a large class of Besov space constraints described by $(\alpha, 1/p)$ and size parameter C. The optimal threshold estimator in each case will depend on (α, p, C): can one give an estimator with optimal or near-optimal properties without needing to specify (α, p, C)? One very simple possibility, already shown in figure 4 and explored at length in Donoho *et al.*(1995), is to use hard or soft thresholding at threshold $\sqrt{2 \log n}$, where n is the number of observations, or wavelet coefficients. This estimator has a remarkably robust near adaptivity—it nearly achieves, up to logarithmic terms, the minimax rate of convergence simultaneously over a wide range of both functional classes *and* error measures—not solely mean squared error.

5   Concluding remarks

Wavelets have enabled the development of theoretical support in statistics for the important notions of sparsity and thresholding.

In contrast, much work in modern curve fitting and regression treats the imposition of *smoothness* as a guiding principle. Wavelets prompt us to think of smoothness as a particular case of a more general principle, namely sparsity of representation. It is in fact sparsity of representation that determines when good estimation is possible.

Pursuing the sparsity idea for functions of more than one variable leads to systems other than literal wavelets, as discussed by Candès & Donoho (this issue).

We may expect to see more use of 'dyadic thinking' in areas of statistics and data analysis that have little to do with wavelets directly. This is likely both in the development of methods, and also as a metaphor in making simpler models for theoretical analysis of other more complex procedures.

6 Acknowledgements

Many thanks to Marc Raimondo for his help in the preparation of figure 6. The author gratefully acknowledges financial support from the National Science Foundation (DMS 9505151).

7 References

Brown, L. D. & Low, M. G. 1996 Asymptotic equivalence of nonparametric regression and white noise. *Ann. Statistics* **3**, 2384–2398.

Cleveland, W. S. 1979 Robust locally weighted regression and smoothing scatterplots. *J. Am. Statist. Assoc.* **74**, 829–836.

Cohen, A., Daubechies, I. & Vial, P. 1993 Wavelets and fast wavelet transform on an interval. *Appl. Comp. Harmonic Analysis* **1**, 54–81.

Daubechies, I. 1992 *Ten lectures on wavelets.* CBMS-NSF Series in Applied Mathematics, no. 61. Philadelphia, PA: SIAM.

Donoho, D. 1993 Unconditional bases are optimal bases for data compression and statistical estimation. *Appl. Comp. Harmonic Analysis* **1**, 100–115.

Donoho, D. L. & Johnstone, I. M. 1994*a* Ideal spatial adaptation via wavelet shrinkage. *Biometrika* **81**, 425–455.

Donoho, D. L. & Johnstone, I. M. 1994*b* Minimax risk over ℓ_p-balls for ℓ_q-error. *Probability Theory Related Fields* **99**, 277–303.

Donoho, D. L. & Johnstone, I. M. 1995 Adapting to unknown smoothness via wavelet shrinkage. *J. Am. Statist. Assoc.* **90**, 1200–1224.

Donoho, D. L., Johnstone, I. M., Kerkyacharian, G. & Picard, D. 1995 Wavelet shrinkage: asymptopia? (With discussion.) *J. R. Statist. Soc.* B **57**, 301–369.

Friedman, J. & Stuetzle, W. 1981 Projection pursuit regression. *J. Am. Statist. Assoc.* **76**, 817–823.

Hotelling, H. 1933 Analysis of a complex of statistical variables into principal components. *J. Educ. Psych.* **24**, 417–441, 498–520.

Huber, P. J. 1981 *Robust statistics.* Wiley.

Ibragimov, I. & Khas'minskii, R. 1981 *Statistical estimation: asymptotic theory.* Springer.

Johnstone, I. M. 1994 Minimax Bayes, asymptotic minimax and sparse wavelet priors. In *Statistical decision theory and related topics* (ed. S. Gupta & J. Berger), vol. V, pp. 303–326. Springer.

Johnstone, I. M. 1998 Function estimation: white noise, sparsity and wavelets. Lecture notes.

Kahane, J., de Leeuw, K. & Katznelson, Y. 1977 Sur les coefficients de fourier des fonctions continues. *Comptes Rendus Acad. Sci. Paris* A **285**, 1001–1003.

Le Cam, L. 1986 *Asymptotic methods in statistical decision theory.* Springer.

Mallat, S. 1998 Applied mathematics meets signal processing. In *Proc. of ICM, 18–27 August 1998, Berlin*. (*Doc. Math.* Extra Volume **I**, 319–338.)

Meyer, Y. 1990 *Ondelettes et Opérateurs. I. Ondelettes. II. Opérateurs de Calderón–Zygmund. III. Opérateurs multilinéaires (with R. Coifman)*. Paris: Hermann. (English translation published by Cambridge University Press.)

Nemirovskii, A. 1985 Nonparametric estimation of smooth regression function. *Izv. Akad. Nauk. SSR Teckhn. Kibernet.* **3**, 50–60 (in Russian). (Transl. in *J. Comput. Syst. Sci.* **23** (1986), 1–11.)

Nemirovskii, A., Polyak, B. & Tsybakov, A. 1985 Rate of convergence of nonparametric estimates of maximum-likelihood type. *Problems Information Transmission* **21**, 258–272.

Nussbaum, M. 1996 Asymptotic equivalence of density estimation and white noise. *Ann. Statist.* **24**, 2399–2430.

Pinsker, M. 1980 Optimal filtering of square integrable signals in Gaussian white noise. *Problems Information Transmission* **16**, 120–133. (Originally published in Russian in *Problemy Peredatsii Informatsii* **16**, 52–68.)

Wald, A. 1950 *Statistical decision functions*. Wiley.

6

Ridgelets: a key to higher-dimensional intermittency?

Emmanuel J. Candès and David L. Donoho

Department of Statistics, Stanford University, Stanford, CA 94305–4065, USA

Abstract

In dimensions two and higher, wavelets can efficiently represent only a small range of the full diversity of interesting behaviour. In effect, wavelets are well adapted for point-like phenomena, whereas in dimensions greater than one, interesting phenomena can be organized along lines, hyperplanes and other non-point-like structures, for which wavelets are poorly adapted.

We discuss in this paper a new subject, ridgelet analysis, which can effectively deal with line-like phenomena in dimension 2, plane-like phenomena in dimension 3 and so on. It encompasses a collection of tools which all begin from the idea of analysis by ridge functions $\psi(u_1 x_1 + \cdots + u_n x_n)$ whose ridge profiles ψ are wavelets, or alternatively from performing a wavelet analysis in the Radon domain. The paper reviews recent work on the continuous ridgelet transform (CRT), ridgelet frames, ridgelet orthonormal bases, ridgelets and edges and describes a new notion of smoothness naturally attached to this new representation.

Keywords: Ridge functions; wavelets; singularities; edges; Radon transform; nonlinear approximation

1 Introduction

This paper is part of a series around the theme—*wavelets: a key to intermittent information?*. The title itself raises a fundamental question; we shall argue that the answer is both *no* and *yes*. We say *no* because wavelets *per se* only address a portion of the intermittency challenge; we intend to make clear how much larger the question is than just the portion which wavelets can face effectively. Roughly speaking, wavelets deal efficiently only with one type of intermittency—singularities at points—and in higher dimensions there are many other kinds of intermittency—singularities along lines, along hyperplanes, etc.—which wavelets do not deal with efficiently. But we also say *yes*, because by using wavelets in a novel way, we have been able to build new systems of representations—ridgelets– which are efficient at many of the tasks where wavelets fail.

In this expository paper, we will primarily focus on the study of objects defined in two-dimensional space since, on one hand, this case already exhibits the main concepts underlying the ridgelet analysis and, on the other hand, it is a very practical setting because of the connection with image analysis. However, we will refer to extensions to higher dimensions wherever it is conceptually straightforward to do so.

1.1   Wavelets and point singularities

To begin, we call the reader's attention to one of the really remarkable facts about wavelet bases. Suppose that we have a function $f(t)$ of a single real variable $t \in [0, 1]$ and that f is smooth apart from a discontinuity at a single point t_0. For example, let $f(t) = t - 1_{\{t > t_0\}}$. In some sense this is a very simple object, and we would like to find an expansion that reveals its simplicity. However, in traditional types of expansions, the representation of this object is quite complicated, involving contributions from many terms. This is so of the Fourier representation; viewing $[0, 1]$ as the circle, we can calculate the appropriate Fourier series on $[0, 1]$; the number of the Fourier coefficients of f exceeding $1/N$ in absolute value exceeds $c \cdot N$ as $N \to \infty$, for some positive constant c. It is true of traditional orthogonal series estimates; an expansion of f in Legendre polynomials has at least $c \cdot N$ coefficients exceeding $1/N$. In stark contrast, in a nice wavelet orthonormal basis (Daubechies 1988), such as the Lemarié–Meyer inhomogeneous periodized wavelet basis, the number of coefficients exceeding $1/N$ in amplitude grows more slowly that N^ρ for any positive ρ. In effect, the singularity at t_0 causes widespread effects throughout the Fourier and Legendre representations; but the singularity causes highly localized or concentrated effects to the wavelet representation. Alternately, we can say that *in analysing an object exhibiting punctuated smoothness, the wavelet coefficients are sparse, while the coefficients of classical transforms are not sparse.*

The potential for sparsity of wavelet representations has had a wide impact, both in theory and in practice. It has a well-understood meaning for nonlinear approximation and for data compression of objects exhibiting punctuated smoothness (Donoho 1993): since the energy associated with the singularity is mostly concentrated in just a few big coefficients, partial reconstruction using a relatively small number of wavelet terms (the terms associated with the biggest wavelets coefficients) can give excellent approximations. The recognition that wavelets deal successfully with functions which are smooth away from singularities has led to a great deal of interest in their applications in image coding, where a great deal of the important structure consists of singularities—namely, edges. Wavelet-based coders have found wide application in various 'niche' data-compression applications, and are now being considered for inclusion in the JPEG-2000 still-picture data-compression standard.

1.2   Singularities along lines

Unfortunately some claims for wavelets have been overstated, and wavelets are sometimes being used for applications well outside their actual domain of expertise. To understand this point requires a more careful look at the notion of *singularity*. A function $f(x)$ of n variables may have singularities of any integer dimension d in the range $0, \ldots, n - 1$. A zero-dimensional singularity is a point of bad behaviour.

A one-dimensional singularity is a curve of bad behaviour. An $(n-1)$-dimensional singularity is a hypersurface of bad behaviour. *Wavelets are fully efficient at dealing with zero-dimensional singularities* only. Unfortunately, in higher dimensions, other kinds of singularities can be present, or even dominant: in typical images, the edges represent one-dimensional singularities, and there are no zero-dimensional singularities to speak of.

To be more concrete, consider the function g supported in the unit square

$$g(x_1, x_2) = 1_{\{x_1+x_2>1/2\}} \, w(x_1, x_2), \quad x \in \mathbb{R}^2, \tag{1.1}$$

where $w(x_1, x_2)$ is a smooth function tending to zero together with its derivatives at the boundary of the unit square. This simple object has a singularity along the line $x_1+x_2 = \frac{1}{2}$. Such an object poses a difficult problem of approximation both for two-dimensional Fourier analysis and for two-dimensional wavelet analysis. Although the object is very simple, its wavelet transform does not decay rapidly: as $N \to \infty$, there are greater than or equal to $c \cdot N$ orthonormal wavelet coefficients exceeding $1/N$ in size. Its bivariate Fourier series does not decay rapidly either: as $N \to \infty$, there are $\geq c \cdot N$ Fourier coefficients exceeding $1/N$ in size. Neither wavelets nor Fourier methods perform really well here. For example, if we used either approach as the basis of transform coders (Donoho 1996), we would have, as a direct corollary of the fact that at least $c \cdot N$ coefficients of g have amplitude $\geq 1/N$, that the number of bits one must retain to achieve a distortion less than or equal to ϵ for wavelet transform coding grows as $\epsilon \to 0$ at least as rapidly as $c \cdot \epsilon^{-1}$, and the number of bits one must retain to achieve a distortion ϵ for Fourier transform coding grows as $\epsilon \to 0$ at least as rapidly as $c \cdot \epsilon^{-1}$.

In effect, wavelets are being used in image data compression although their theoretical properties are not nearly as favourable as one might have imagined, given the degree of attention they have received.

The concept of intermittency does not have a universal acceptance. We now take the liberty of identifying this concept as a situation where objects of interest are typically smooth apart from occasional singularities on, say, a set of measure zero. From this point of view we can say that wavelets have a role to play in dealing with a particular kind of intermittency—unusual behaviour at one point (or occasional points)—but not with every kind of intermittency; in dimension two they already fail when asked to deal efficiently with unusual behaviour on a line.

We are entitled here to say that wavelets 'fail' because we know of representing systems which, in a precise sense, can succeed in dealing with unusual behaviour on a line.

1.3   Ridgelet analysis

In this paper we describe a recently developed approach to problems of functional representation—*ridgelet analysis*. Ridgelet analysis makes available representations of functions by superpositions of *ridge functions* or by simple elements that are in some way related to ridge functions $r(a_1 x_1 + \cdots + a_n x_n)$; these are functions of

n variables, constant along hyperplanes $a_1x_1 + \cdots + a_nx_n = c$; the graph of such a function in dimension two looks like a 'ridge'. The terminology 'ridge function' arose first in tomography (Logan & Shepp 1975), and ridgelet analysis makes use of a key tomographic concept, the Radon transform.

However, multiscale ideas, as found in the work of Littlewood & Paley or Calderòn (Meyer 1990) and culminating in wavelet theory, also appear as a crucial tool in the story. From wavelet theory, ridgelet analysis borrows the localization idea: fine-scale ridgelets are concentrated near hyperplanes at all possible locations and orientations.

As an example of what this family of ideas can do, consider the function g of (1.1). It will turn out that there are ridgelet expansions—by frames and even by orthonormal sets—having the property that the number of coefficients exceeding $1/N$ in amplitude grows more slowly that N^ρ for any positive ρ. In effect, the singularity in g across the line $x_1 + x_2 = \frac{1}{2}$ has widespread effects in the Fourier and wavelet representation, but the singularity causes highly concentrated effects in the ridgelet representation. Moreover, a ridgelet transform coding method, based on scalar quantization and run-length coding, can code such objects with a bit length that grows more slowly as $\epsilon \to 0$ than any fractional power of ϵ^{-1}. Hence ridgelets do for linear singularities in dimension two what wavelets did for point singularities in dimension one—they provide an extremely sparse representation; neither wavelets nor Fourier can manage a similar feat in representing linear singularities in dimension two.

1.4   Ridgelets and ridge functions

The ability of ridgelets to give a sparse analysis of singularities is just one point of entry into our topic. Another interesting entry point is provided by the connection of ridgelet analysis with the theory of approximation by superpositions of ridge functions. Since the 1970s, it has been proposed that superpositions of ridge functions could offer interesting alternatives to standard methods of multivariate approximation. Friedman & Stuetzle (1981) introduced into statistics the topic of 'projection pursuit regression', specifically suggesting that by such means one might perhaps evade the curse of dimensionality as suffered by then-typical methods of function approximation. Approximation by superpositions of ridge functions acquired further interest in the late 1980s under the guise of approximation by single-hidden-layer feedforward neural nets. In such neural nets, one considers the m-term approximation

$$f(x_1, \ldots, x_n) \approx \sum_{i=1}^{m} c_i \sigma(a_{i,1}x_1 + \cdots + a_{i,n}x_n).$$

Celebrated results in the neural-nets literature include Cybenko's (1989) result that every nice function of n-variables can be approximated arbitrarily well in a suitable norm by a sequence of such m-term approximations, and results of Barron (1993) and Jones (1992) that describe function classes and algorithms under which such m-term approximations converge at given rates, including specific situations in which the rates do not worsen with increasing dimension.

Ridgelet analysis provides an alternate approach to obtaining approximations by superpositions of ridge functions; one which is quantitative, constructive and stable. Roughly speaking, the earlier theory of m-term ridge-function approximations assures us only of the *existence* of superpositions with prescribed features; the theory of ridgelet analysis, growing as it does out of wavelets and computational harmonic analysis, goes to a new level, and gives a particular way to build an approximation which is both constructive and stable. It also gives theoretical insights, previously unavailable, about those objects which can be well represented by ridge functions.

2   The continuous ridgelet transform

The (continuous) ridgelet transform in $\mathbb{R}^2$ can be defined as follows (Candès 1999). Pick a smooth univariate function $\psi : \mathbb{R} \to \mathbb{R}$ with sufficient decay and vanishing mean, $\int \psi(t)\,dt = 0$. For each $a > 0$, each $b \in \mathbb{R}$ and each $\theta \in [0, 2\pi)$, define the bivariate function $\psi_{a,b,\theta} : \mathbb{R}^2 \to \mathbb{R}^2$ by

$$\psi_{a,b,\theta}(x) = a^{-1/2} \cdot \psi((\cos\theta x_1 + \sin\theta x_2 - b)/a).$$

This function is constant along 'ridges' $\cos\theta x_1 + \sin\theta x_2 = \text{const}$ Transverse to these ridges it is a wavelet; hence the name *ridgelet*. Given an integrable bivariate function $f(x)$, define its ridgelet coefficients

$$\mathcal{R}_f(a, b, \theta) = \int \bar{\psi}_{a,b,\theta}(x) f(x)\,dx.$$

Our hypotheses on ψ guarantee that $\int |\hat{\psi}(\lambda)|^2 \lambda^{-2}\,d\lambda < \infty$, and we suppose further that ψ is normalized so that

$$\int |\hat{\psi}(\lambda)|^2 \lambda^{-2}\,d\lambda = 1.$$

Candès (1999) proves the exact reconstruction formula

$$f(x) = \int_0^{2\pi} \int_{-\infty}^{\infty} \int_0^{\infty} \mathcal{R}_f(a, b, \theta)\psi_{a,b,\theta}(x)\, \frac{da}{a^3}\,db\,\frac{d\theta}{4\pi}$$

valid a.e. for functions which are both integrable and square integrable. This shows that 'any' function may be written as a superposition of 'ridge' functions. Such integral representations have been independently discovered by Murata (1996). In addition, our representation is stable, as we have a Parseval relation:

$$\int |f(x)|^2\,dx = \int_0^{2\pi} \int_{-\infty}^{\infty} \int_0^{\infty} |\mathcal{R}_f(a, b, \theta)|^2\, \frac{da}{a^3}\,db\,\frac{d\theta}{4\pi}.$$

(This relation is, however, absent from Murata's papers.) This approach generalizes to any dimension. Given a ψ obeying

$$\int |\hat{\psi}(\lambda)|^2 \lambda^{-n} \, d\lambda = 1,$$

define $\psi_{a,b,u}(x) = \psi((u'x - b)/a)/\sqrt{a}$ and $\mathcal{R}f(a, b, \theta) = \langle f, \psi_{a,b,u} \rangle$. Then there is an n-dimensional reconstruction formula

$$f = c_n \iiint \mathcal{R}f(a, b, u)\psi_{a,b,u}(x) \frac{da}{a^{n+1}} \, db \, du,$$

with du the uniform measure on the sphere; and a Parseval relation

$$\|f\|^2_{L^2(\mathbb{R}^n)} = c_n \iiint |\mathcal{R}f(a, b, \theta)|^2 \frac{da}{a^{n+1}} \, db \, du.$$

2.1   Relation to Radon transform

The continuous ridgelet transform is intimately connected with the Radon transformation (an excellent reference for the Radon transform is Helgason (1986)). If we put

$$Rf(u, t) = \int f(x)\delta(u'x - t) \, dx$$

for the integral of f over the hyperplane $u'x = t$, then $\mathcal{R}f(a, b, u) = \langle \psi_{a,b}, Rf(u, \cdot) \rangle$, where $\psi_{a,b}(t) = \psi((t - b)/a)/\sqrt{a}$ is a one-dimensional wavelet. Hence the Ridgelet transform is precisely the application of a one-dimensional wavelet transform to the slices of the Radon transform where u is constant and t is varying.

2.2   An example

Let g be the mutilated Gaussian

$$g(x_1, x_2) = 1_{\{x_2 > 0\}} e^{-x_1^2 - x_2^2}, \quad x \in \mathbb{R}^2. \tag{2.1}$$

This is discontinuous along the line $x_2 = 0$ and smooth away from that line. One can calculate immediately the Radon transform of such a function; it is

$$(Rg)(t, \theta) = e^{-t^2} \bar{\Phi}(-t \sin\theta/|\cos\theta|) \quad t \in \mathbb{R}, \quad \theta \in [0, 2\pi], \tag{2.2}$$

where

$$\bar{\Phi}(v) \equiv \int_v^\infty e^{-u^2} \, du.$$

We can get immediate insight into the form of the CRT from this formula. Remember that the wavelet transform $\langle \psi_{a,b}, e^{-t^2} \cdot \bar{\Phi}(-t \sin\theta/|\cos\theta|) \rangle$ needs to be computed. Effectively, the Gaussian window e^{-t^2} makes little difference; it is smooth and of rapid decay, so it does little of interest; in effect the object of real interest to us is $\langle \psi_{a,b}, \bar{\Phi}(-s(\theta)t)\rangle$, where $s(\theta) = \sin\theta/|\cos\theta|$. Define then $W(a,b) = \langle \psi_{a,b}, \bar{\Phi}(-t)\rangle$; this is the wavelet transform of a smooth sigmoidal function. By the scale-invariance of the wavelet transform,

$$\langle \psi_{a,b}, \bar{\Phi}(-s(\theta)t)\rangle = W(s(\theta)a, s(\theta)b) \cdot |s(\theta)|^{-1/2}, \quad \text{for } \theta \in (0,\pi)$$

and, of course, a similar relationship holds for $(\pi, 2\pi)$. In short, for a caricature of $R_f(a,b,\theta)$, we have, for each fixed θ a function of a and b which is a simple rescaling of the wavelet transform of $\bar{\Phi}$ as function of θ. This rescaling is smooth and gentle away from $\theta = \frac{1}{2}\pi$ and $\theta = \frac{3}{2}\pi$, where it has singularities.

We remark that in a certain sense the CRT of g is sparse; if we use a sufficiently nice wavelet, such as a Meyer wavelet, the CRT belongs to $L^p((da/a^3)\,db\,d\theta)$ for every $p > 0$. This is a fancy way of saying that the CRT decays rapidly as one moves either spatially away from $b = 0$ or $\theta \in \{\frac{1}{2}\pi, \frac{3}{2}\pi\}$ as one goes to fine scales $a \to 0$.

3 Discrete ridgelet transform: frames

It is important for applications that one obtains a discrete representation using ridgelets. Typical discrete representations include expansions in orthonormal bases. Here we describe an expansion in two dimensions by frames (see also Candès (1999), where the case for all dimensions $n \geq 2$ is treated).

We now develop a formula for the CRT of f using the Fourier domain. Obviously, with $\hat{f}$ denoting Fourier transform,

$$R_f(a,b,\theta) = \frac{1}{2\pi}\int \bar{\hat{\psi}}_{a,b,\theta}(\xi)\hat{f}(\xi)\,d\xi,$$

where $\hat{\psi}_{a,b,\theta}(\xi)$ is interpreted as a distribution supported on the radial line in the frequency plane. Letting $\xi(\lambda,\theta) = (\lambda\cos(\theta), \lambda\sin(\theta))$ we may write

$$R_f(a,b,\theta) = \frac{1}{2\pi}\int_{-\infty}^{\infty} a^{1/2}\bar{\hat{\psi}}(a\lambda)e^{-e\lambda b}\hat{f}(\xi(\lambda,\theta))\,d\lambda. \tag{3.1}$$

This says that the CRT is obtainable by integrating the weighted Fourier transform $w_{a,b}(\xi)\hat{f}(\xi)$ along a radial line in the frequency domain, with weight $w_{a,b}(\xi)$ given by

$$a^{1/2}\bar{\hat{\psi}}(a|\xi|)$$

times a complex exponential in $e^{-e\lambda b}$. Alternatively, we can see that the function of b (with a and θ considered fixed), $\rho_{a,\theta}(b) = R_f(a,b,\theta)$, satisfies

$$\rho_{a,\theta}(b) = \mathcal{F}_1^{-1}\{\hat{\rho}_{a,\theta}(\lambda)\},$$

where $\mathcal{F}_1$ stands for the one-dimensional Fourier transform, and

$$\hat{\rho}_{a,\theta}(\lambda) = a^{1/2}\bar{\hat{\psi}}(a\lambda)\hat{f}(\xi(\lambda,\theta)), \quad -\infty < \lambda < \infty$$

is the restriction of $w_{a,0}(\xi)\hat{f}(\xi)$ to the radial line. Hence, conceptually, the CRT at a certain scale a and angle θ can be obtained by the following steps:

(1) two-dimensional Fourier transform, obtaining $\hat{f}(\xi)$;

(2) radial windowing, obtaining $w_{a,0}(\xi)\hat{f}(\xi)$, say; and

(3) one-dimensional inverse Fourier transform along radial lines, obtaining $\rho_{a,\theta}(b)$, for all $b \in \mathbb{R}$.

We are interested in finding a method for sampling $(a_j, b_{j,k}, \theta_{j,\ell})$ so that we obtain frame bounds, i.e. so we have equivalence:

$$\sum_{j,k,\ell} |\mathcal{R}_f(a_j, b_k, \theta_{j,\ell})|^2 \asymp \iiint |\mathcal{R}_f(a, b, \theta)|^2 \frac{da}{a^3} db d\theta. \qquad (3.2)$$

To simplify our exposition, we will suppose that $\hat{\psi}(\lambda) = 1_{\{1 \le |\xi| \le 2\}}$ although the frame result holds for a large class of ψ as exposed in Candès (1999). Guided by the Littlewood–Paley and the wavelet theories, the scale a and location parameter b are discretized dyadically, as $a_j = a_0 2^j$ and $b_{j,k} = 2\pi k 2^{-j}$. Following (3.1) the ridgelet coefficients may be written as

$$\mathcal{R}_f(a_j, b_{j,k}, \theta) = \frac{1}{2\pi} 2^{-j/2} \int_{2^j \le |\lambda| \le 2^{j+1}} e^{-e\lambda 2\pi 2^{-j}} \hat{f}(\xi(\lambda, \theta)) d\lambda,$$

and hence, the Plancherel theorem gives

$$\sum_k |\mathcal{R}_f(a_j, b_{j,k}, \theta)|^2 = \frac{1}{\sqrt{2\pi}} \int_{2^j \le |\lambda| \le 2^{j+1}} |w_{2^j,0}|^2 |\hat{f}(\xi(\lambda, \theta))|^2 d\lambda.$$

In short, at a fixed scale and angular location, the sum of squares of ridgelet coefficients across a varying spatial location amounts to integrating the square of the Fourier transform along a dyadic segment.

Discretizing the angular variable θ amounts to performing a sampling of such segment-integrals from which the integral of $|\hat{f}(\xi)|^2$ over the whole frequency domain needs to be inferred. This is not possible without support constraints on f, as functions f can always be constructed with $f(x)$ having slow decay as $|x| \to \infty$ so that $\hat{f}$ will vanish on a collection of disjoint segments without being identically zero. However, under a support restriction, so that f is supported inside the unit disc (or any other compact set), the integrals over the segments can provide sufficient information to infer $\int |\hat{f}(\xi)|^2 d\xi$.

Indeed, under a support constraint, the Fourier transform $\hat{f}(\xi)$ is a band-limited function, and over 'cells' of appropriate size can only display very banal behaviour.

If we sample once per cell, we will capture enough of the behaviour of this object to be in a position to infer the size of the function from those samples. The solution found by Candès (1999) is to sample something like the following with increasing angular resolution at increasingly fine scales:

$$\theta_{j,\ell} = 2\pi \ell 2^{-j}.$$

This strategy gives the equivalence (3.2). It then follows that the collection

$$\{ 2^{j/2} \psi (2^j (x_1 \cos(\theta_{j,\ell}) + x_2 \sin(\theta_{j,\ell}) - 2\pi k 2^{-j})) \}_{(j \ge j_0, \ell, k)}$$

is a frame for the unit disc; for any f supported in the disk with finite L^2 norm,

$$\sum_{j,k,l} |\langle \psi_{a_j, b_{j,k}, \theta_{j,l}}, f \rangle|^2 \asymp \|f\|^2.$$

The construction generalizes to any dimension n; in two dimensions, the discretization involves the sampling of angles from the circle and in n dimensions the sampling of angles from the unit sphere. The angular variable is also sampled at increasing resolution so that at scale j the discretized set is a net of nearly equispaced points at a distance of order 2^{-j} (see Candès (1999) for details).

The existence of frame bounds implies, by soft analysis, that there are 'dual ridgelets' $\tilde{\psi}_{j,k,\ell}$ so that

$$f = \sum_{j,k,\ell} \langle f, \tilde{\psi}_{j,k,\ell} \rangle \psi_{j,k,\ell} \text{ and } f = \sum_{j,k,\ell} \langle f, \psi_{j,k,\ell} \rangle \tilde{\psi}_{j,k,\ell},$$

with equality in a an L^2 sense, and so that

$$\sum_{j,k,\ell} |\langle f, \tilde{\psi}_{j,k,\ell} \rangle|^2 \asymp \sum_{j,k,\ell} |\langle f, \psi_{j,k,\ell} \rangle|^2 \asymp \|f\|^2_{L^2}.$$

At the moment, only qualitative properties of the dual ridgelets $\tilde{\psi}_{j,k,\ell}$ are known; for example there are no closed-form expressions for their structure.

4 Orthonormal ridgelets in dimension 2

Donoho (1998) had the idea to broaden somewhat the notion of a ridgelet, to allow the possibility of systems obeying certain frequency/angle localization properties, and showed that if we allow this broader notion, then it becomes possible to have orthonormal ridgelets whose elements can be specified in closed form. Such a system can be defined as follows: let $(\psi_{j,k}(t) : j \in \mathbb{Z}, k \in \mathbb{Z})$ be an orthonormal basis of Meyer wavelets for $L^2(\mathbb{R})$ (Lemarié & Meyer 1986) and let

$$(w^0_{i_0,\ell}(\theta), \ell = 0, \dots, 2^{i_0} - 1; \ w^1_{i,\ell}(\theta), i \ge i_0, \ell = 0, \dots, 2^i - 1)$$

be an orthonormal basis for $L^2[0, 2\pi)$ made of periodized Lemarié scaling functions $w^0_{i_0,\ell}$ at level i_0 and periodized Meyer wavelets $w^1_{i,\ell}$ at levels $i \geq i_0$. (We suppose a particular normalization of these functions). Let $\hat{\psi}_{j,k}(\omega)$ denote the Fourier transform of $\psi_{j,k}(t)$, and define ridgelets $\rho_\lambda(x)$, $\lambda = (j, k; i, \ell, \varepsilon)$ as functions of $x \in \mathbb{R}^2$ using the frequency-domain definition

$$\hat{\rho}_\lambda(\xi) = \tfrac{1}{2}|\xi|^{-1/2}(\hat{\psi}_{j,k}(|\xi|)w^\varepsilon_{i,\ell}(\theta) + \hat{\psi}_{j,k}(-|\xi|)w^\varepsilon_{i,\ell}(\theta + \pi)). \qquad (4.1)$$

Here the indices run as follows: $j, k \in \mathbb{Z}$, $\ell = 0, \ldots, 2^{i-1} - 1$; $i \geq i_0, i \geq j$. Notice the restrictions on the range of ℓ and on i. Let Λ denote the set of all such indices λ. It turns out that $(\rho_\lambda)_{\lambda \in \Lambda}$ is a complete orthonormal system for $L^2(\mathbb{R}^2)$.

In the present form the system is not visibly related to ridgelets as defined earlier, but two connections can be exhibited. First, define a fractionally differentiated Meyer wavelet:

$$\psi^+_{j,k}(t) = \frac{1}{2\pi} \int_{-\infty}^\infty |\omega|^{1/2} \hat{\psi}_{j,k}(\omega) e^{e\omega t} \, d\omega.$$

Then for $x = (x_1, x_2) \in \mathbb{R}^2$,

$$\rho_\lambda(x) = \frac{1}{4\pi} \int_0^{2\pi} \psi^+_{j,k}(x_1 \cos\theta + x_2 \sin\theta) w^\varepsilon_{i,\ell}(\theta) \, d\theta. \qquad (4.2)$$

Each $\psi^+_{j,k}(x_1 \cos\theta + x_2 \sin\theta)$ is a *ridge* function of $x \in \mathbb{R}^2$, i.e. a function of the form $r(x_1 \cos\theta + x_2 \sin\theta)$. Therefore ρ_λ is obtained by 'averaging' ridge functions with ridge angles θ localized near $\theta_{i,\ell} = 2\pi\ell/2^i$. A second connection comes by considering the sampling scheme underlying ridgelet frames as described in Section 3. This scheme says that one should sample behaviour along line segments and that those segments should be spaced in the angular variable proportional to the scale 2^{-j} of the wavelet index. The orthonormal ridgelet system consists of elements which are organized angularly in just such a fashion; the elements $\hat{\rho}_\lambda$ are localized 'near' such line segments because the wavelets $w^\varepsilon_{i,\ell}(\theta)$ are localized 'near' specific points $\theta_{i,\ell}$.

Orthonormal ridgelet analysis can be viewed as a kind of wavelet analysis in the Radon domain; if we let $Rf(\theta, t)$ denote the Radon transform and if we let $\tau_\lambda(t, \theta)$ denote the function $\tfrac{1}{2}(\psi^+_{j,k}(t)w^\varepsilon_{i,l}(\theta) + \psi^+_{j,k}(-t)w^\varepsilon_{i,l}(\theta + \pi))$, the $(\tau_\lambda : \lambda \in \Lambda)$ give a system of antipodally symmetrized non-orthogonal tensor wavelets. The ridgelet coefficients α_λ are given by analysis of the Radon transform via $\alpha_\lambda = [Rf, \tau_\lambda]$. This means that the ridgelet coefficients contain within them information about the smoothness in t and θ of the Radon transform. In particular, if the Radon transform exhibits a certain degree of smoothness, we can immediately see that the ridgelet coefficients exhibit a corresponding rate of decay.

5 Ridgelet synthesis of linear singularities

Consider again the Gaussian-windowed half-space (2.1). The CRT of this object is sparse, which suggests that a discrete ridgelet series can be made which gives a sparse representation of g. This can be seen in two ways.

5.1 Using dual frames

It can be shown that there exist constructive and simple approximations using dual frames (which are not pure ridge functions) which achieve any desired rate of approximation on compact sets (Candès 1998, ch. 5). Indeed, let A be compact and ψ_i be a ridgelet frame for $L_2(A)$. Out of the exact series

$$g = \sum_i \langle g, \psi_i \rangle \tilde{\psi}_i, \qquad (5.1)$$

extract the m-term approximation $\tilde{g}_m$ where one only keeps the dual-ridgelet terms corresponding to the m largest ridgelet coefficients $\langle g, \psi_i \rangle$; then the approximant $\tilde{g}_m$ achieves the rate

$$\| g - \tilde{g}_m \|_{L_2(A)} \leq C_r m^{-r} \quad \text{for any } r > 0,$$

provided, say, ψ is a nice function whose Fourier transform is supported away from 0 (like the Meyer wavelet). The result generalizes to any dimension n and is not limited to the Gaussian window. The argument behind this fact is the sparsity of the ridgelet coefficient sequence; each ridgelet coefficient $\langle \psi_{j,k}, Rg(\theta_{j,\ell}, \cdot) \rangle$ being the one-dimensional wavelet coefficient of the Radon transform $Rg(\theta_{j,\ell}, \cdot)$—for fixed θ. From the relation $Rg(\theta, t) = e^{-t^2} \bar{\Phi}(-t \cdot \sin\theta / |\cos\theta|)$, it is easy to see that the coefficients $\langle f, \psi_{a,\theta,b} \rangle$ decay rapidly as θ and/or b move away from the singularities

$$(\theta = \tfrac{1}{2}\pi, \ t = 0) \text{ and } (\theta = \tfrac{3}{2}\pi, \ t = 0).$$

5.2 Using orthonormal ridgelets

Donoho (1998) shows that the orthonormal ridgelet coefficients of g belong to ℓ^p for every $p > 0$. This means that if we form an m-term approximation by selecting the m terms with the m-largest coefficients, the reconstruction $f_m = \sum_{i=1}^m \alpha_{\lambda_i} \rho_{\lambda_i}$ has any desired rate of approximation.

The argument for the orthonormal ridgelet approximation goes as follows. Because orthonormal ridgelet expansion amounts to a special wavelet expansion in the Radon domain, the question reduces to considering the sparsity of the wavelet coefficients of the Radon transform of g. Now, the Radon transform of g, as indicated above, will have singularities of order 0 (discontinuities) at $(t = 0, \theta = \tfrac{1}{2}\pi)$ and at $(t = 0, \theta = \tfrac{3}{2}\pi)$. Away from these points the Radon transform is infinitely differentiable, uniformly so, outside any neighbourhood of the singularities. If we 'zoom in' to fine scales on one of the singularities and make a smooth change of coordinates, the picture we see is that of a function $S(u, v) = |v|^{-1/2}\sigma(u/|v|)$ for a certain smooth bounded function $\sigma(\cdot)$. The wavelet coefficients of such an object are sparse.

6   Ridgelet analysis of ridge functions

Although ridge functions are not in L^2, the continuous ridgelet transform of a ridge function $f = r(x_1 \cos\theta_0 + x_2 \sin\theta_0)$ makes sense; if the ridge profile r is bounded, the transform can be obtained in a distributional sense and obeys

$$(\mathcal{R}_f)(a, b, \theta) = \delta(\theta - \theta_0) \cdot (Wr(a, b)). \tag{6.1}$$

Thus, the transform is *perfectly localized* to the slice $\theta = \theta_0$ of the precise ridge direction and it amounts to the one-dimensional wavelet transform of the profile function there. This exceptional degree of concentration suggests that ridge functions ought to have very sparse representations by discrete ridgelet systems and that a high rate of approximation can be obtained via m-term ridgelet approximations to such ridge functions using simple thresholding. This can be verified in two ways.

6.1   Using dual ridgelets

Suppose that the ridge profile r is supported in the interval $[-1, 1]$ and obeys a sparsity condition on the wavelet coefficients in a nice wavelet basis: the coefficient sequence $\beta \in w\ell_p$ $(p < 2)$. Then the best m-term one-dimensional wavelet approximation to r has an $L_2[-1, 1]$ convergence rate of order $m^{-(1/p-1/2)}$. There exist approximations by superpositions of m dual ridgelets (which are not pure ridge functions) which achieve the $L^2(A)$ rate of approximation $m^{-(1/p-1/2)}$, where A is now the unit disc (Candès 1998, ch. 5 and 7). Such approximants can be constructed by selecting the m terms out of the series (5.1) corresponding to the m largest coefficients.

6.2   Using orthonormal ridgelets

A key point about orthonormal ridgelets is that they are not only in $L^2(R^2)$, but also in $L^1(R^2)$; hence the integral defining orthonormal ridgelet coefficients makes sense for every bounded ridge function. Let the ridge profile $r(t)$ belong to the homogeneous Besov space $\dot{B}^s_{p,p}(\mathbb{R})$, where $s = 1/p$. This means that the best one-dimensional m-term wavelet approximation to r has an $L^\infty(\mathbb{R})$ convergence rate of $m^{-(s-1/p)}$.

Consider now the rate of convergence of thresholded ridgelet expansions. Let $\bar{\eta}_\delta(y, x) = y1_{\{y\cdot x>\delta\}}$ be a thresholding function with a second 'scaling' argument allowing for adjustment of the threshold. For a bounded function f, with

$$\bar{m}(\delta) = \sum_\Lambda 1_{\{|\langle f, \rho_\lambda\rangle| > \delta/\|\rho_\lambda\|_{L^\infty(D)}\}}$$

finite, set

$$\tilde{f}_\delta = \sum_\Lambda \eta_\delta^{(2)}(\langle f, \rho_\lambda \rangle, \|\rho_\lambda\|_{L^\infty(D)}) \rho_\lambda.$$

In effect, thresholding is driven by the interaction between the size of a coefficient and the 'effect' of the corresponding basis function inside the unit disc.

Let $r_\theta(x)$ denote the corresponding ridge function of $x \in \mathbb{R}^2$. Let $\bar{f}_{m(\delta)}$ be the $\bar{m}(\delta)$-term orthonormal ridgelet approximation to the ridge function f. Then

$$\|f - \bar{f}_m\|_{L^\infty(D)} \leq C \cdot m^{-(s-1/p)}, \quad m \to \infty. \tag{6.2}$$

In effect, this result is ideal, as it gives the same rate $m^{-(s-1/p)}$ we could hope to obtain by knowing that the underlying approximand was a ridge function in a specific direction and exploiting that information fully—even though the ridgelet thresholding does not 'know' or 'exploit' such information.

These results suggest that dual ridgelet frames and orthonormal ridgelets, although *not* ridge functions, can play the same role in approximation as pure ridge functions. More precisely, suppose an arbitrary function f is well-approximated by a sequence of m-term superpositions of ridge functions; it seems that f should also be well approximated by m-term superpositions from discrete ridgelet systems.

7 Ridge spaces

An important fact about wavelets is their relationship to two special families of functional spaces—the Besov spaces and the Triebel spaces. Taken together, these families of spaces include an important collection of classical functional spaces, such as L^2 spaces, L^p spaces, Sobolev spaces, Hölder spaces and so on. Wavelets provide a special basis for such spaces (an unconditional basis) (Meyer 1990) and provide near-optimal approximations to elements taken from functional balls of such spaces.

With the existence of a new family of transforms, we have the possibility to ask: what are the spaces that these transforms are most naturally associated to? Candès (1998) defines a family of spaces $R^s_{p,q}$—'ridge spaces'—which consist of functions f with ridgelet coefficients obeying certain constraints:

$$\|f\|_{\dot{R}^s_{p,q}} = \left(\int \left[\int |\mathcal{R}_f(a, \theta, b)|^p \, db \, d\theta \right]^{q/p} \frac{da}{a^{q(s+1)+1}} \right)^{1/q}$$

and similarly for higher dimensions where $d\theta$ is replaced by the uniform measure on the sphere and the scale factor $a^{q(s+1)+1}$ by $a^{q(s+n/2)+1}$. (The above display corresponds to the homogeneous ridge spaces (see Candès (1998) for a corresponding inhomogeneous version).) Although the definition looks rather internal, it is possible to give an external characterization of such spaces because of the intimate relationship between the ridgelet analysis and the wavelet analysis of the Radon transform $Rf(u, t)$. In fact, letting $p = q$, one can check that

$$\|f\|^p_{\dot{R}^s_{p,p}} \asymp \text{Ave}_u \|Rf(u, \cdot)\|^p_{\dot{B}^{s+(n-1)/2}_{p,p}},$$

where the notation $\dot{B}^{s+(n-1)/2}_{p,p}$ stands for the usual one-dimensional homogeneous Besov norm. From this characterization, it is clear that s is a smoothness parameter and that both parameters p, q serve to measure smoothness. Here, smoothness has to be understood in a non-classical way; we are not talking about the local behaviour of a function but rather about its behaviour near lines (or if one is in dimension $n > 2$, near hyperplanes).

To capture the essence of such spaces, let us return to our original mutilated Gaussian example, (2.1), generalized to dimension n:

$$g(x_1, \ldots, x_n) = 1_{\{x_n > 0\}} e^{-(x_1^2 + \cdots + x_n^2)}.$$

From a classical point of view, in any dimension, this object has barely one derivative (in an L_1 sense) meaning that its first derivative is a singular measure, the singularity being supported on the plane $\{x_n = 0\}$. However, under our new definition, this same object is quite smooth and in fact its regularity increases as the dimension increases, as explained in Candès (1998). What do typical elements of these new spaces look like? The mutilated Gaussian is a typical element of $\dot{R}^s_{1,\infty}$ for $s \leq 1 + \frac{1}{2}(n - 1)$.

For classical Besov spaces, Meyer (1990) tells us that typical elements of $\dot{B}^1_{1,1}$, for instance, are bumps of various scales and at various locations and that the latter space is nothing else than the collection of convex combinations of those bumps (bump algebra). An analogous observation can be made for the ridge spaces (Candès 1998, ch. 4). On the real line, a normalized *point singularity* σ of degree zero, say, is a smooth function away from the origin that may or may not have a pathological behaviour at the origin: that is, we want $|\sigma(t)| \leq 1$ and for a few derivatives $|d^m \sigma(t)/dt^m| \leq |t|^{-m}$ for $t \neq 0$ and $m \leq M$. As an example we have the Heaviside $1_{\{x > 0\}}$, or a smoothly windowed version of the Heaviside. Next, out of a one-dimensional point singularity σ, we create a *ridge singularity* $\sigma(u'x - b)$, where u is a unit vector and b a scalar, and consider the set of functions arising as convex combinations of such ridge singularities:

$$\mathcal{S} = \left\{ f(x) = \sum_i a_i \sigma_i(u'_i x - b_i), \ \sum_i |a_i| \leq 1 \right\}.$$

Then, if we look at objects restricted to the unit ball, the membership of an object in $\mathcal{S}$ is essentially equivalent to a statement about the norm of this object in the norm $\dot{R}^s_{p,q}$ for appropriate (s, p, q). More precisely, we have the following double inclusion:

$$R^{1+(n-1)/2}_{1,1}(C_1) \subset \mathcal{S} \subset R^{1+(n-1)/2}_{1,\infty}(C_2), \tag{7.1}$$

saying that compactly supported objects with $R_{1,1}^{1+(n-1)/2}$ norm not exceeding C_1 are convex combinations of ridge singularities, and that every such convex combination has a bounded $\dot{R}_{1,\infty}^{1+(n-1)/2}$ norm.

It follows from this characterization that ridge spaces model very special conormal objects: objects that are singular across a collection of hyperplanes and smooth elsewhere, where there might be an arbitrary number of hyperplanes in all possible spatial orientations and/or locations.

Earlier, we claimed that ridgelets were naturally associated with the representation of ridge spaces. In fact ridgelets provide near-optimal approximations to elements of these spaces, in much the same way that wavelets provide near-optimal approximations to elements of Besov spaces. For instance, we know that the L_2 error of approximation to a mutilated Gaussian by an m-term linear combination of dual-ridgelets decays more rapidly than m^{-r} for any $r > 0$; the space $R_{1,\infty}^{1+(n-1)/2}$ being more or less the convex hull of such mutilated smooth objects, it is natural to guess that ridgelets provide the right dictionary to use for approximating these spaces.

We can make this more precise. Suppose we are given a dictionary $\mathcal{D} = \{g_\lambda, \lambda \in \Lambda\}$ and that we are interested in the L_2 approximation of a generic class of functions $\mathcal{F}$ out of finite linear combinations of elements of $\mathcal{D}$. For a function f and dictionary $\mathcal{D}$, we define its m-term approximation error by

$$d_m(f, \mathcal{D}) \equiv \inf_{(\alpha_i)_{i=1}^m} \inf_{(\lambda_i)_{i=1}^m} \left\| f - \sum_{i=1}^m \alpha_i g_{\lambda_i} \right\|,$$

and measure the quality of approximation of the class $\mathcal{F}$ using m selected elements of $\mathcal{D}$ by

$$d_m(\mathcal{F}, \mathcal{D}) \equiv \sup_{f \in \mathcal{F}} d_m(f, \mathcal{D})$$

(the worst case error over $\mathcal{F}$). Then, let us consider the class $\mathcal{F}$ of functions whose $R_{p,q}^s$-norm is bounded by some constant C (that will be denoted $R_{p,q}^s(C)$), to be approximated in the metric of $L_2(A)$ for some compact set A. We impose the additional restriction $s > n(1/p - \frac{1}{2})_+$ to guarantee that our class belongs to L_2 also. Then, Candès (1998, ch. 5) shows that no reasonable dictionary would give a better rate of approximation than $m^{-s/d}$: that is, for any reasonable dictionary,

$$d_m(R_{p,q}^s(C), \mathcal{D}) \geq K m^{-s/d}.$$

On the other hand, thresholding the ridgelet expansion gives the optimal rate of approximation. Namely, if $|\alpha|_{(m)}$ denotes the mth largest amplitude among the $(|\alpha_i|)$, the m-term series

$$\tilde{f}_m = \sum_i \alpha_i \mathbb{1}_{\{|\alpha_i| \geq |\alpha|_{(m)}\}} \tilde{\psi}_i$$

produced by thresholding at $|\alpha|_{(m)}$ achieves the optimal rate

$$\sup_{f \in R_{p,q}^s(C)} \| f - \tilde{f}_m \|_{L_2(A)} \leq K' m^{-s/d},$$

for some constant $K' = K'(A, C, s, p, q)$.

The result says that we have an asymptotically near-optimal procedure for binary encoding elements of $R^s_{p,q}(C)$: let $L(\epsilon, R^s_{p,q}(C))$ be the minimum number of bits necessary to store in a lossy encoding–decoding system in order to be sure that the decoded reconstruction of every $f \in R^s_{p,q}(C)$ will be accurate to within ϵ (in an L_2 sense). Then, a coder–decoder based on simple uniform quantization (depending on ϵ) of the coefficients α_i followed by simple run-length coding achieves both a distortion smaller than ϵ and a code length that is optimal up to multiplicative factors like $\log(\epsilon^{-1})$ (Donoho 1996).

8   Ridgelets and curves

As we have said earlier, wavelets are in some sense adapted to zero-dimensional singularities, whereas ridgelets are adapted to higher-dimensional singularities; or more precisely, singularities on curves in dimension two, singularities on surfaces in dimension three, and singularities on $(n-1)$-dimensional hypersurfaces in dimension n. Unfortunately, the task that ridgelets must face is somewhat more difficult than the task which wavelets must face, since zero-dimensional singularities are inherently simpler objects than higher-dimensional singularities. In effect, zero-dimensional singularities are all the same—points—while a one-dimensional singularity—lying along a one-dimensional set—can be curved or straight. *Ridgelets are specially adapted only to straight singularities.*

One way to see this is to look at the CRT of a curved singularity. Again in dimension $n = 2$, consider the object $g' = e^{-x_1^2 - x_2^2} \cdot 1_{\{x_2 > x_1^2\}}$. Qualitatively, it is not hard to see that the Radon transform of such an object has a singularity along a curve, and not just at a point: that is, in the Radon domain, there is a smooth curve $t_0(\theta)$ so that in a neighbourhood of $(t_0(\theta), \theta)$, we have $Rg(t, \theta) \sim w(\theta)(t - t_0(\theta))_+^{1/2}$ for some smooth function w. When we take the wavelet transform in t along each fixed value of θ, we will find that the transform is not nearly as sparse as it was with g.

One can adapt to this situation by the method of localization, which has been frequently used, for example, in time-frequency analysis. We divide the domain in question into squares, and smoothly localize the function into smooth pieces supported on or near those squares either by partition of unity or by smooth orthonormal windowing. We then apply ridgelet methods to each piece. The idea is that, at sufficiently fine scale, a curving singularity looks straight, and so ridgelet analysis—appropriately localized—works well in such cases.

9   Discussion

Because of space limitation, the situation in higher dimensions and the structure of fast ridgelet transform algorithms for lower dimensions, for example, have not been mentioned in this paper. Information on these and related topics can be found in the references below.

10 References

Barron, A. R. 1993 Universal approximation bounds for superpositions of a sigmoidal function. *IEEE Trans. Inform. Theory* **39**, 930–945.

Candès, E. J. 1998 Ridgelets: theory and applications. PhD thesis, Department of Statistics, Stanford University.

Candès, E. J. 1999 Harmonic analysis of neural networks. *Appl. Comput. Harmon. Analysis* **6**(2), 197–218.

Cybenko, G. 1989 Approximation by superpositions of a sigmoidal function. *Math. Control Signals Systems* **2**, 303–314.

Daubechies, I. 1988 Orthonormal bases of compactly supported wavelets. *Commun. Pure Appl. Math.* **41**, 909–996.

Donoho, D. L. 1993 Unconditional bases are optimal bases for data compression and for statistical estimation. *Appl. Comput. Harmon. Analysis* **1**, 100–115.

Donoho, D. L. 1996 Unconditional bases and bit-level compression. *Appl. Comput. Harmon. Analysis* **3**, 388–392.

Donoho, D. L. 1998 Orthonormal ridgelets and linear singularities. Report no. 1998-19, Department of Statistics, Stanford University.

Friedman, J. H. & Stuetzle, W. 1981 Projection pursuit regression. *J. Am. Statist. Ass.* **76**, 817–823.

Helgason, S. 1986 *Groups and geometric analysis*. New York: Academic.

Jones, L. K. 1992 A simple lemma on greedy approximation in Hilbert space and convergence rates for projection pursuit regression and neural network training. *Ann. Statist.* **20**, 608–613.

Lemarié, P. G. & Meyer, Y. 1986 Ondelettes et bases Hilbertiennes. *Rev. Mat. Iberoamericana* **2**, 1–18.

Logan, B. F. & Shepp, L. A. 1975 Optimal reconstruction of a function from its projections. *Duke Math. Jl* **42**, 645–659.

Meyer, Y. 1990 *Ondelettes et opérateurs*. Paris: Hermann.

Murata, N. 1996 An integral representation of functions using three-layered networks and their approximation bounds. *Neural Networks* **9**, 947–956.

7

Wavelets in time-series analysis

Guy P. Nason *Department of Mathematics, University of Bristol, University Walk, Bristol BS8 1TW, UK*

Rainer von Sachs
Institute of Statistics, Catholic University of Louvain, Louvain-la-Neuve, Belgium

Abstract

This article reviews the role of wavelets in statistical time-series analysis. We survey work that emphasizes scale, such as estimation of variance, and the scale exponent of processes with a specific scale behaviour, such as $1/f$ processes. We present some of our own work on locally stationary wavelet (LSW) processes, which model both stationary and some kinds of non-stationary processes. Analysis of time-series assuming the LSW model permits identification of an evolutionary wavelet spectrum (EWS) that quantifies the variation in a time-series over a particular scale and at a particular time. We address estimation of the EWS and show how our methodology reveals phenomena of interest in an infant electrocardiogram series.

Keywords: Allan variance; locally stationary time-series; long-memory processes; time-scale analysis; wavelet processes; wavelet spectrum

1 Introduction

Reviewing the role of wavelets in statistical time-series analysis (TSA) appears to be quite an impossible task. For one thing, wavelets have become so popular that such a review could never be exhaustive. Another, more pertinent, reason is that there is no such thing as *one* statistical time-series *analysis*, as the very many different fields encompassed by TSA are, in fact, so different that the choice of a particular methodology must naturally vary from area to area. Examples for this are numerous: think about the fundamentally different goals of treating comparatively short correlated biomedical time-series to explain, for example, the impact of a set of explanatory variables on a response variable, or of analysing huge inhomogeneous datasets in sound, speech or electrical engineering, or, finally, building models for a better understanding and possibly prediction of financial time-series data.

Hence, here we can only touch upon some aspects of where and why it can be advantageous to use wavelet methods in some areas of statistical TSA. We consider the common situation where the trend and (co-)variation of autocorrelated data are to be modelled and estimated. Classically, the time-series data are assumed to be *stationary*: their characterizing quantities behave homogeneously over time. Later in this paper, we shall consider situations where some controlled deviation from stationarity is allowed.

We stress that this article does not provide an exhaustive review of this area. In particular, we shall not go into detail concerning the time-series aspects of wavelet denoising, nor the role of wavelets in *deterministic* time-series, but briefly summarize these next. Later on, we shall concentrate on seeing how wavelets can provide information about the *scale* behaviour of a series: i.e. what is happening at different scales in the series. This type of analysis, as opposed to a frequency analysis, can be appropriate for certain types of time-series as well as provide interpretable and insightful information that is easy to convey to the non-specialist.

Wavelet denoising.   Silverman (this issue) and Johnstone (this issue) describe the method and rationale behind *wavelet shrinkage*, a general technique for curve denoising problems. The specific connection with the analysis of correlated time-series data $\{Y_t\}$ is that in a general regression-like model,

$$Y_t = m(X_t) + \sigma(X_t)\epsilon_t, \quad t = 1, \ldots, T,$$

non-parametric estimation of the trend function m and/or the function σ, which measures the variability of the data, can be performed in exactly the same framework of nonlinear wavelet shrinkage as for the original simple situation of Gaussian i.i.d. data (see Silverman, this issue). In typical biomedical applications, the errors (and, hence, the Y_t themselves) need be neither uncorrelated nor even stationary (for one example, in the case of a deterministic equidistant design, $\{X_t\} = \{t/T\}$ (see von Sachs & MacGibbon 2000)). In financial time-series analysis, where the choice of $\{X_t\} = \{(Y_{t-1}, \ldots, Y_{t-p})\}$ leads to a non-parametric autoregressive model (of order p), the task is estimation of a conditional mean and variance under the assumption of *heteroscedastic* stationary errors (see, for example, Hoffmann (1999) for the case $p = 1$). Both examples, but particularly the second one, call for *localized* methods of estimation. Problems of this sort typically show regimes of comparatively smooth behaviour, which, from time to time, may be disrupted by break points or other discontinuities. A subordinate task would then be to detect and estimate the precise location of these break points, and here wavelet methods again prove useful. Finally, for this little review, we mention that wavelet shrinkage can be used for the estimation of spectral densities of stationary time-series (Neumann 1996; Gao 1997) and of time-varying spectra using localized periodogram-based estimators (von Sachs & Schneider 1996; Neumann & von Sachs 1997).

Deterministic time-series.   A completely different use of wavelets in statistical time-series analysis was motivated by how wavelets originally entered the field of (deterministic) time–frequency analysis (see, for example, Rioul & Vetterli 1991; or Flandrin 1998). Here, wavelets, being *time-scale* representation methods, deliver a tool complementary to both classical and localized (e.g. windowed) Fourier analyses. We will focus next on this aspect for *stochastic* signals.

Stochastic time-series.   Recently, the search for localized time-scale representations of stochastic correlated signals led to both analysing and synthesizing (i.e. modelling) mainly *non-stationary* processes with wavelets or wavelet-like bases. By non-stationarity we mean here two different types of deviation from stationarity: first, but not foremost, we will address, in Section 2, wavelet analyses of certain long-memory processes that show a specific (global or local) scale behaviour, including the well-known $1/f$ (or power law) processes.   However, our main emphasis is on signals with a possibly time-changing probability distribution or characterizing quantities. Overall, from the modelling point of view, wavelets offer clear advantages mainly for the two types of non-stationary data mentioned above.

Statistical time-scale analysis performs a statistical wavelet *spectral analysis* of time-series analogous to the classical *Fourier* spectral analysis, where the *second-order* structure of the series such as variance, autocovariance (dependence structure within the series) and spectral densities are to be estimated (see Priestley 1981). The analyses depend fundamentally on processes possessing a representation in terms of random coefficients with respect to some localizing basis: in classical spectral analysis, the Fourier basis, which is perfectly localized in frequency; in time-scale analysis, a basis that is localized in time and scale. Then, the respective second-order quantities of interest (e.g. variance or autocovariance) can be represented by a superposition of the (Fourier or the wavelet) spectra. Our specific model, in Section 3 *b*, uses a particular set of basis functions: discrete non-decimated wavelets.

However, before turning to scale- *and* time-localized models, the next section reviews the basic ideas about using wavelets for the *analysis* of statistical phenomena with characteristic behaviour living on certain global scales. For example, a variance decomposition on a scale-by-scale basis has considerable appeal to scientists who think about physical phenomena in terms of variation operating over a range of different scales (a classical example being $1/f$ processes). We will review a very simple example to demonstrate how insights from *analysis* can be used to derive models for the *synthesis* of stochastic processes.

Both in analysis and synthesis it is, of course, possible to *localize* these scale-specific phenomena in time as well.   For the specific example of $1/f$ processes, we refer to Gonçalvès & Flandrin (1993) and more recent work of, for example, Wang *et al.*(1997), and the overview on wavelet analysis, estimation and synthesis of scaling data by Abry *et al.*(1999). The common paradigm of all these approaches is the separation of the scale on which the process data are sampled from the scale(s) where the respective behaviour is observed.

2   Estimation of 'global' scale behaviour

This section gives examples on estimating the scale characteristics of processes that do not show a location dependency. In fact, we restrict our discussion to the utility of wavelets for the analysis and synthesis of long-memory processes. We begin with a brief introduction to scalograms using the example of the *Allan variance*, originally developed by Allan (1966) as a time-domain measure of frequency stability in high-frequency oscillators (McCoy & Walden 1996). Then we turn, more specifically, to the problem of estimation of the scale exponent of $1/f$ processes, in the specific case of fractional Brownian motion (fBm).

2.1   Long-memory processes, Allan and wavelet variance

For pedagogical reasons, we will concentrate on Percival & Guttorp (1994), one of the earlier papers in the vast literature in this field, and also, for a simple exposition, we will only concentrate on the Haar wavelet although other wavelets may be equally, if not more, useful.

Consider a stretch of length T of a given zero mean stochastic process $\{X_t\}_{t \in Z}$. The *Allan variance* $\sigma_X^2(\tau)$ at a particular scale $\tau \in Z$ is a measure of how averages, over windows of length τ, change from one period to the next. If

$$\bar{X}_t(\tau) = \frac{1}{\tau} \sum_{n=0}^{\tau-1} X_{t-n},$$

then

$$\sigma_X^2(\tau) := \tfrac{1}{2} E(|\bar{X}_t(\tau) - \bar{X}_{t-\tau}(\tau)|^2). \tag{2.1}$$

In order to have a meaningful quantity independent of time t, X_t must be stationary, or at least have stationary increments of order 1.

In fact, the Allan variance turns out to be proportional to the Haar wavelet variance, another 'scale variance' based on the standard discrete Haar wavelet transform as follows. Let $\{\hat{d}_{jk}\}$ denote the (empirical) wavelet coefficients of the signal $\{X_t\}_{t=0,\dots,T-1}$, where we deviate from standard notation in that our scales $j = -1, \dots, -\log_2(T)$ become more negative the coarser the level of the transform, and, hence, the location index k runs from 0 to $T/2^{-j} - 1$. This proportionality can easily be observed by writing the Haar coefficients as successive averages of the data with filters $1/\sqrt{2}$ and $-1/\sqrt{2}$ (which are the high-pass filter coefficients $\{g_k\}$ of the Haar transform (see Silverman, this issue)). For example, for $j = -1$ at scale $\tau_j = 2^{-j-1} = 1$,

$$\hat{d}_{-1,k} = (1/\sqrt{2})(X_{2k+1} - X_{2k}), \quad k = 0, \dots, T/2 - 1, \tag{2.2}$$

and it is easy to see that $\text{var}\{\hat{d}_{-1,k}\} = \sigma_X^2(1)$. More generally,

$$\text{var}\{\hat{d}_{jk}\} = E\hat{d}_{jk}^2 = \tau_j \sigma_X^2(\tau_j). \tag{2.3}$$

Motivated by this last equation, an unbiased estimator for the Allan variance, the so-called 'non-overlapped' estimator, is the appropriately normalized sum of the squared wavelet coefficients:

$$\hat{\sigma}_X^2(\tau_j) := \frac{2}{T} \sum_{k=0}^{T/2^{-j}-1} \hat{d}_{jk}^2. \tag{2.4}$$

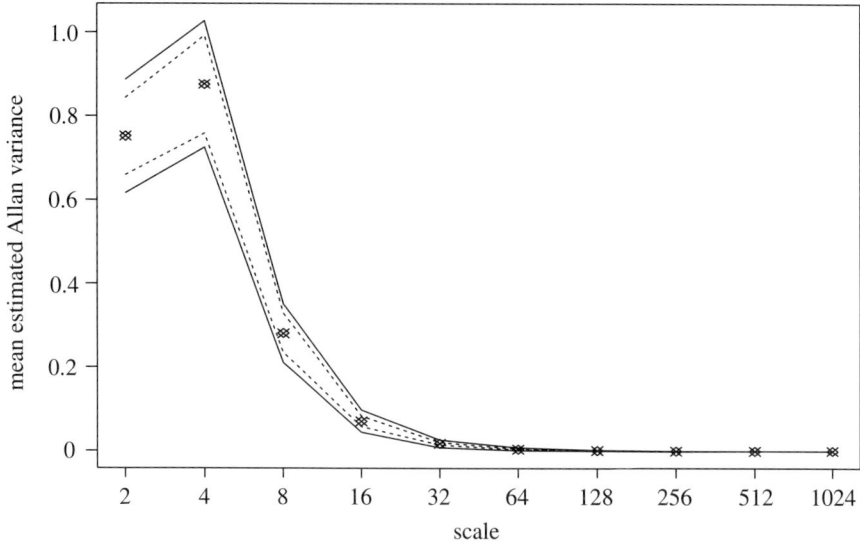

Fig. 1: Mean of estimated Allan variances over 1000 simulations of the MA(3) process computed using Haar wavelet variances (see text). The mean of the estimates is shown by the symbols, the mean plus or minus twice the standard deviation of the estimate is shown by lines. For the 'non-overlap' estimator the symbols are × with solid error lines; for the 'maximal-overlap' estimator the symbols are ⋄ and the error lines are dotted. The theoretical values of the Allan variance at scales 2, 4 and 8 are $\frac{3}{4}$, $\frac{7}{8}$ and $\frac{3}{6}$ 128 respectively.

This estimator has the property that each datapoint X_t contributes to exactly one coefficient $\hat{d}_{jk}$. Again only considering the finest scale, $j = -1$, formula (2.4) can be written in terms of the data as

$$\hat{\sigma}_X^2(1) = \frac{1}{T} \sum_{k=0}^{T/2-1} (X_{2k+1} - X_{2k})^2.$$

We observe immediately that one can improve upon the above estimator by summing over not just $T/2$ values of these time-series differences but over all $T - 1$ possible ones. The resulting estimator will clearly have a smaller variance and also possesses independence with respect to the choice of the origin of the series $\{X_t\}$. This 'maximal-overlap' estimator, denoted by $\tilde{\sigma}_X^2(\tau_j)$, is based on the 'non-decimated' (Haar) wavelet transform (NDWT), with wavelet coefficients d_{jk} (see the appendix for a description). The NDWT amounts to a uniform sampling in k instead of the inclusion of subsampling or decimation in each step of the standard discrete (decimated) wavelet transform (DWT).

Figure 1 shows an estimator for the Allan variance of the MA(3) process

$$X_t^{(2)} = \tfrac{1}{2}(\varepsilon_t + \varepsilon_{t-1} - \varepsilon_{t-2} - \varepsilon_{t-3}),$$

with standard Gaussian white noise ε_t, using both the 'non-overlap' and 'maximal-overlap' estimators. This MA process is one of a class that we will meet in Section 3. The figure was produced by simulating 1024 observations from the process and computing the estimated Allan variance. This simulation procedure was repeated 1000 times and the figure actually shows the mean of all the estimated Allan variances with lines showing the accuracy of the estimates. It is clear that the 'maximal-overlap' estimator has a smaller variance for a wide range of scales τ_j. It is clear from the formula of the MA(3) process that it 'operates' over scale 4, and, thus, the Allan variance at scale 4 ($j = -2$) is largest in figure 1. Further, the Allan variance indicates that there is variation at scale 2; this is not surprising because the process formula clearly links quantities over that scale as well. However, at scale 8 and larger scales, the process has insignificant variation, and so the Allan variance decays for large scales. Using (2.3) and the orthogonality of the DWT, it is easy to see that the Allan variance for standard white noise is

$$\sigma^2(\tau_j) = 1/\tau_j = 2^{j+1}, \quad j < 0.$$

More general wavelets could be used in place of Haar in the wavelet variance estimators given above. In any case, the use of the NDWT will be beneficial as is clear from the example above.

Why is the concept of a 'scale variance' useful at all? The 'scale variance' permits a new decomposition of the process variance which is different (but related) to the classical spectral decomposition using the (Fourier) spectral density. That is, suppose we have a stationary process $\{X_t\}$, then we can decompose its total variance var$\{X_t\}$ into quantities that measure the fluctuation separately scale by scale:

$$\text{var}\{X_t\} = \tfrac{1}{2} \sum_{j=-\infty}^{-1} \sigma_X^2(\tau_j) = \sum_j \text{var}\{\tilde{d}_{jk}\}/2\tau_j. \tag{2.5}$$

Here, on the right-hand side, the scale-dependent quantities play the role of a 'wavelet spectrum' $S_j := \text{var}\{\tilde{d}_{jk}\}/2\tau_j$, which, in the stationary case, is independent of time k. For white noise, $S_j = 2^j$. We will define a time-dependent wavelet spectrum in (3.4) for more general non-stationary processes in Section 3.

The Allan (or wavelet) variance is of additional interest for the study of long-memory processes. These are processes in which the autocorrelation of the process decays at a very slow rate, such that it is possible for effects to persist over long time-scales (see Beran 1994). Consider a 'fractional Gaussian noise' (fGn) process with self-similarity parameter $\tfrac{1}{2} < H < 1$, or, more precisely, the first-order increments $\{Y_t\}$ of an fBm $B = B_H$ (with $B_0 = 0$). This process is characterized by having normally distributed stationary and self-similar increments

$$Y_t := \frac{B_{s+t} - B_s}{t} \sim \frac{B_t - B_0}{t} \sim |t|^{H-1} B_1 \sim N(0, \sigma^2|t|^{2H-2})$$

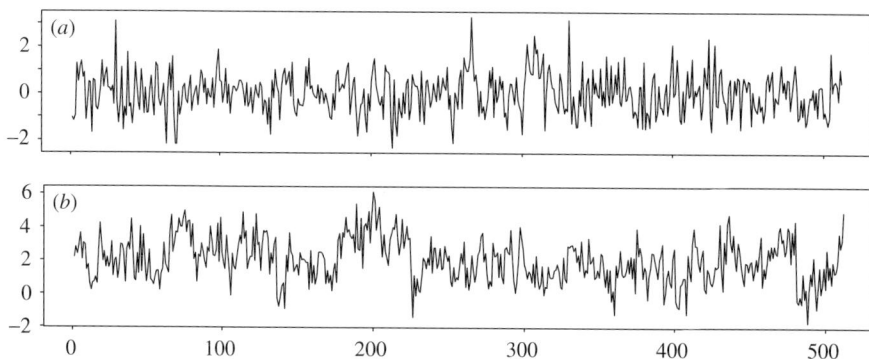

Fig. 2: (*a*) Realization of an fGn process with very little 'long memory'. (*b*) Realization of an fGn process with evident long memory. In (*b*) you can 'see' a 'slow oscillation' underlying the process.

(see, for example, Abry *et al.* 1995). Figure 2 shows sample paths for two different simulated fGn processes. Try and guess which one has the long-memory before you study the caption!

It can be shown that the Allan variance of $\{Y_t\}$ follows a power law, i.e.

$$\sigma_Y^2(\tau) = L(\tau)|\tau|^{2H-2} \quad |\tau| \gg t_0,$$

where $L(\cdot)$ is a slowly varying function for $\tau \to \infty$. Hence, a plot of $\log\{\sigma_Y^2(\tau)\}$ versus $\log(\tau)$ (or, more precisely, a least-squares or maximum-likelihood estimate of this log-linear relationship based on one of the estimators for σ_Y^2) can reveal an estimator of the parameter H, and, in general, for large enough τ one can observe to a good approximation a line with slope $2H - 2$. Here we now see that it can be useful to use wavelets other than Haar when investigating a potential relationship because different wavelets cover slightly different frequency ranges and possess differing phase behaviour.

Some example applications can be found in Percival & Guttorp (1994) on the analysis of the scales of variation within vertical ocean shear measurements, which obey a power-law process over certain ranges. Serroukh *et al.*(1998) investigate the wavelet variance of the surface albedo of pack ice, which happens to be a strongly non-Gaussian series. This paper also derives statistical properties of the estimators used for obtaining confidence intervals.

2.2 Wavelet spectral analysis of $1/f$ processes

We now slightly change our point of view, and instead of variances over certain scales we now examine the more general quantity: the spectral density or spectrum. The spectrum is the Fourier transform of the autocovariance function of a stationary process. For its proper definition in case of the (non-stationary) fBm, we again must use the fGn, which has a power-law spectral density as follows,

$$f_Y(\omega) = L_f(\omega)|\omega|^{1-2H}, \quad 0 < |\omega| \ll t_0^{-1},$$

where, again, $L_f(\cdot)$ is a slowly varying function, now for $\omega \to 0$. If $H > \frac{1}{2}$, we observe a singularity in zero frequency that is, again, an indicator for a strongly correlated time-series, i.e. one with long memory.

As before, from the statistical point of view, there is a linear relation between $\log\{f_Y(\omega)\}$ and $\log(|\omega|)$ for small enough $|\omega|$, which allows us to base an estimator for the spectral exponent $\alpha = 1 - 2H$ on an estimator for the spectrum f_Y. This will, in fact, be one of the estimators $\hat{\sigma}_Y^2$ or $\tilde{\sigma}_Y^2$ for the wavelet variance σ_Y^2, which is related to the spectrum f_Y by the equation (2.6) below. We summarize Section 2 of Abry *et al.*(1995) who clarify why wavelet is superior to traditional Fourier spectral analysis for power law processes. The two methods may be compared by examining the expectation of the wavelet variance estimator, $\hat{\sigma}_Y^2(\tau_j)$, and the expectation of the average of short-time Fourier periodograms over segments of equal length of $\{Y_t\}$. The Fourier estimator is constructed in a similar way to the wavelet variance estimator, except that Gabor-like basis functions (appropriately weighted exponentials) instead of wavelets are used. In other words, the Fourier estimator is the time marginal of a particular bilinear time–frequency distribution, the *spectrogram*, which is the squared modulus of a Gabor or short-time Fourier transform. In this context, the wavelet variance estimator is the time marginal of the *scalogram*, i.e. the squared wavelet coefficients. The properties of the two estimators are different because of the way that energy is distributed over the two different configurations of atoms (Fourier, rectangular equal-area boxes centred at equispaced nodes; wavelets, the famous constant-Q tiling with centres located on the usual wavelet hierarchy).

The expectation of the time marginal of the spectrogram can be written as the convolution of the Fourier spectrum $f_Y(\omega)$ of $\{Y_t\}$ with the squared Fourier transform of the moving window. Similarly, the expectation of the time marginal of the scalogram, i.e. of $\hat{\sigma}_Y^2(\tau_j)$ (for reasons of simplicity we only refer to the inferior estimated based on the DWT), is the convolution of $f_Y(\omega)$ with the squared modulus of the Fourier transform $\hat{\psi}_{jk}(\omega)$ of the wavelets in the DWT, i.e.

$$\sigma_Y^2(\tau_j) = E\hat{\sigma}_Y^2(\tau_j) = \frac{2}{T}\sum_k E\hat{d}_{jk}^2 = \tau_j^{-1}\int_{-\pi}^{\pi} f_Y(\omega)|\hat{\psi}_{jk}(\omega)|^2\,d\omega. \qquad (2.6)$$

Abry *et al.*(1995) show foremost that in a log–log relationship of equation (2.6), the bias for estimating the spectral parameter α of f_Y becomes frequency *independent* when using averaged scalograms instead of averaged spectrograms. This is a consequence of the fact that in the Fourier domain, wavelets scale multiplicatively with respect to frequency, a property that the fixed-window spectrograms do not enjoy. Further considerations in Abry *et al.*(1995), such as those pertaining to the efficiency of the estimators, support the wavelet-based approach for these processes.

Equation (2.6) also helps to further interpret the variation of the considered estimators of $\sigma_Y^2(\tau_j)$ in the example given in figure 1. As the spectrum $f_Y(\omega)$ of the

MA(3) process of this example has its power concentrated near to high frequencies, and as the variance of these estimators are approximately proportional to the square of their mean, it is clear from the integral in (2.6) that this variance increases with frequency, i.e. if we go to finer scales in the plots of figure 1.

Further examples for processes with such a singular power-law behaviour near zero frequency can be found (see Flandrin 1998), e.g. in the areas of atmospheric turbulence (see, for example, Farge, this issue), hydrology, geophysical and financial data, and telecommunications traffic (Abry *et al.*1999), to name but a few. Whitcher *et al.*(1998) detect, test and estimate time-series variance changes. Their work can identify the scale at which the change occurs. Generalizations of these kind of tests to time-varying autocovariance and spectral density for short-memory non-stationary processes can be found in von Sachs & Neumann (2000).

In Section 3 we discuss ideas of how to localize both the analysis and synthesis of the global scale behaviour discussed in this section.

2.3 Synthesis of long-memory processes using wavelets

In the above considerations on *analysis* of $1/f$ processes, we saw that wavelets formed a key role. In reverse, it is not surprising that they are useful also for *synthesis*, i.e. in the theory of modelling $1/f$ processes (see, again, Abry *et al.*1999). Indeed, it is possible to, for example, simulate $1/f$ processes using wavelets (see Wornell & Oppenheim 1992). One method for simulating fGn is given by McCoy & Walden (1996) as follows:

(a) compute the variances, S_j, of the required fGn processes by integrating its spectrum over dyadic intervals $[-2^j, -2^{j-1}] \cup [2^{j-1}, 2^j]$;

(b) for each scale j, draw a sequence of 2^{-j} independent and identically distributed normal random variables d_{jk};

(c) apply the inverse DWT to the $\{d_{jk}\}$ coefficients to obtain an approximate realization of a $1/f$ process.

Figure 2 shows two realizations from fGn processes using the McCoy & Walden (1996) methodology.

3 Wavelet processes: a particular time-scale model

3.1 Local stationarity

Suppose that we have a time-series $\{X_t\}_{t \in \mathbb{Z}}$, and that we wish to estimate the variance $\sigma_t^2 = \text{var}(X_t)$ over time. If the series is stationary, then σ_t^2 will be constant and equal to σ^2 and we can use the usual sum of squared deviations from the mean estimator on a single stretch of the observed time-series $X_1, \ldots, X_T$. As more data become available (as T increases), the estimate of the variance $\hat{\sigma}^2$ improves. Alternatively, suppose that we know that the variance of the series changes at each time point t, i.e.

assume that $\text{var}(X_t) = \sigma_t^2$ for all $t \in \mathbb{Z}$ where none of the σ_t^2 are the same. Here the series is non-stationary and we do not have much hope in obtaining a good estimate of σ_t^2 since the *only* information we can obtain about σ_t^2 comes from the single X_t. As a third alternative, suppose the variance of a time-series changes *slowly* as a function of time t. Then the variance around a particular time t^* could be estimated by pooling information from X_t close to t^*.

A similar situation occurs if the long-memory parameter H in the previous section changes over time, i.e. $H = H(t)$, then the Allan variance would also change over time:

$$\sigma_Y^2(t, \tau) = E(|\bar{Y}_t(\tau) - \bar{Y}_{t-\tau}(\tau)|^2) \sim \tau^{2H(t)-2}.$$

For a series with such a structure, we would hopefully observe

$$\tilde{d}_{j,k}^2 \approx \tilde{d}_{j,k+1}^2,$$

for the NDWT coefficients. To estimate the Allan variance we would need to construct local averages. In other words, we would not sum over *all* empirical NDWT coefficients but would perform adaptive averaging of the $\tilde{d}_{jk}^2$ over k for fixed scale j.

Time-series whose statistical properties are slowly varying over time are called *locally stationary*. Loosely speaking, if you examine them at close range they appear to be stationary and if you can collect enough data in their region of local stationarity then you can obtain sensible estimates for their statistical properties, such as variance (or autocovariance or the frequency spectrum).

One possibility for modelling time-series such as these is to *assume* that

$$\text{var}(\tilde{d}_{j,k}) \approx \text{var}(\tilde{d}_{j,k+1}), \tag{3.1}$$

so that we have some chance of identifying/estimating coefficients from one realization of a time-series. More generally, an early idea for 'local stationarity', due to Silverman (1957), proposes

$$\text{cov}(X_t, X_s) = c(t, s) \approx m(\tfrac{1}{2}(s + t))\gamma(s - t) := m(k)\gamma(\tau), \tag{3.2}$$

where $k = \frac{1}{2}(s + t)$ and $\tau = t - s$. This model says that the covariance behaves locally as a typical stationary autocovariance but then varies from place to place depending on k. The Silverman model reflects our own wavelet-specific model given in (3.5) except that ours decomposes γ over scales using a wavelet-like basis. Other early important work in this area can be found in Page (1952) and Priestley (1965). More recently, Dahlhaus (1997) introduced an interesting model that poses estimation of time-series statistical properties (such as variance) as a curve estimation problem (which bestows great advantages when considering the performance of estimators because it assumes a *unique* spectrum).

Figure 3 shows an example of a time-series that is not stationary. It shows the electrocardiogram (ECG) recording of a 66-day-old infant. There are a number of interesting scientific and medical issues concerning such ECG data, e.g. building

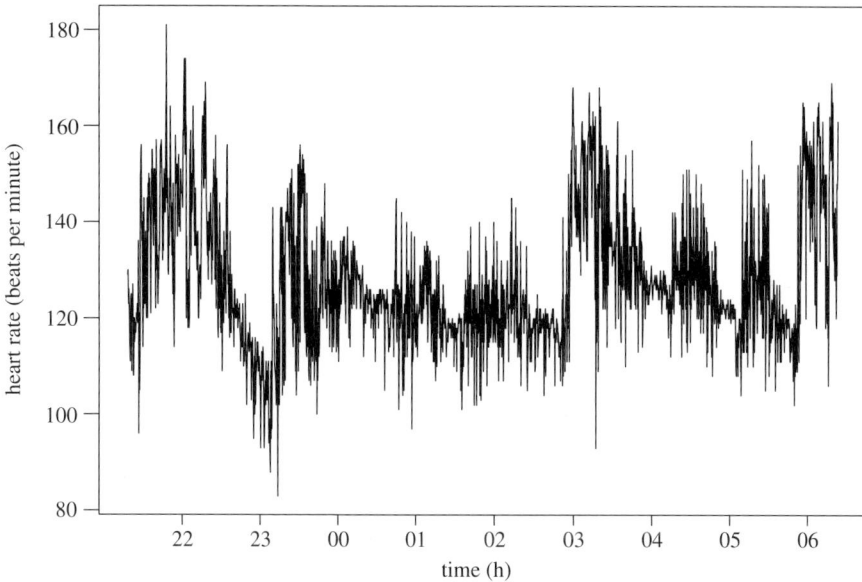

Fig. 3: ECG recording of a 66-day-old infant. Series is sampled at $\frac{1}{16}$ Hz and is recorded from 21:17:59 to 06:27:18. There are $T = 2048$ observations.

and interpreting models between ECG and other covariates such as infant sleep state (see Nason *et al.*1999). However, for the purposes of this article, we shall confine ourselves to examining how the variance of the series changes as a function of time and scale. Further analyses of this sort can be found in Nason *et al.*(1998).

3.2 Locally stationary wavelet processes

3.2.1 *The processes model*

A time-domain model for encapsulating localized scale activity was proposed by Nason *et al.*(1999). They define the *locally stationary wavelet* (LSW) process by

$$X_t = \sum_{j=-J}^{-1} \sum_{k\in\mathbb{Z}} w_{j,k}\psi_{jk}(t)\xi_{jk}, \quad \text{for } t = 0,\ldots,T-1, \tag{3.3}$$

where the $\{\xi_{jk}\}$ are mutually orthogonal zero mean random variables, the $\psi_{jk}(t)$ are discrete non-decimated wavelets (as described in the appendix), and the $w_{j,k}$ are amplitudes that quantify the energy contribution to the process at scales j and location k. Informally, the process X_t is built out of *wavelets with random amplitudes*. The LSW construction is similar in some ways to the well-known construction of stationary processes out of sinusoids with random amplitudes.

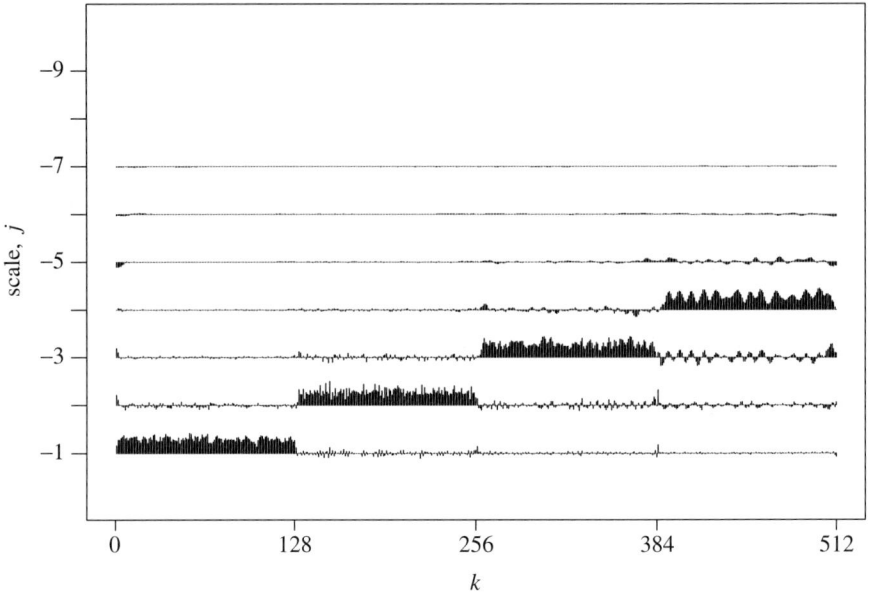

Fig. 4: EWS estimate using non-decimated Haar wavelet transform. (Simulation generated) plot showing (estimated) EWS for concatenated Haar process. The horizontal axis shows 'normal' time k but could be labelled in rescaled time $z = k/512$.

3.2.2 *Evolutionary wavelet spectrum and local variance*

To quantify how the size of $w_{j,k}$ changes over time, we embed our model (3.3) into the Dahlhaus (1997) framework. To model locally stationary processes we use our assumption (3.1) to insist that $w_{j,k}^2 \approx w_{j,k+1}^2$, which forces $w_{j,k}^2$ to change slowly over k. Stationary processes can be included in this model by ensuring that $w_{j,k}^2$ is constant with respect to k. A convenient measure of the variation of $w_{j,k}^2$ is obtained by introducing rescaled time, $z \in (0, 1)$, and defining the *evolutionary wavelet spectrum* (EWS) by

$$S_j(z) = S_j(k/T) \approx w_{j,k}^2, \tag{3.4}$$

for $k = 0, \ldots, T - 1$. In the stationary case we lose the dependence on z (k) and obtain the 'wavelet spectrum' or Allan variance given just after formula (2.5). One can see that as more time-series observations are collected (as T increases), one obtains more information about $S_j(z)$ on a grid of values k/T, which makes the estimation of $S_j(z)$ a standard statistical problem.

The important thing to remember about the EWS is that

> $S_j(z)$ *quantifies the contribution to process variance at scale j and time z.*

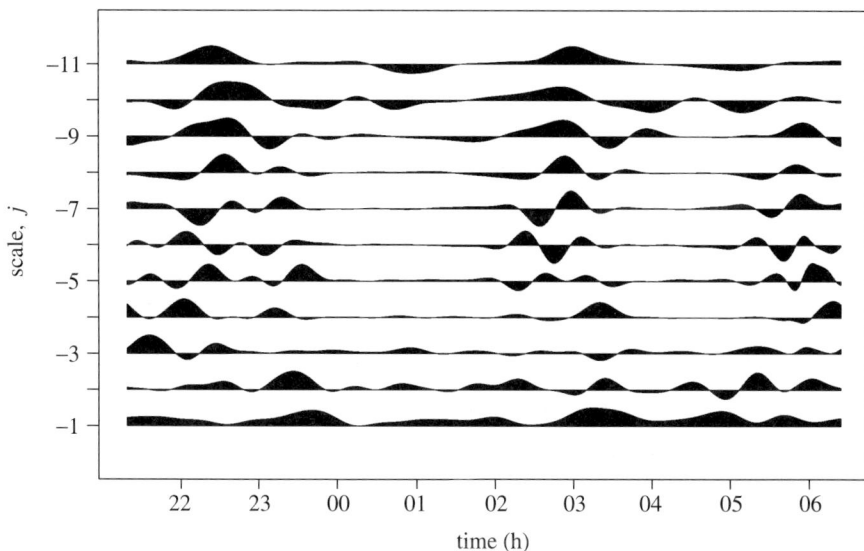

Fig. 5: EWS estimate using non-decimated Haar wavelet transform. Estimate of the evolutionary spectrum $S_j(z)$ for infant ECG series shown in figure 3 (just as figure 4 shows for the simulated concatenated Haar process). The original time axis is shown although it is really rescaled time in the interval $(0, 1)$.

In other words, a large value of $S_j(z)$ indicates that there is a large amount of oscillatory power operating at scale j around location z. For examples of this, see figures 4 and 5.

For a non-stationary process, we would expect the variance of the process to vary over time and so we would expect our model to exhibit a time-localized version of (2.5), i.e. something like

$$\text{var}(X_k) = \sum_j w_{j,k}^2,$$

or, more precisely, (3.7) below. If one takes our process model (3.3) and forms the autocovariance of X_t, then one (asymptotically) obtains an expression in terms of the *autocorrelation* function of the *wavelets*. Let $c(z, \tau)$ define this *localized autocovariance*,

$$\lim_{T \to \infty} \text{cov}(X_{[zT]-\tau}, X_{[zT]+\tau}) = c(z, \tau) = \sum_{j=-\infty}^{-1} S_j(z)\Psi_j(\tau), \qquad (3.5)$$

where $\Psi_j(\tau)$ is the autocorrelation function of the discrete non-decimated wavelets defined by (for Haar ⊓⊔ * ⊓⊔ = ⩕)

$$\Psi_j(\tau) = \sum_{k=-\infty}^{\infty} \psi_{jk}(0)\psi_{jk}(\tau), \tag{3.6}$$

for $j < 0$. The representation in (3.3) is not unique because of the nature of the overdetermined non-decimated wavelet system. However, the autocovariance representation in (3.5) is unique. Relation (3.5) is reminiscent of the Silverman (1957) idea with $S_j(\cdot)$ playing the role of $m(\cdot)$ and $\Psi_j(\tau)$ the role of $\gamma(\tau)$, except that our model separates the behaviour over scales. Relation (3.5) also allows us to define the *localized variance* at time z by

$$v(z) = c(z, 0) = \sum_{j=-\infty}^{-1} S_j(z), \tag{3.7}$$

i.e. the promised localized version of (2.5), note that $\Psi_j(0)$ is always 1 for all scales j. For stationary series, the localized autocovariance collapses to the usual autocovariance $c(\tau)$.

For estimation of the EWS we implement the idea of local averaging expressed in Section 3 a. The EWS may be estimated by a corrected *wavelet periodogram*, which is formed by taking the NDWT coefficients, $\tilde{d}_{jk}$, of the sample realization, squaring them, and then performing a correction to disentangle the effect of the overdetermined NDWT. Like the classical periodogram (see Priestley 1981), the corrected wavelet periodogram is a noisy estimator of the EWS and needs to be smoothed to provide good estimates. The smoothing could be carried out by any number of methods, but, since we wish to be able to capture sharp changes in the EWS, we adopt wavelet shrinkage techniques (Donoho *et al.*1995; Coifman & Donoho 1995). Figure 5 shows a smoothed corrected wavelet periodogram for the infant ECG data. For further detailed analyses on this dataset see Nason *et al.*(1998).

3.2.3 *Motivating example: Haar MA processes*

The MA(1) process,
$$X_t^{(1)} = (\varepsilon_t - \varepsilon_{t-1})/\sqrt{2},$$

is an LSW process as in (3.3), where the amplitudes are equal to 1 for $j = -1$ and zero otherwise, and the constructing wavelets ψ_{jk} are Haar non-decimated wavelets as given in the appendix. The autocovariance function of $X_t^{(1)}$ is $c(\tau) = 1, -\frac{1}{2}, 0$ for $\tau = 0, \pm 1$, otherwise, and this c is precisely the finest scale Haar autocorrelation wavelet $\Psi_{-1}(\tau)$. So this special Haar MA process satisfies (3.5) with $S_{-j}(z) = 0$ for all $j < -1$ and $S_{-1}(z) = 1$ (and this agrees with (3.4) since the amplitudes are 1 for $j = -1$ only, and zero otherwise). We already met the MA process $X_t^{(2)}$ in Section 2 (its Allan variance was plotted in figure 1). By the same argument, its autocovariance function is, this time, the next-finest scale Haar autocorrelation wavelet $\Psi_{-2}(\tau)$.

Similarly, we can continue in this way defining the rth order Haar MA($2^r - 1$) process $X_t^{(r)}$, which has $\Psi_{-r}(\tau)$ for its autocovariance function for integers $r > 0$. Each of the Haar MA processes is stationary, but we can construct a non-stationary process by concatenating the Haar MA processes. For example, suppose we take 128 observations from each of $X_t^{(1)}$, $X_t^{(2)}$, $X_t^{(3)}$ and $X_t^{(4)}$ and concatenate them (a realization from such a process is shown in Nason et al.(1998)). As a time-series of 512 observations, the process will not be stationary. The Haar MA processes have $S_{-j}(z) = 0$ for $-j \neq r$ and all z, and $S_{-r}(z) = 1$ for $z \in ([r-1]/4, r/4)$, a plot of which appears in figure 4 (remember $z = k/512$ is rescaled time).

The plot clearly shows that from time 1 to 128, the Haar MA(1) process is active with variation active at scale -1 (scale $2^{-j} = 2$), then, at time 128, the MA(1) process, $X_t^{(1)}$, changes to the MA(3) process, $X_t^{(2)}$, until time 256, and so on.

Indeed, the section from 128 to 256 ($z \in (\frac{1}{4}, \frac{1}{2})$) should be compared with figure 1, which shows the Allan variance of $X_t^{(2)}$, which would be equivalent to averaging over the 128–256 time period. However, the EWS plot above does not show any power at scales $j = -1$ ($\tau_j = 2$) and $j = -3$ ($\tau_j = 8$) unlike the Allan variance plot. The absence of power in the EWS plot is because of the disentanglement mentioned in (ii).

3.2.4 Application to infant ECG

Figure 5 shows an estimate of the evolutionary wavelet spectrum for the infant ECG data. It can be useful to aggregate information over scales. Figure 6 displays the variation in the ECG series at fine scales. The fine-scale plot was obtained by summing over contributions from scales -1 to -4 in the smoothed corrected wavelet periodogram (ca. 32 s, 1, 2 and 4 min time-scales). The dashed line in figure 6 shows another covariate: the sleep state of the infant as judged by a trained observer. It is clear that there is a correlation between the ECG and the sleep state that can be exploited. There is no real clear link between the sleep state and the EWS at coarse and medium scales. However, a plot of an estimate of the localized variance v (not shown here) gives some idea of overall sleep behaviour.

The EWS is a useful tool in that it provides insight into the time-scale behaviour of a time-series (much in the same way as a periodogram gives information about power in a stationary time-series at different frequencies). However, the EWS has additional uses: Nason et al.(1999) have also used the EWS to build models between sleep state (expensive and intrusive to measure) and ECG (cheap and easy), which allow estimates of future sleep state values to be predicted from future ECG values.

4 Acknowledgements

We thank P. Fleming, A. Sawczenko and J. Young of the Institute of Child Health, Royal Hospital for Sick Children, Bristol, for supplying the ECG data. G.P.N. was supported in part by EPSRC grants GR/K70236 and GR/M10229, and R.v.S. by DFG grant Sa 474/3-1 and by EPSRC Visiting Fellowship GR/L52673.

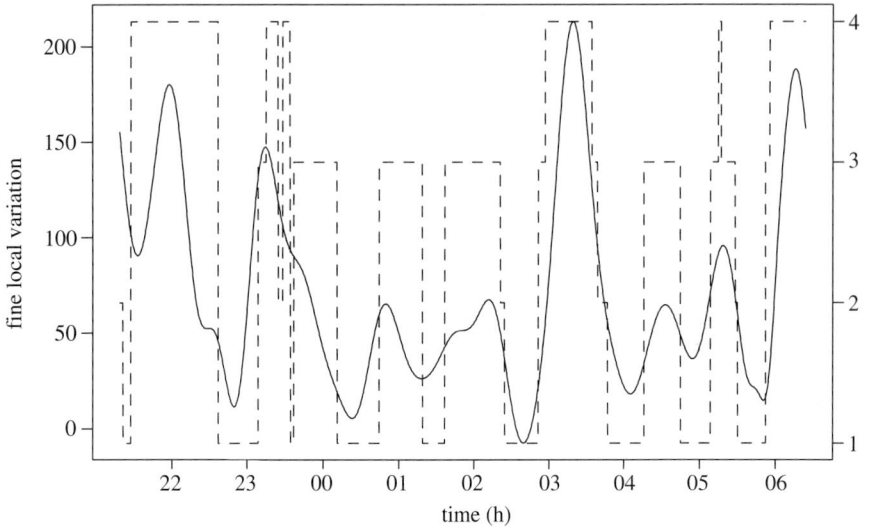

Fig. 6: Estimate of local variation at fine scales for infant ECG series shown in figure 3 (solid line). The dashed line shows the sleep state of the infant (visually determined by a trained observer) ranging from 1 (quiet sleep), 2 (between 1 and 3), 3 (active sleep), 4 (awake). The original time axis is shown although z is really rescaled time in the interval $(0, 1)$.

Appendix A. Discrete non-decimated wavelets

Discrete non-decimated wavelets are merely the vectors of 'filter' coefficients that appear in the matrix representation of the DWT. We shall define those that appear at scale $-j$ to be ψ_j. So, for example, with Haar

$$\psi_{-1} = \left(\frac{1}{\sqrt{2}}, -\frac{1}{\sqrt{2}} \right), \qquad \psi_{-2} = (\tfrac{1}{2}, \tfrac{1}{2}, -\tfrac{1}{2}, -\tfrac{1}{2}),$$

and so on for other scales. The ψ_j for $j = -1, \ldots, -J$ can be obtained for any Daubechies compactly supported wavelet using the formulae

$$\tilde{h}_{j-1,n} = \sum_k h_{n-2k} \tilde{h}_{j,k}, \qquad \psi_{j,n} = \sum_k g_{n-2k} \tilde{h}_{j,k},$$

where $\{h_k\}$ and $\{g_k\}$ are the usual Daubechies quadrature mirror filters, and

$$\tilde{h}_{-1,k} = h_k.$$

For non-decimated wavelets, $\psi_{jk}(\tau)$ is the kth element of the vector $\psi_{j(k-\tau)}$, i.e. ψ_{jk} shifted by integers τ. The key point for *non-decimated* discrete wavelets is that they can be shifted to any location and not just by shifts of 2^{-j} (as in the DWT). Hence, non-decimated discrete wavelets are no longer orthogonal but are an overcomplete collection of shifted vectors. The NDWT can be computed through a fast algorithm (similar to the Mallat pyramid algorithm described in Silverman, this issue), which takes a computational effort of order $n \log n$ for a dataset of length n. See Nason & Silverman (1995) for a more detailed description of the NDWT.

5 References

Abry, P., Gonçalvès, P. & Flandrin, P. 1995 Wavelets, spectrum analysis and $1/f$ processes. In *Wavelets and statistics* (ed. A. Antoniadis & G. Oppenheim). Springer Lecture Notes in Statistics, no. 103, pp. 15–29.

Abry, P., Flandrin, P., Taqqu, M. & Veitch, D. 1999 Wavelets for the analysis, estimation and synthesis of scaling data. In *Self similar network traffic analysis and performance evaluation* (ed. K. Park & W. Willinger). Wiley.

Allan, D. W. 1966 Statistics of atomic frequency clocks. *Proc. IEEE* **31**, 221–230.

Beran, J. 1994 *Statistics for long-memory processes*. London: Chapman and Hall.

Coifman, R. R. & Donoho, D. L. 1995 Translation-invariant denoising. In *Wavelets and statistics* (ed. A. Antoniadis & G. Oppenheim). Springer Lecture Notes in Statistics, no. 103, pp. 125–150.

Dahlhaus, R. 1997 Fitting time series models to nonstationary processes. *Ann. Statist.* **25**, 1–37.

Donoho, D. L., Johnstone, I. M., Kerkyacharian, G. & Picard, D. 1995 Wavelet shrinkage: asymptopia? (with discussion). *J. R. Statist. Soc.* B **57**, 301–369.

Flandrin, P. 1998 *Time–frequency/time-scale analysis*. Wavelet Analysis and its Applications, vol. 10. Academic.

Gao, H.-Ye. 1997 Choice of thresholds for wavelet shrinkage estimate of the spectrum. *J. Time Series Analysis* **18**, 231–251.

Gonçalvès, P. & Flandrin, P. 1993 Bilinear time-scale analysis applied to local scaling exponents estimation. In *Progress in wavelet analysis and applications* (ed. Y. Meyer & S. Roques), pp. 271–276. Gif-sur-Yvette: Editions Frontières.

Hoffmann, M. 1999 On nonparametric estimation in nonlinear AR(1)-models. *Statist. Prob. Lett.* **44**(1), 29–45.

McCoy, E. J. & Walden, A. T. 1996 Wavelet analysis and synthesis of stationary long-memory processes. *J. Computational Graphical Statist.* **5**, 26–56.

Nason, G. P. & Silverman, B. W. 1995 The stationary wavelet transform and some statistical applications. In *Wavelets and statistics* (ed. A. Antoniadis & G. Oppenheim). Springer Lecture Notes in Statistics, no. 103, pp. 281–300.

Nason, G. P., von Sachs, R. & Kroisandt, G. 1998 Wavelet processes and adaptive estimation of the evolutionary wavelet spectrum. Discussion Paper 98/22, Institut de Statistique, UCL, Louvain-la-Neuve.

Nason, G. P., Sapatinas, T. & Sawczenko, A. 1999 Statistical modelling of time series using non-decimated wavelet representations. Preprint, University of Bristol.

Neumann, M. 1996 Spectral density estimation via nonlinear wavelet methods for stationary non-Gaussian time series. *J. Time Series Analysis* **17**, 601–633.

Wavelets in time-series analysis

Neumann, M. & von Sachs, R. 1997 Wavelet thresholding in anisotropic function classes and application to adaptive estimation of evolutionary spectra. *Ann. Statist.* **25**, 38–76.

Page, C. H. 1952 Instantaneous power spectra. *J. Appl. Phys.* **23**, 103–106.

Percival, D. B. & Guttorp, P. 1994 Long-memory processes, the Allan variance and wavelets. In *Wavelets in geophysics* (ed. E. Foufoula-Georgiou & P. Kumar), pp. 325–344. Academic.

Priestley, M. B. 1965 Evolutionary spectra and non-stationary processes. *J. R. Statist. Soc.* B **27**, 204–237.

Priestley, M. B. 1981 *Spectral analysis and time series.* Academic.

Rioul, O. & Vetterli, M. 1991 Wavelets and signal processing. *IEEE Sig. Proc. Mag.* **8**, 14–38.

Serroukh, A., Walden, A. T. & Percival, D. B. 1998 Statistical properties of the wavelet variance estimator for non-Gaussian/non-linear time series. Statistics Section Technical Report, TR-98-03. Department of Mathematics, Imperial College, London.

Silverman, R. A. 1957 Locally stationary random processes. *IRE Trans. Inform. Theory* **IT-3**, 182–187.

von Sachs, R. & MacGibbon, B. 2000 Nonparametric curve estimation by wavelet thresholding with locally stationary errors. *Scandinavian J. Stats* **27** (to appear).

von Sachs, R. & Neumann, M. 2000 A wavelet-based test for stationarity. *J. Time Series Analysis* **21** (to appear).

von Sachs, R. & Schneider, K. 1996 Wavelet smoothing of evolutionary spectra by non-linear thresholding. *Appl. Comp. Harmonic Analysis* **3**, 268–282.

Wang, Y., Cavanaugh, J. & Song, Ch. 1997 Self-similarity index estimation via wavelets for locally self-similar processes. Preprint, Department of Statistics, University of Missouri.

Whitcher, B., Byers, S. D., Guttorp, P. & Percival, D. B. 1998 Testing for homogeneity of variance in time series: long memory, wavelets and the Nile river. (Submitted.)

Wornell, G. W. & Oppenheim, A. V. 1992 Estimation of fractal signals from noisy measurements using wavelets. *IEEE Trans. Sig. Proc.* **40**, 611–623.

8

Wavelets, vision and the statistics of natural scenes

D. J. Field

Uris Hall, Cornell University, Ithaca, NY 14853, USA

Abstract

The processing of spatial information by the visual system shows a number of similarities to the wavelet transforms that have become popular in applied mathematics. Over the last decade, a range of studies has focused on the question of 'why' the visual system would evolve this strategy of coding spatial information. One such approach has focused on the relationship between the visual code and the statistics of natural scenes under the assumption that the visual system has evolved this strategy as a means of optimizing the representation of its visual environment. This paper reviews some of this literature and looks at some of the statistical properties of natural scenes that allow this code to be efficient. It is argued that such wavelet codes are efficient because they increase the independence of the vectors' outputs (i.e. they increase the independence of the responses of the visual neurons) by finding the sparse structure available in the input. Studies with neural networks that attempt to maximize the 'sparsity' of the representation have been shown to produce vectors (neural receptive fields) that have many of the properties of a wavelet representation. It is argued that the visual environment has the appropriate sparse structure to make this sparse output possible. It is argued that these sparse/independent representations make it computationally easier to detect and represent the higher-order structure present in complex environmental data.

Keywords: wavelet; vision; independent components analysis; natural scenes

1 Introduction

Over the last decade, the wavelet transform in its various incarnations has grown to be a highly popular means of analysis, with a wide range of applications in processing natural signals. Although there is some debate regarding who developed the first wavelet transform, most of the claims of priority apply to only this century. In this paper, we consider wavelet-like transforms that pre-date these recent studies by possibly as much as several hundred million years. These wavelet-like transforms are found within the sensory systems of most vertebrates and probably a number of invertebrates. The most widely studied of these is the mammalian visual system. This paper focuses on recent work exploring the visual system's response to spatial patterns and on recent theories of 'why' the visual system would use this strategy for coding its visual environment. Much of this work has concentrated on the relationship between the mathematical relationships within environment stimuli (e.g. the statistics of natural scenes), and these wavelet-like properties of the visual system's code (see, for example, Atick 1992; Atick & Redlich 1990, 1992; Field 1987, 1989, 1993, 1994; Olshausen & Field 1996; Hancock *et al.*1992; Ruderman 1994; Shouval *et al.*1997; Srinivisan *et al.*1982). The first section begins by looking at the visual system's 'wavelet-like' transform of spatial information. We then look at some of the statistical regularities found in natural images and their relationship to the properties of the visual transform. In particular we will review research that suggests that the particulars of this coding strategy produces a nearly optimal sparse/independent transform of natural scenes. Finally, we look at a neural-network approach that attempts to search for efficient representations of natural scenes, and results in a 'wavelet-like' representation with many similarities to that found in the visual cortex.

2   The mammalian visual system

Although there are a number of differences between the visual systems of different mammals, there are a considerable number of similarities, especially in the representation of spatial information. The most extensively studied systems are those of the cat and monkey and it is studies on these animals that provide the basis of much of our knowledge about visual coding. The acuity of the cat is significantly lower than that of the monkey, but within the range of sensitivities covered by these visual systems (i.e. the spatial frequency range), the methods by which spatial information is processed follow a number of similar rules. The area that we will be considering is a region at the back of the brain referred to as the primary visual cortex (area V1). This area is the principal projection area for visual information and consists of neurons that receive input from neurons in the eye (via an area called the lateral geniculate nucleus (LGN)).

The behaviour of these neurons is measured by placing an electrode near the cell body and recording small voltage changes in the neuron's membrane. The neuron produces a response 'spike' or a series of spikes when the visual system is presented with the appropriate stimulus. Hubel & Wiesel (1962) were the first to provide a spatial mapping of the response properties of these neurons. By moving stimuli (e.g. spots, lines, edges, etc.) in front of the animal, they found that the neurons would respond when an appropriate stimulus was presented at a particular region in the visual field of the animal. The map describing the response region of the cell is referred to as the 'receptive field'. Figure 1 shows examples of the types of receptive fields that are obtained from these neurons. If a spot of light is shown within the receptive field, then the cell may either increase its firing rate (excitation) or decrease its firing rate (inhibition) depending on the region. The neurons in the primary visual cortex are described as 'simple cells' and are marked by elongated excitatory regions (causing an increase in the number of spikes) and inhibitory regions (causing a decrease in the number of spikes) as shown in figure 1.

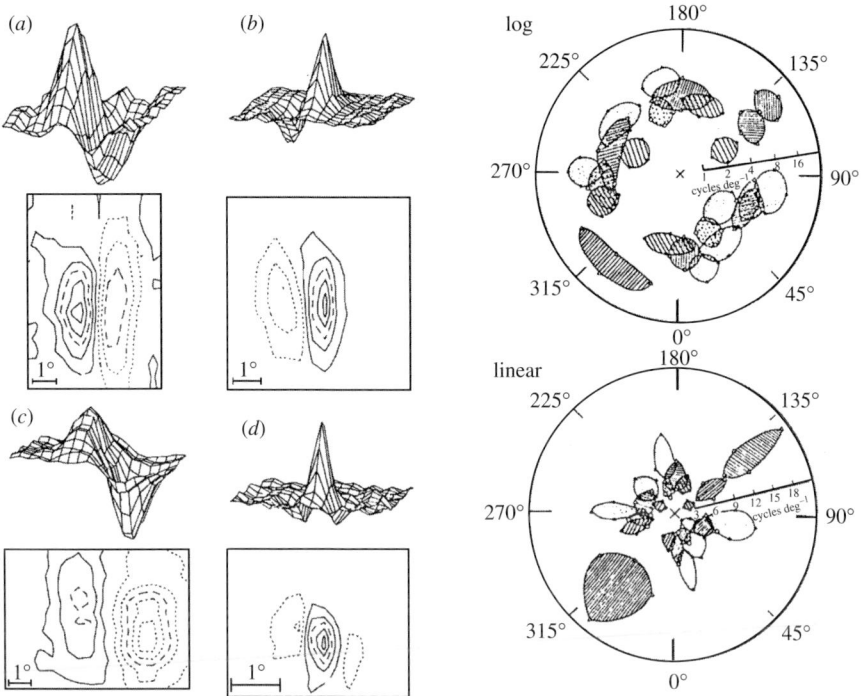

Fig. 1: Results from two laboratories looking at the receptive field profiles of cortical simple cells in the cat. On the left are results derived from Jones & Palmer (1987) showing the two-dimensional receptive field profiles of X cortical simple cells. The data on the right show results from DeValois *et al.*(1982) that represent the spatial frequency tuning of a variety of different cortical cells when plotted on both a log (*a*) and a linear frequency plot (*b*). Although there is significant variability, bandwidths increase with increasing frequency (i.e. on the linear axis (*b*)). Therefore, when bandwidths are plotted on the log axis (*a*), they remain roughly constant at different frequencies.

With diffuse illumination or a line placed horizontally across the receptive field, the excitation and inhibition will typically cancel and the cell will not respond. Different neurons respond to different positions within the visual field. Furthermore, at any given position in the visual field, different neurons have receptive fields orientated at different angles and show a variety of sizes. Thus, the entire visual field is covered by receptive fields that vary in size and orientation. Neurons with receptive fields like the one above were described by Hubel & Weisel (1962) as 'simple cells' and were distinguished from other types of neurons in the primary visual cortex referred to as 'complex' and 'hyper-complex'. (The principal difference is that these neurons show a higher degree of spatial nonlinearity.)

Throughout the 1960s and 70s there was considerable discussion of how to describe these receptive field profiles and what the function of these neurons might be. Early accounts described these cells as edge and bar detectors and it was suggested that the visual code was analogous to algorithms performing a local operation like edge detection (Marr & Hildreth 1980). In opposition to this way of thinking were those that used the terms of linear systems theory (Campbell & Robson 1968). In the latter case, the selectivity of cells was described in terms of their tuning to orientation and spatial frequency (see, for example, Blakemore & Campbell 1969). It was not until 1980 (Marcelja 1980) that the functions describing these receptive fields were considered in terms of Gabor's 'theory of communication' (Gabor 1946). Marcelja noted that the functions proposed by Gabor to analyse time-varying signals showed a number of interesting similarities to the receptive fields of cortical neurons.

Marcelja's suggestion was that the profile described by the line-weighting function appeared to be well described by a Gaussian modulated sinusoid:

$$f(x) = \sin(2\pi k x + \theta) e^{-x^2/2\sigma^2}. \tag{2.1}$$

This function, now referred to as a 'Gabor function', has served as a model of cortical neurons by a wide variety of visual scientists. Early tests of this notion showed that such functions did indeed provide an excellent fit to the receptive fields of cortical neurons (Webster & DeValois 1985; Field & Tolhurst 1986; Jones & Palmer 1987). Daugman (1985) and Watson (1983) generalized Gabor's notion to the two dimensions of space where the two-dimensional basis function is described as the product of a two-dimensional Gaussian and a sinusoid. Although Jones & Palmer (1987) found that the full two-dimensional receptive field profiles were well described by this two-dimensional 'Gabor function', other studies (Hawken & Parker 1987) have found that other types of functions (e.g. sum of Gaussians) may provide a better fit (see also Stork & Wilson 1990). Although some of the differences between these various models may prove to be important, the differences are not large. All of the basis functions proposed involve descriptions in terms of orientated functions that are well localized in both space and frequency. However, as shown in figure 1, there is considerable variability in the receptive field shapes across neurons, and no single basis set will likely capture all of this variability. There is also significant variability in receptive field bandwidths. For example, the bandwidths of cortical cells average around 1.4 octaves (width at half height) but bandwidths less than 1.0 or greater than 2.0 octaves are found (Tolhurst & Thompson 1982; DeValois *et al.*1982).

2.1   Wavelet-like transforms

When these cortical codes were first converted to mathematical representations (see, for example, Watson 1991; Kulikowski *et al.*1982; Daugman 1988), they were known as Gabor transforms, self-similar Gabor transforms or log–Gabor transforms (see, for example, Field 1987). More recently, with the popularity of

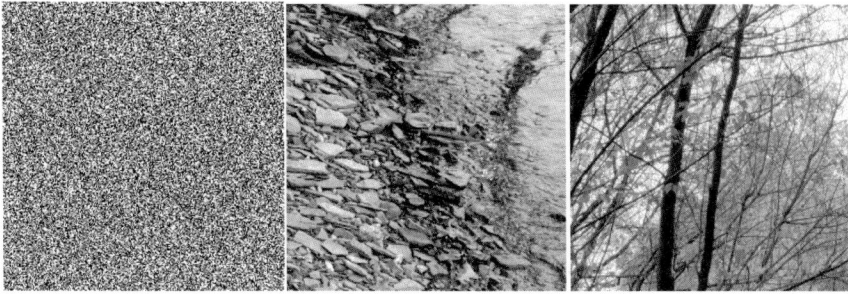

Fig. 2: Noise versus natural scenes (see text).

the wavelet ideas, these transforms have come to be known as wavelet, or wavelet-like transforms. However, in most of these transforms the basis vectors are not orthogonal. Furthermore, it is also common that these functions be truncated in both space and frequency (e.g. a fixed window size). Finally, these 'transforms' may not be in a one-to-one relation to the numbers of pixels, and instead may be overcomplete with more basis vectors than dimensionality of the data (see, for example, Olshausen & Field 1997). In visual research, most of these aspects of the transform are not crucial to the questions that are addressed. Only in cases where there is an attempt to reconstruct the inputs do issues of orthogonality and critical sampling become a major issue.

3   Image transforms and the statistics of natural scenes

There are various ways to describe the statistical redundancy in a dataset. One approach is to consider the nth-order statistical dependencies among the basis vectors. This works well when the basis vectors have 'all or none' outputs like letters, where the frequency of occurrence can be defined by a single number. However, for real-valued vectors that show a continuous output, it can often be useful to consider a description of images in terms of a 'state-space' where the axes of the space represent the intensities of the pixels of the image. For any n-pixel image, one requires an n-dimensional space to represent the set of all possible images. Every possible image (e.g. a particular face, tree, etc.) is represented in terms of its unique location in the space. The white-noise patterns like that shown in figure 2 (i.e. a pattern with random pixel intensities), represent random locations in that n-dimensional space. It is probably obvious that probability of generating anything resembling a natural scene from images with random pixel intensities will be extremely low. This suggests that in this state-space of all possible scenes, the region of the space occupied by natural scenes is also extremely low (Field 1994).

Just as any image can be represented as a point in the state-space of possible images, it is also possible to describe the response of any particular visual neuron in terms of the region of the state-space that can produce a response. If the neuron's response is linear, then we can treat it as a vector projecting from the origin into the state-space, and its response is simply the projection of the point representing the image against this vector. In reality, visual neurons show a variety of very interesting and important nonlinearities. However, it is argued that treating the visual cells as linear, to a first approximation, provides a means of exploring the relative advantages of different response properties (e.g. orientation tuning, spatial frequency tuning, localization, etc.).

In addition, with the state-space description, any orthonormal transform, such as the Fourier transform, is simply a rotation in the state-space (Field 1994). Although wavelet transforms may be orthogonal (see, for example, Adelson *et al.*1987; Daubechies 1988), the wavelet transforms used by the visual system and in the analyses that follow are neither orthogonal nor normal. Nonetheless, we can treat the visual code to a first approximation as a rotation of the coordinate system. As long as the total number of vectors remains constant, such a rotation will not change the entropy or the redundancy of the overall representation (i.e. the relative density of the space remains constant). The question then becomes, why the visual system would evolve this particular rotation. Or more specifically, what is it about the population of natural scenes that would make this particular rotation useful? Several theories have been proposed and the following sections will consider two of the principal theories as well as a more general approach referred to as independent components analysis (ICA).

4   The goal of visual coding

Why and when are wavelet codes effective? And what is the reason that the wavelet-like transform would evolve within the mammalian visual system? Some of the early theories of sensory coding were developed by Barlow (1961), who suggested that one of the principal goals should be to reduce the redundancy of the representation. Field (1994) contrasted two approaches to transforming redundancy. We will discuss these below and follow this with a discussion of ICA that has gained considerable attention.

Figure 3 shows two examples of two-dimensional datasets and the effects that a particular transform (e.g. a rotation) has on the outputs. In figure 3*a*, the data are correlated and the collection of data forms a Gaussian ellipse. The figure shows a rotation to align the vectors with the correlations (i.e. the axes of the ellipse). After the rotation, there exist no correlations in the data; however, the basis vectors now have unequal variance. In this new coordinate system, most of the variance in the data can be represented with only a single vector (A′). Removing B′ from the code produces only minimal loss in the description of the data. This rotation of the coordinate system to allow the vectors to be aligned with the principal axes of the data is what is achieved with a process called principal component analysis (PCA)— sometimes called the Karhonen–Lo'eve transform. The method provides a means of

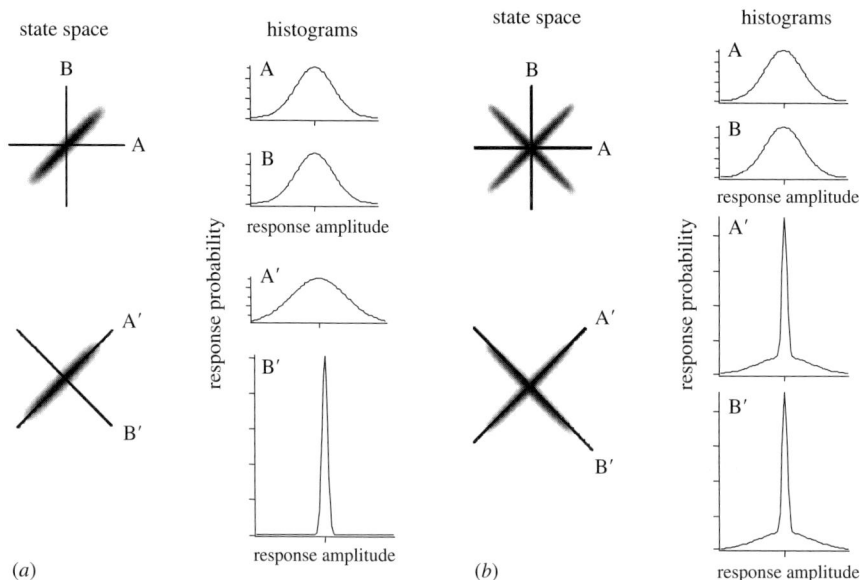

Fig. 3: State-spaces and projections of two populations of two-dimensional data: (*a*) data that are correlated; (*b*) data that are not correlated but are sparse.

compressing high-dimensional data onto a subset of vectors.

4.1 Principal components and the amplitude spectra of natural scenes

An interesting and important idea involves PCA when the statistics of a dataset are stationary. Stationarity implies that over the population of images in the dataset (e.g. all natural scenes), the statistics at one location are no different than at any other location:

$$\text{across all images } P(x_i \mid x_{i+1}, x_{i+2}, \ldots) = P(x_j \mid x_{j+1}, x_{j+2}, \ldots) \quad \forall\, i \text{ and } j.$$
$$(4.1)$$

This is a fairly reasonable assumption with natural scenes since it implies that there are no 'special' locations in an image where the statistics tend to be different (e.g. the camera does not have a preferred direction). It should be noted that stationarity is not a description of the presence or lack of local features in an image. Rather, stationarity implies that over the population all features have the same probability of occurring in one location versus another. When the statistics of an image set are stationary, the real and imaginary amplitudes of the Fourier coefficients of the image must all be uncorrelated with each other (see, for example, Field 1989). This means that when the statistics of a dataset are stationary then all the redundancy reflected in the correlations between pixels is captured by the amplitude spectra of the data. This should not be surprising since the Fourier transform of the autocorrelation function is the power spectrum. Therefore, with stationary statistics, the amplitude spectrum describes the principal axes (i.e. the principal components) of the data in the state-space (Pratt 1978). With stationary data, the phase spectra of the data are irrelevant to the directions of the principal axes.

As noted previously (Field 1987), an image that is scale invariant will have a well-ordered amplitude spectrum. For a two-dimensional image, the amplitudes will fall inversely with frequency (i.e. power falls as a k^{-2}, where k is the spatial frequency). Natural scenes have been shown to have spectra that fall as roughly k^{-2} (see, for example, Burton & Moorhead 1987; Field 1987, 1993; Tolhurst *et al.*1992). If we accept that the statistics of natural images are stationary, then the k^{-1} amplitude spectrum provides a complete description of the pairwise correlations in natural scenes. The amplitude spectrum certainly does not provide a complete description of the redundancy in natural scenes, but it does describe the relative amplitudes of the principal axes.

A number of recent studies have discussed the similarities between the principal components of natural scenes and the receptive fields of cells in the visual pathway (Bossomaier & Snyder 1986; Atick & Redlich 1990, 1992; Atick 1992; Hancock *et al.*1992; Intrator 1992), and there have been a number of studies that have shown that, under the right constraints, units in competitive networks can develop large orientated receptive fields (see, for example, Lehky & Sejnowski 1990; Linsker 1988; Intrator 1992). However, the PCA will not produce wavelet-like transforms, since they depend only on the amplitude spectrum. Since the phases are unconstrained, the resulting functions will not be localized and therefore the sizes can not scale with frequency (Field 1994).

To account for the localized self-similar aspects of the wavelet coding, it has been argued that one must go beyond this second-order structure as described by the amplitude spectrum and the principal components (Field 1987, 1993, 1994; Bell & Sejnowski 1997). However, does an understanding of the amplitude spectrum provide any insights into the visual system's wavelet code? Field (1987) argued that if the peak spatial frequency sensitivity of the wavelet bases is constant, then the average response magnitude will be flat in the presence of images with $1/f$ amplitude spectra. Brady & Field (1995) and Field & Brady (1997) propose that this model provides a reasonable account of the sensitivity of neurons and has some support from visual neurophysiology (Croner & Kaplan 1995). In these models the vector magnitude increases with frequency, reaching a maximum around 25 cycles per degree. Such an approach explains why a white-noise pattern, like that shown on the left in figure 2, appears to be dominated by high frequencies when the spectrum is actually flat (Field & Brady 1997).

Atick & Redlich (1990, 1992) have suggested that the spatial frequency tuning of retinal ganglion cells is well matched to the combination of amplitude spectra of natural scenes and high-frequency quantal limitations found in the natural environment. They have stressed the importance that the role of the noise plays in limiting information processing by the visual system and have effectively argued that the fall-off in frequency sensitivity of individual neurons and the system as a whole is due to the decrease in signal-to-noise at these higher frequencies.

Since the principal components conform to the Fourier coefficients for natural scenes, and since the amplitudes of the Fourier coefficients fall with increasing frequency, removing the lowest amplitude principal components of natural scenes effectively removes the high spatial frequencies. Removing the high spatial frequencies is the most effective means of reducing dimensionality of the representation with minimal loss in entropy. This is exactly what occurs in the early stages of the visual system. The number of photoreceptors in the human eye is approximately 120 million and this is reduced to approximately 1 million fibres in the optic nerve. This compression is achieved almost entirely by discarding the high spatial frequencies in the visual periphery. Only the fovea codes the highest spatial frequencies with eye movements, allowing this high-acuity region to be directed towards points of interest.

Therefore, it is argued that the visual system does perform compression of the spatial information and this is possible because of the correlations in natural scenes. However, the two insights one gains from this approach are (1) in understanding the spatial frequency cut-off (especially in the visual periphery); and (2) in understanding the relative sensitivity of visual neurons as a function of spatial frequency. To account for the wavelet-like properties of localization, spatial frequency tuning and self-similarity found in the visual cortex, we must consider statistics beyond the pairwise correlations.

4.2 Discovering sparse structure

How does the presence of sparse localized structure modify the state-space? Field (1994) has suggested the simplified state-space shown above in figure 3*b* to characterize sparse structure. In this particular example the data are not correlated. However, the data are redundant since the state-space is not filled uniformly. One might think of these data as containing two kinds of structure: pixels that are positively correlated and pixels that are negatively correlated. This is generally true of neighbouring pixels in images which have been 'whitened' to remove the pairwise correlations. If a pixel has a non-zero value, the neighbouring pixel also is likely to have a non-zero value, but the polarity of the value cannot be predicted since the pixel values are uncorrelated.

The same transformation performed as before (i.e. a rotation) produces a marked change in the histograms of the basis functions A' and B'. This particular dataset allows a 'sparse' response output. Although the variance of each basis function remains constant, the histogram describing the output of each basis function has changed considerably. After the transformation, vector A' is high or vector B' is high but they are rarely high at the same time. The histograms of each vector show a dramatic change. Relative to a normal distribution, there is a higher probability of

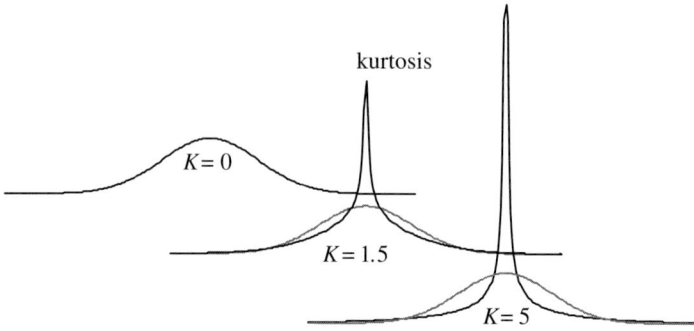

Fig. 4: Non-Gaussian distributions in the direction of increasing kurtosis.

low magnitude and a higher probability of a high magnitude, but a reduction in the probability of a midlevel magnitude.

This change in shape can be represented in terms of the kurtosis of the distribution where the kurtosis is defined as the fourth moment according to

$$K = \frac{1}{n} \sum \left[\frac{(x - \bar{x})^4}{\sigma^4} \right]^{-3}. \tag{4.2}$$

Figure 4 provides an example of distributions with various degrees of kurtosis. In a sparse code, any given input can be described by only a subset of cells, but that subset changes from input to input. Since only a small number of vectors describe any given image, any particular vector should have a high probability of no activity (when other vectors describe the image) and a higher probability of a large response (when the vector is part of the family of vectors describing the image). Thus, a sparse code should have response distributions with high kurtosis.

As we move to higher dimensions (e.g. images with a larger number of pixels), we might consider the case where only one basis vector is active at a time (e.g. vector 1 or vector 2 or vector 3 $\cdots$):

$$ax_1 \cup ax_2 \cup ax_3 \cup ax_4 \cdots . \tag{4.3}$$

In this case, each image can be described by a single vector and the number of images equals the number of vectors. However, this is a rather extreme case and is certainly an unreasonable description of most datasets, especially natural scenes.

When we go to higher dimensions, there exist a wide range of possible shapes that allow sparse coding. Overall, the shape describing the probability distribution of natural scenes must be such that any location can be described by a subset of vectors, but the shape requires the full set of vectors to describe the entire population of images (i.e. shape requires the full dimensionality of the space):

$$\text{image} = \sum_i^n a V_i, \quad \text{where } n < m, \tag{4.4}$$

where m is the number of dimensions required to represent all images in the population (e.g. all natural scenes).

For example, with a three-pixel image, where only two pixels are non-zero at a time, it is possible to have

$$(ax_1 + bx_2) \cup (ax_2 + bx_3) \cup (ax_1 + bx_3). \tag{4.5}$$

This state-space consists of three orthogonal planes. By choosing vectors aligned with the planes (e.g. x_1, x_2, x_3), it is possible to have a code in which only two vectors are non-zero for any input. Of course, for high-dimensional data like natural scenes these low-dimensional examples are too simplistic and more interesting geometries (e.g. conic surfaces) have been proposed (Field 1994). The basic proposal is that there exist directions in the state-space (i.e. features) that are more probable than others. And the direction of this higher-density region is not found by looking at the pairwise correlations in the image. The wavelet transform does not reduce the number of dimensions needed to code the population of natural scenes. It reduces only the number of dimensions needed to code a particular instance of a natural scene. As Donoho (1993) has argued, the wavelet transform can provide an optimal representation when the data consist of an arbitrary number of singularities (e.g, transition points). Here, it is proposed that the signature of a sparse code is found in the kurtosis of the response distribution (Field 1994). A high kurtosis signifies that a large proportion of the cells are inactive (low variance) with a small proportion of the cells describing the contents of the image (high variance). However, an effective sparse code is not determined solely by the data or solely by the vectors but by the relation between the data and the vectors.

4.3 Sparse structure in natural scenes

Is the visual systems code optimally sparse in response to natural scenes? First, it should be noted that we are modelling the visual system with linear codes. Real visual neurons have a number of important nonlinearities which include a threshold (i.e. the output cannot go below a particular value: the cell cannot go below a zero firing rate). Several studies suggest that cells with properties similar to those in the mammalian visual cortex will show high kurtosis in response to natural scenes. In Field (1987), visual codes with a range of different bandwidths were studied to determine how populations of cells would respond when presented with natural scenes. It was found that when the parameters of the visual code matched the properties of simple cells in the mammalian visual cortex, a small proportion of cells could describe a high proportion of the variance in a given image. When the parameters of the code differed from those of the mammalian visual system, the response histograms for any given image were more equally distributed. The published response histograms by both Zetzsche (1990) & Daugman (1988) also suggest that codes based on the properties of the mammalian visual system will show positive kurtosis in response to natural scenes. Burt & Adelson (1983) noted that the histograms of their 'Laplacian pyramids' showed a concentration near zero when presented with their images and suggested that this property could be used for an efficient coding strategy.

Field (1989, 1993) demonstrated that the bandwidths of cortical cells were well matched to the degree of phase alignment across scale in natural scenes. Because edges are rarely very straight in natural scenes, the orientation and position of any given edge will typically shift in position and orientation across scale (i.e. across spatial frequency). In natural scenes, the degree of predictability is around the 1–2 octave range, which is why cortical neurons have bandwidths in the 1–2 octave range. This is also the reason that this wavelet-like code is sparse when presented with natural scenes. Field (1994) looked at the kurtosis of the histograms of various wavelet codes in the presence of natural scenes, and found that the kurtosis (sparsity) peaked when the wavelet transforms used bandwidths in this range of 1–2 octaves.

Recently, however, more direct tests have been developed. If the wavelet-like code used by the visual system is near to optimal in its sparse response to natural scenes, then a gradient decent algorithm like a neural network, which attempts to optimize this response, should develop receptive fields that are wavelet-like. The following work explores this idea.

5   Neural networks and independent coding

There is no known analytic solution to the problem of finding the most independent representation for a complex dataset like natural scenes. However, recently, number studies using neural networks have attempted to find relatively independent solutions using gradient descent techniques (Olshausen & Field 1996, 1997; Bell & Sejnowski 1997). Some of these studies describe their approach as independent components analysis (ICA). This author believes that the such a description is a poor use of the term, since most complex datasets are not likely to have independent components, and the current techniques search only for specific forms of independence. For example, in some of these studies, there is an assumption that the most independent solution must necessarily have vectors that are completely decorrelated. By forcing this particular form of redundancy, one is limited to solutions that are orthogonal in the whitened space. 'Sphering' refers to the process of rotating to the principal axes and adjusting the gain of the vectors to produce equal variance. All rotations in this whitened spaced will certainly be orthogonal—but the optimal rotation in this space is not guaranteed to be the most independent possible.

Olshausen & Field (1996, 1997) describe networks which search for one particular form of independence (sparse codes) by searching for a non-Gaussian response histogram. There are two competing components of the network. One component attempts to reconstruct the input with the available vectors and produces small modifications in the vectors to minimize the error in the reconstruction. A second component imposes a cost function that attempts to push the shape of the histogram away from Gaussian towards higher kurtosis. The main point to note regarding the cost function is that it is nonlinear with the gradient change with

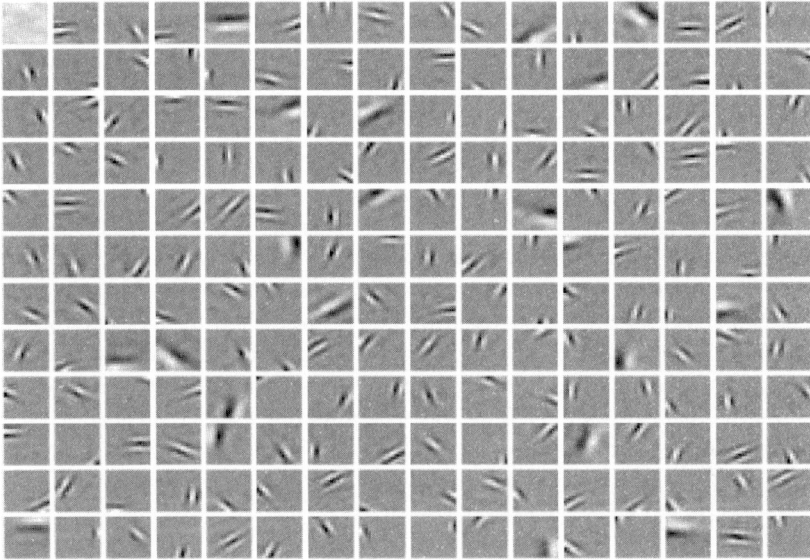

Fig. 5: Results from Olshausen & Field (1996) who used a neural network to search for a sparse representation of natural scenes. Each template represents one vector of the population.

response magnitude. What this does is reduce the magnitude of the low-magnitude vectors more than it reduces the magnitude of the high-magnitude vectors. Overall, the network attempts to find a method of reconstructing any given input with a few high magnitude vectors—although the vectors involved in the reconstruction are allowed to change from input to input. An example of the results of the network are shown in figure 5 (Olshausen & Field 1997). It should also be noted that this particular network allows non-orthogonal solutions by allowing inhibition of the output vectors. With this particular nonlinearity, it also turns out that the code can be more sparse, if one allows an overcomplete basis set (more vectors than dimensions/pixels in the data). Similar results have been obtained by other studies (see, for example, Bell & Sejnowski 1997). Although there is some debate as to whether such a solution is more or less independent than the results of Olshausen & Field (1997), the results are globally similar, producing localized orientated vectors.

The results shown in figure 5 have a number of similarities to the wavelet-like transforms found in the mammalian primary visual cortex. The results suggest that a possible reason for this transform by the visual system is that it reduces statistical dependencies and allows the firing of any particular cell to provide maximal information about the image.

One of the criticisms of this approach is that for a biological system, a sparse code has a serious disadvantage. If a given cell is providing maximal information about a particular feature, and is not shared with other cells, then what happens should that cell die? This is actually one of the advantages to the locally competitive overcomplete codes described by Olshausen & Field (1997). The output is quite sparse, but the loss of any given cell will not result in a loss of the ability to completely recapture the input. However, with a one-to-one 'critically sampled' wavelet transform, the removal of the information provided by one basis vector would make it impossible to completely reconstruct the input.

A second criticism of this approach is that such networks are not biologically plausible. Most of the networks discussed above rely on some measure of the response magnitudes (i.e. the histogram) of all the cells (vectors) in the code. These sorts of global measures would be quite difficult to calculate with known physiology. Secondly, these networks typically attempt to reconstruct the input, and use the error in the reconstruction to modify the weights of the network. Again, this error is a global measure and, even though the network might be able to calculate the error locally, plausibility is in question.

5.1   Nonlinear decorrelation

As noted earlier, it is possible to calculate the principal components with a Hebbian network that can be made biologically plausible. Unfortunately, if the network is linear, the networks are sensitive to only pairwise correlations and do not produce wavelet-like receptive fields unless the relative sizes and positions of the fields are directly imposed. However, the addition of nonlinear weights can allow the network to become sensitive to structure beyond the pairwise correlations (Foldiak 1990; Bienenstock *et al.*1982). Foldiak (1990) demonstrated with relatively restricted stimuli that a combination of Hebbian and anti-Hebbian can learn a sparse code.

Can a biologically plausible network produce a wavelet-like code similar to the results shown above? Field & Millman (1999) have developed a network with Hebbian and anti-Hebbian learning rules with similarities to that of Foldiak (1990). Such a network can produce results similar to Olshausen & Field (1996) when the correct threshold is applied to the output. The method of learning is relatively straightforward. For any given stimulus (a patch of a natural scene), an output is associated to each vector by calculating the inner product of the vector with that image patch. A nonlinear threshold is then imposed on each of the outputs and the learning algorithm is applied only to those vectors exceeding the threshold value. In the learning algorithm, each vector above this threshold is compared with every other vector above threshold. For every pair, the vector with the larger output becomes more like the input (Hebbian learning) and the vector with the smaller output becomes less like the input (anti-Hebbian learning). The results are comparable to those shown in figure 5.

Why can this network learn sparse codes? Figure 6 demonstrates what the imposition of a threshold does in the presence of sparse data like that shown in figure 3*b*. There will be no correlations in the original data, so the principal axes will not describe the axes of the data. However, by using a threshold to break up the

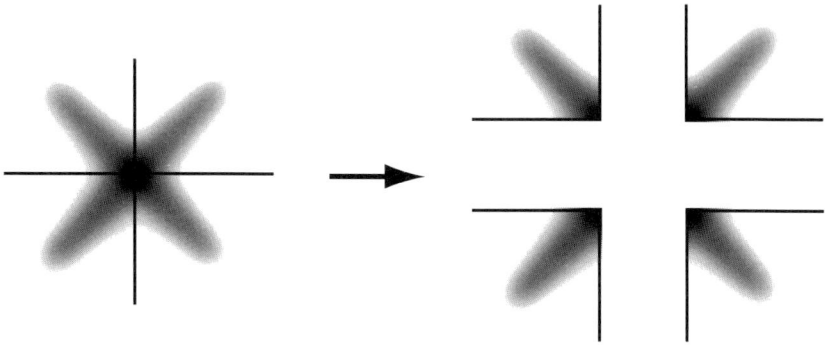

Fig. 6: On the left is an example of the state-space of data (x, y) that are sparse and have no correlations and therefore no principal components. However, when the data are split into quadrants by using vectors that allow only non-zero values $(x, -x, y, -y)$, the resultant data are correlated. Networks with appropriate Hebbian and anti-Hebbian learning rules can now learn the axes of these data.

quadrants of the data, the correlations can now provide the sparse axes. However, one should note that the two-dimensional data now require four dimensions. If the vectors are limited to positive values, as shown in this case, then one needs twice the number of vectors to cover the full dimensionality of the space. Increasing the threshold to higher levels allows the network to search for non-orthogonal solutions. It should be noted that these networks are searching for the high-density regions of the state-space. In this two-dimensional example, the high density is treated as a spike, but as noted earlier, it is probably more likely that we are dealing with high-dimensional surfaces, given that the relative positions of features are smoothly continuous across the image. Nonetheless, since we cannot assume that the structure of these sparse features is orthogonal, networks that allow non-orthogonal solutions are likely to find more efficient solutions.

6 Overview

This paper explored the possible reasons why the mammalian visual system might evolve a wavelet-like code for representing the natural environment. It was argued that this particular wavelet representation is extremely well matched to the particular statistical structure of our natural visual environment. It is argued that, in general, wavelet codes are effective because they match the sparse localized oriented band-limited structure that exists in most natural data. The result of such a coding strategy is that the activity of any particular cell will be relatively independent of the activity of other cells. It is presumed that this assists the sensory system in finding more complex structure remaining in the input. It is argued here that the use of wavelets in sensory systems is not to compress data, but to aid in extracting this complex structure. The wavelet representation is only a first step, although an excellent first step.

Complex natural data can take on many forms, and we should not presume that the wavelet transform is an optimal representation for all of these forms. Eventually, we may find many ways of fine tuning our transforms to match both the individual needs of the user and the statistics of the input. And we may find, in many cases, that our own sensory systems have found the solution first.

7   Acknowledgements

D.J.F. was supported by NIH Grant MH50588.

8   References

Adelson, E. H., Simoncelli, E. & Hingorani, R. 1987 Orthogonal pyramid transforms for image coding. *SPIE Visual Commun. Image Processing* II **845**, 50–58.

Atick, J. J. 1992 Could information theory provide an ecological theory of sensory processing. *Network* **3**, 213–251.

Atick, J. J. & Redlich, A. N. 1990 Towards a theory of early visual processing. *Neural Computation* **4**, 196–210.

Atick, J. J. & Redlich, A. N. 1992 What does the retina know about natural scenes? *Neural Computation* **4**, 449–572.

Barlow, H. B. 1961 The coding of sensory messages. *Current problems in animal behavior*. Cambridge University Press.

Bell, A. J. & Sejnowski, T. J. 1997 The independent components of natural scenes are edge filters. *Vision Res.* **37**, 3327–3338.

Bienenstock, E. L., Cooper, L. N. & Monro, P. W. 1982 Theory for the development of neuron selectivity: orientation selectivity and binocular interaction in visual cortex. *J. Neurosci.* **128**, 3139–3161.

Blakemore, C. & Campbell, F. W. 1969 On the existence of neurones in the human visual system selectively sensitive to the orientation and size of retinal images. *J. Physiol.* **203**, 237–260.

Bossomaier, T. & Snyder, A. W. 1986 Why spatial frequency processing in the visual cortex? *Vision Res.* **26**, 1307–1309.

Brady, N. & Field, D. J. 1995 What's constant in contrast constancy: the effects of scaling on the perceived contrast of bandpass patterns. *Vision Res.* **35**, 739–756.

Burt, P. J. & Adelson, E. H. 1983 The Laplacian pyramid as a compact image code. *IEEE Trans. Commun.* **31**, 532–540.

Burton, G. J. & Moorehead, I. R. 1987 Color and spatial structure in natural scenes. *Appl. Optics* **26**, 157–170.

Campbell, F. W. & Robson, J. G. 1968 Application of Fourier analysis to the visibility of gratings. *J. Physiol.* **197**, 551–556.

Croner, L. J. & Kaplan, E. 1995 Receptive fields of P and M ganglion cells across the primate retina. *Vision Res.* **35**, 7–24.

Daubechies, I. 1988 Orthonormal bases of compactly supported wavelets. *Commun. Pure Appl. Math.* **41**, 909–996.

Daugman, J. G. 1988 Complete discrete 2-D Gabor transforms by neural networks for image analysis and compression. *IEEE Trans. on Acoustics, Speech and Signal Processing* **36**, 1169–1179.

DeValois, R. L., Albrecht, D. G. & Thorell, L. G. 1982 Spatial frequency selectivity of cells in macaque visual cortex. *Vision Res.* **22**, 545–559.

Donoho, D. L. 1993 Unconditional bases are optimal bases for data compression and for statistical estimation. *Appl. Computational Harmonic Analysis* **1**, 100–115.

Field, D. J. 1987 Relations between the statistics of natural images and the response properties of cortical cells. *J. Opt. Soc. Am.* **4**, 2379–2394.

Field, D. J. 1989 What the statistics of natural images tell us about visual coding. *Proc. SPIE* **1077**, 269–276.

Field, D. J. 1993 Scale-invariance and self-similar 'wavelet' transforms: an analysis of natural scenes and mammalian visual systems. In *Wavelets, fractals and Fourier transforms* (ed. M. Farge, J. Hunt & J. C. Vassilicos). Oxford University Press.

Field, D. J. 1994 What is the goal of sensory coding? *Neural Computation* **6**, 559–601.

Field, D. J. & Brady, N. 1997 Visual sensitivity, blur and the sources of variability in the amplitude spectra of natural scenes. *Vision Res.* **37**, 3367–3383.

Field, D. J. & Millman, T. J. 1999 A biologically plausible nonlinear decorrelation network can learn sparse codes. (In preparation.)

Field, D. J. & Tolhurst, D. J. 1986 The structure and symmetry of simple-cell receptive field profiles in the cat's visual cortex. *Proc. R. Soc. Lond.* B **228**, 379–400.

Foldiak, P. 1990 Forming sparse representations by local anti-Hebbian learning. *Biol. Cybernetics* **64**, 165–170.

Gabor, D. 1946 Theory of communication. *J. IEE, Lond.* **93**, 429–457.

Hancock, P. J., Baddeley, R. J. & Smith, L. S. 1992 The principal components of natural images. *Network* **3**, 61–70.

Hawken, M. J. & Parker, A. J. 1987 Spatial properties of neurons in the monkey striate cortex. *Proc. R. Soc. Lond.* B **231**, 251–288.

Hubel, D. H. & Wiesel, T. N. 1962 Receptive fields, binocular interaction and functional architecture in the cat's striate cortex. *J. Physiol.* **160**, 106–154.

Intrator, N. 1992 Feature extraction using an unsupervised neural network. *Neural Computation* **4**, 98–107.

Jones, J. & Palmer, L. 1987 An evaluation of the two-dimensional Gabor filter model of simple receptive fields in cat striate cortex. *J. Neurophysiol.* **58**, 1233–1258.

Kulikowski, J. J., Marcelja, S. & Bishop, P. O. 1982 Theory of spatial position and spatial frequency relations in the receptive fields of simple cells in the visual cortex. *Biol. Cybernetics* **43**, 187–198.

Lehky, S. R. & Sejnowski, T. J. 1990 Network model of shape-from-shading: neural function arises from both receptive and projective receptive fields. *Nature* **333**, 452–454.

Linsker, R. 1988 Self-organization in a perceptual network. *Computer* **21**, 105–117.

Marcelja, S. 1980 Mathematical description of the responses of simple cortical cells. *J. Opt. Soc. Am.* **70**, 1297–1300.

Marr, D. & Hildreth, E. 1980 Theory of edge detection. *Proc. R. Soc. Lond.* B **207**, 187–217.

Olshausen, B. A. & Field, D. J. 1996 Emergence of simple-cell receptive field properties by learning a sparse code for natural images. *Nature* **381**, 607–609.

Olshausen, B. A. & Field, D. J. 1997 Sparse coding with an overcomplete basis set: a strategy employed by V1? *Vision Res.* **37**, 3311–3325.

Pratt, W. K. 1978 *Digital image processing.* New York: Wiley.

Ruderman, D. L. 1994 The statistics of natural images. *Network* **5**, 517–548.

Shouval, H., Intrator, N. & Cooper, L. 1997 BCM network develops orientation selectivity and ocular dominance in natural scene environment. *Vision Res.* **37**, 3339–3342.

Srinivasan, M. V., Laughlin, S. B. & Dubs, A. 1982 Predictive coding: a fresh view of inhibition in the retina. *Proc. R. Soc. Lond.* B **216**, 427–459.

Stork, D. G. & Wilson, H. R. 1990 Do Gabor functions provide appropriate descriptions of visual cortical receptive fields? *J. Opt. Soc. Am.* A **7**, 1362–1373.

Tolhurst, D. J. & Thompson, I. D. 1982 On the variety of spatial frequency selectivities shown by neurons in area 17 of the cat. *Proc. R. Soc. Lond.* B **213**, 183–199.

Tolhurst, D. J., Tadmor, Y. & Tang, C. 1992 The amplitude spectra of natural images. *Ophthalmic Physiol. Opt.* **12**, 229–232.

Watson, A. B. 1983 Detection and recognition of simple spatial forms. In *Physical and biological processing of images* (ed. O. J. Braddick & A. C. Slade). Berlin: Springer.

Watson, A. B. 1991 Multidimensional pyramids in vision and video. In *Representations of vision* (ed. A. Gorea). Cambridge University Press.

Webster, M. A. & DeValois, R. L. 1985 Relationship between spatial-frequency and orientation tuning of striate-cortex cells. *J. Opt. Soc. Am.* A **2**, 1124–1132.

Zetzsche, C. 1990 Sparse coding: the link between low level vision and associative memory. In *Parallel processing in neural systems and computers* (ed. R. Eckmiller, G. Hartmann & G. Hauske) Amsterdam: North-Holland.

9

Image processing with complex wavelets

Nick Kingsbury

*Signal Processing Group, Department of Engineering, University of Cambridge,
Cambridge CB2 1PZ, UK*

ngk@eng.cam.ac.uk

Abstract

We first review how wavelets may be used for multi-resolution image processing, describing the filter-bank implementation of the discrete wavelet transform (DWT) and how it may be extended via separable filtering for processing images and other multi-dimensional signals. We then show that the condition for inversion of the DWT (perfect reconstruction) forces many commonly used wavelets to be similar in shape, and that this shape produces severe shift dependence (variation of DWT coefficient energy at any given scale with shift of the input signal). It is also shown that separable filtering with the DWT prevents the transform from providing directionally selective filters for diagonal image features.

Complex wavelets can provide both shift invariance and good directional selectivity, with only modest increases in signal redundancy and computation load. However, development of a complex wavelet transform (CWT) with perfect reconstruction and good filter characteristics has proved difficult until recently. We now propose the dual-tree CWT as a solution to this problem, yielding a transform with attractive properties for a range of signal and image processing applications, including motion estimation, denoising, texture analysis and synthesis, and object segmentation.

Keywords: image processing; wavelets; shift invariance; directional filters; perfect reconstruction; complex filters

1 Introduction

In this paper we consider how wavelets may be used for image processing. To date, there has been considerable interest in wavelets for image compression, and they are now commonly used by researchers for this purpose, even though the main international standards still use the discrete cosine transform (DCT). However, for image processing tasks, other than compression, the take-up of wavelets has been less enthusiastic. Here we analyse possible reasons for this and present some new ways to use wavelets that offer significant advantages.

A good review of wavelets and their application to compression may be found in Rioul & Vetterli (1991) and in-depth coverage is given in the book by Vetterli

& Kovacevic (1995). An issue of the Proceedings of the IEEE (Kovacevic & Daubechies 1996) has been devoted to wavelets and includes many very readable articles by leading experts.

In Section 2 of this paper we introduce the basic discrete wavelet filter tree and show how it may be used to decompose multi-dimensional signals. In Section 3 we show some typical wavelets and illustrate the similar shapes of those that all satisfy the perfect reconstruction constraints. Unfortunately, as explained in Section 4, discrete wavelet decompositions based on these typical wavelets suffer from two main problems that hamper their use for many image analysis and reconstruction tasks as follows.

(i) Lack of *shift invariance*, which means that small shifts in the input signal can cause major variations in the distribution of energy between wavelet transform coefficients at different scales.

(ii) Poor *directional selectivity* for diagonal features, because the wavelet filters are separable and real.

Complex wavelets are shown, in Section 5, to overcome these two key problems by introducing limited redundancy into the transform. However, a further problem arises here because perfect reconstruction becomes difficult to achieve for complex wavelet decompositions beyond level 1, when the input to each level becomes complex. To overcome this, we have recently developed the dual-tree complex wavelet transform (DT CWT), which allows perfect reconstruction while still providing the other advantages of complex wavelets. This is described in Section 6, and then Section 7 discusses constraints on the DT CWT filter characteristics, required in order to achieve close approximations to shift invariance. Finally, in Section 8, we suggest possible application areas for the DT CWT, including an example of image denoising.

2   The wavelet tree for multi-dimensional signals

For one-dimensional signals, the conventional discrete wavelet transform (DWT) may be regarded as equivalent to filtering the input signal with a bank of bandpass filters, whose impulse responses are all approximately given by scaled versions of a *mother wavelet*. The scaling factor between adjacent filters is usually 2:1, leading to octave bandwidths and centre frequencies that are one octave apart. At the coarsest scale, a lowpass filter is also required to represent the lowest frequencies of the signal. The outputs of the filters are usually maximally decimated so that the number of DWT output samples equals the number of input samples and the transform is invertible. The octave-band DWT is most efficiently implemented by the *dyadic wavelet decomposition tree* of Mallat (1989), a cascade of two-band perfect-reconstruction filter banks, shown in figure 1a. Because of the decimation by two at

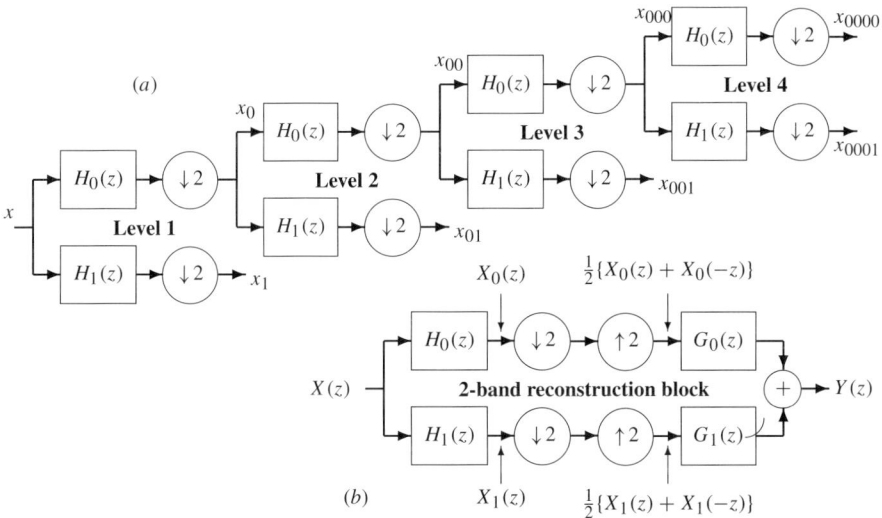

Fig. 1: (*a*) Four-level binary wavelet tree of lowpass filters, H_0, and highpass filters, H_1; and (*b*) the two-band reconstruction block, used to achieve perfect reconstruction from an inverse tree employing filters, G_0 and G_1.

each stage, the total output sample rate equals the input sample rate and there is *no redundancy* in the transform.

In order to reconstruct the signal, a pair of reconstruction filters G_0 and G_1 are used in the arrangement of figure 1*b*, and usually the filters are designed such that the z transform of the output signal $Y(z)$ is identical to that of the input signal $X(z)$. This is known as the condition for *perfect reconstruction*. Hence, in figure 1*a*, x_{000} may be reconstructed from x_{0000} and x_{0001}; and then x_{00} from x_{000} and x_{001}; and so on back to x, using an inverse tree of filters, $G_{...}(z)$.

If the input signal is two dimensional, the binary tree may be extended into a quad-tree structure, as shown for two levels in figure 2. In a separable implementation, each level of the quad-tree comprises two stages of filtering: the first stage typically filters the rows of the image to generate a pair of horizontal lowpass and highpass subimages; and then the second stage filters the columns of each of these to produce four subimages, $x_0, \ldots, x_3$, each one-quarter of the area of x. The lowpass subimage, x_0, is similar to the original but smaller, and is typically decomposed by further levels of the two-dimensional transform.

The filtering in figure 2 is *separable* since it is performed separately in the row and column directions. This is usually the most efficient way to perform two-dimensional filtering. The technique may be extended, straightforwardly, to more than two dimensions, by applying filters to each dimension in turn. For *m*

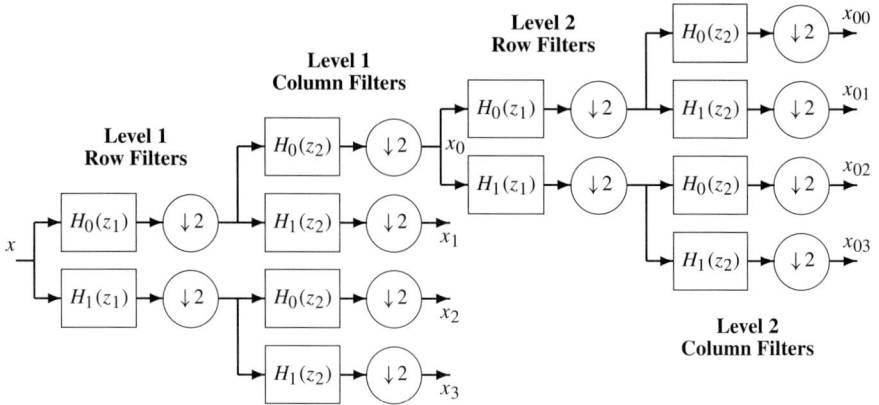

Fig. 2: Two levels of the quaternary separable wavelet tree, normally used for two-dimensional signals. Row filtering is denoted by the z transform parameter z_1, and column filtering by z_2.

dimensions, the number of sub-bands (subimages) at each level increases to 2^m (including the low-frequency sub-band, passed on to the next level).

3   Common wavelets with perfect reconstruction

The art of finding good wavelets lies in the design of the set of filters, $\{H_0, H_1, G_0, G_1\}$ from figure 1b, to achieve various trade-offs between (signal-dependent) spatial and frequency domain characteristics while satisfying the perfect-reconstruction (PR) condition. We now briefly consider this design process.

In figure 1b, multi-rate filter analysis shows that

$$
\begin{aligned}
Y(z) &= \tfrac{1}{2}\{X_0(z) + X_0(-z)\}G_0(z) + \tfrac{1}{2}\{X_1(z) + X_1(-z)\}G_1(z) \\
&= \tfrac{1}{2}X(z)\{H_0(z)G_0(z) + H_1(z)G_1(z)\} \\
&\quad + \tfrac{1}{2}X(-z)\{H_0(-z)G_0(z) + H_1(-z)G_1(z)\}.
\end{aligned}
\tag{3.1}
$$

The first PR condition requires aliasing cancellation and forces the above term in $X(-z)$ to be zero. Hence, $H_0(-z)G_0(z) + H_1(-z)G_1(z) = 0$, which can be achieved if

$$
H_1(z) = z^{-k}G_0(-z) \text{ and } G_1(z) = z^k H_0(-z),
\tag{3.2}
$$

where k must be odd (usually $k = \pm 1$).

The second PR condition is that the transfer function from $X(z)$ to $Y(z)$ should be unity; i.e. $H_0(z)G_0(z) + H_1(z)G_1(z) = 2$. If we define a product filter $P(z) = H_0(z)G_0(z)$ and substitute the results from (3.2), then this condition becomes:

$$H_0(z)G_0(z) + H_1(z)G_1(z) = P(z) + P(-z) = 2. \qquad (3.3)$$

Since the odd powers of z in $P(z)$ cancel with those in $P(-z)$, this requires that $p_0 = 1$ and that $p_n = 0$ for all n even and non-zero.

$P(z)$ is the transfer function of the lowpass branch in figure 1b (excluding the effects of the decimator and interpolator), and $P(-z)$ is that of the highpass branch. For image processing applications, $P(z)$ should be zero phase in order to minimize distortions when the wavelet coefficients are modified in any way; so to obtain PR it must be of the form:

$$P(z) = \cdots + p_5 z^5 + p_3 z^3 + p_1 z + 1 + p_1 z^{-1} + p_3 z^{-3} + p_5 z^{-5} + \cdots . \qquad (3.4)$$

To simplify the tasks of choosing $P(z)$, based on the zero-phase symmetry we usually transform $P(z)$ into $P_t(Z)$ such that:

$$P(z) = P_t(Z) = 1 + p_{t,1} Z + p_{t,3} Z^3 + p_{t,5} Z^5 + \cdots , \quad \text{where } Z = \tfrac{1}{2}(z + z^{-1}). \qquad (3.5)$$

If T_s is the sampling period, the frequency response is given by $z = e^{j\omega T_s}$, and therefore by $Z = \cos(\omega T_s)$. To obtain smooth wavelets (after many levels of decomposition), Daubechies (1990) has shown that $H_0(z)$ and $G_0(z)$ should have a number of zeros at $z = -1$ ($\omega T_s = \pi$), so $P_t(Z)$ needs zeros at $Z = -1$. In general, more zeros at $Z = -1$ (and, hence, at $z = -1$) produce smoother wavelets.

Figure 3 shows five different wavelets satisfying the above equations. They are shown after four levels of decomposition. On the left are the wavelets formed from the analysis filters, H, and on the right are the wavelets from the reconstruction filters, G. These filters may be swapped, so we have chosen the smoother wavelets for reconstruction, since, for many applications (particularly compression), this gives the least visible distortions.

The LeGall (3,5)-tap wavelet is designed by choosing $P_t(Z)$ to be third order, and constraining the maximum number (two) of zeros to be at $Z = -1$, consistent with the even-order term(s) in equation (3.5) being zero. This gives filters, H_0 and G_0, which are five and three taps long, respectively. The Antonini (7,9)-tap wavelet is obtained in the same way, when $P_t(Z)$ is seventh order. In this case there are four zeros at $Z = -1$. This wavelet is one of the most favoured by image-compression researchers and is used in the FBI fingerprint-compression system.

Both of these wavelets are constrained such that all the H and G filters are linear-phase finite impulse-response filters. Unfortunately, this means that the reconstruction filters cannot have the same frequency responses as the analysis filters. Wavelets of this type are known as *biorthogonal*.

For the frequency responses to be the same, the wavelets must be *orthogonal*. The Daubechies (8,8)-tap wavelet is an orthogonal wavelet, obtained from the same

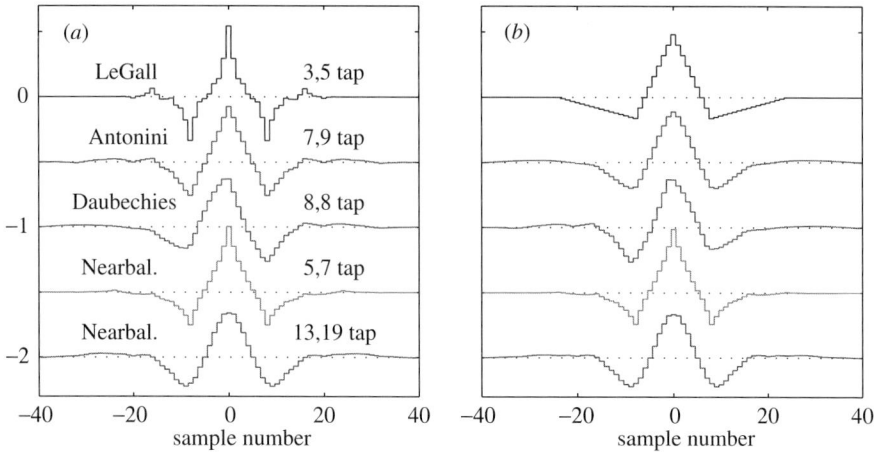

Fig. 3: Comparison of some common wavelets. (*a*) Level 4 analysis wavelets. (*b*) Level 4 reconstruction wavelets.

seventh-order $P_t(Z)$ as the Antonini wavelet, but where the factors of z in $P(z)$ are grouped into reciprocal pairs that are divided equally between H_0 and G_0 so as to provide the best approximation to linear-phase responses.

The final two wavelets in figure 3 are linear-phase wavelets that have frequency responses that are *nearly balanced* between analysis and reconstruction, and are, therefore, nearly orthogonal. To achieve this condition, we increase the order of $P_t(Z)$ by two, without adding further zeros at $Z = -1$, and hence provide an additional design parameter that can be used to obtain approximate balance.

We find (Tay & Kingsbury 1993) that the following factorization of a fifth-order $P_t(Z)$ produces good balance:

$$P_t(Z) = \tfrac{1}{5}0(50 + 41Z - 15Z^2 - 6Z^3) \cdot \tfrac{1}{7}(7 + 5Z - 2Z^2) = H_{0t}(Z) \cdot G_{0t}(Z).$$
$$(3.6)$$

Using the transformation $Z = \tfrac{1}{2}(z + z^{-1})$ gives the simplest near-balanced wavelets with five and seven tap filters. These wavelets are well balanced but have quite sharp cusps, as can be seen in the fourth row of figure 3.

Smoother near-balanced wavelets may be obtained by employing a higher-order transformation from Z to z, such as

$$Z = pz^3 + (\tfrac{1}{2} - p)(z + z^{-1}) + pz^{-3}.$$

Four zeros on the unit circle near $z = -1$ are achieved for each zero, $Z = -1$, if $p = -\tfrac{3}{2}2$. When substituted into $P_t(Z)$, this gives relatively high-order filters with 13 and 19 taps (although two taps of each filter are zero and they do have an efficient ladder implementation). The final row of figure 3 shows the improved smoothness of these wavelets.

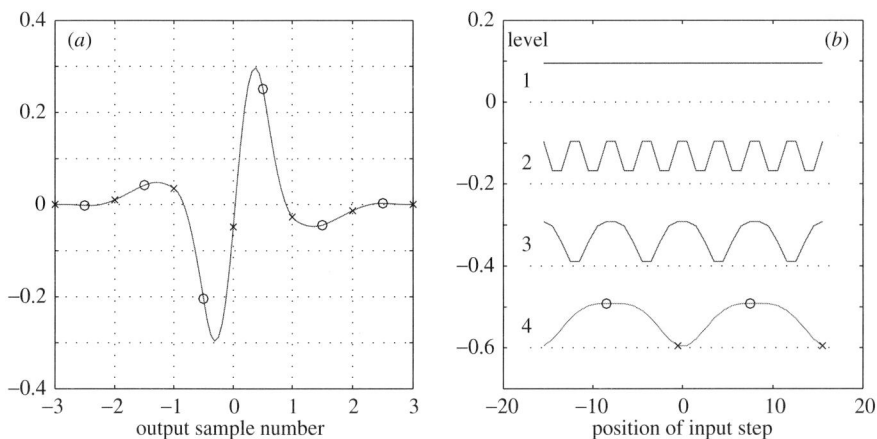

Fig. 4: (*a*) Step response at level 4 of the Antonini (7,9)-tap wavelet; and (*b*) its shift dependence, showing the variation of energy in each level of wavelet coefficients, for a unit step input as the position of the input step is shifted. The zero lines (shown dotted) for each level have been offset for clarity.

4 Some problems with common wavelets

4.1 Shift dependence

Figure 3 shows a strong similarity between the shapes of various wavelets. This is because PR constrains each filter in figure 1 to be approximately a half-band filter. This causes aliasing and results in severe *shift dependence* of the wavelet transform.

When we analyse the Fourier spectrum of a signal, we expect the energy in each frequency bin to be invariant to any shifts of the input in time or space. It would be desirable if wavelet transforms behaved similarly, but, unfortunately, real wavelet transforms, even though they have perfect reconstruction properties, do not provide energy shift invariance separately at each level.

Consider a step function input signal, analysed with the DWT using Antonini (7,9)-tap filters. The step response at wavelet level 4 is shown in figure 4*a*, assuming that wavelet coefficients are computed at the full input sampling rate. In practice, they are computed at $\frac{1}{16}$ of this rate, yielding samples at points such as those of the crosses in figure 4*a*. If the input step is shifted relative to the output sampling grid, then this is equivalent to sampling the step response with a different horizontal offset; e.g. for an offset of eight input samples, we obtain samples at the circles in figure 4*a*.

Now, comparing the total energy of the samples at the crosses (which are all quite small in amplitude) with the energy of the samples at the circles (two of which are rather large), we find a large energy fluctuation (over 24:1). This illustrates a significant drawback to using the standard DWT as a tool for analysing signals;

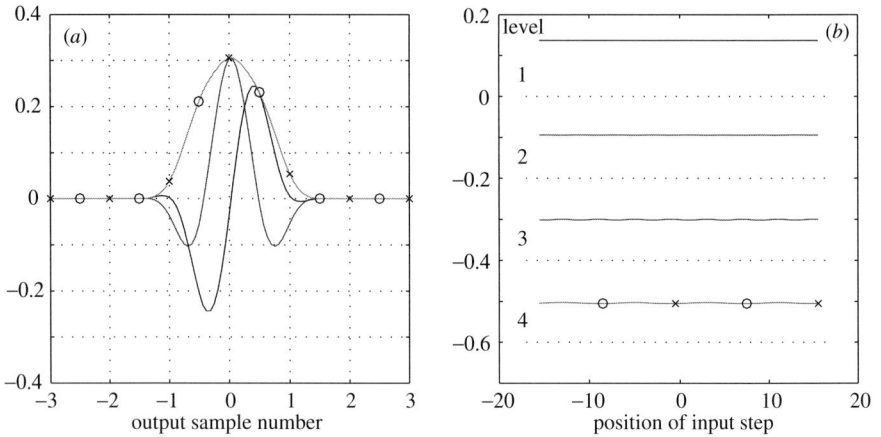

Fig. 5: (*a*) Step response at level 4 of complex wavelet (*a*). (*b*) Shift dependence of output energy at levels 1 to 4, showing the variation of energy in each level of wavelet coefficient for a unit step input as the position of the input step is shifted.

the energy distribution between the various wavelet levels depends critically on the *position* of key features of the signal relative to the wavelet subsampling grid. Ideally, we would like it to depend on just the features themselves.

This problem is illustrated more generally in figure 4*b*, which shows how the total energy at each wavelet level varies as the input step is shifted. The period of each variation equals the subsampling period at that level, and the crosses and circles in figure 4*b* show the energies at level 4 corresponding to the shift positions shown by the equivalent symbols in figure 4*a*.

Hence we conclude that, due to this shift dependence, real DWTs are unlikely to give consistent results when used to detect key features in images. This problem is caused by aliasing due to subsampling at each wavelet level. It can, therefore, be avoided by not decimating the DWT outputs. However, this produces considerable data redundancy because each subimage at each wavelet level is the same size as the input image. This is often known as the undecimated wavelet transform, but, in addition to being computationally inefficient (even when using the 'à trous' algorithm), this does not solve the other main wavelet problem, discussed below.

4.2   Poor directional selectivity

In figure 2, separable filtering of the rows and columns of an image produces four subimages at each level. The Lo–Hi and Hi–Lo bandpass subimages (e.g. x_1 and x_2) can select mainly horizontal or vertical edges, respectively, but the Hi–Hi subimage (x_3) contains components from diagonal features of either orientation. This means that the separable real DWT has *poor directional selectivity*.

One way of explaining this is that real highpass row filters select both positive

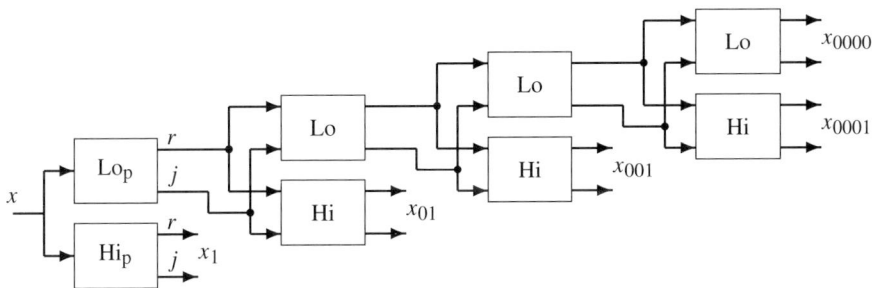

Fig. 6: Four levels of the complex wavelet tree for a real one-dimensional input signal x. The real and imaginary parts (r and j) of the inputs and outputs are shown separately. Where there is only one input to a block, it is a real signal.

and negative horizontal high frequencies, while real highpass column filters select both positive and negative vertical high frequencies. Hence, the combined Hi–Hi filter must have passbands in all four quadrants of the two-dimensional frequency plane. On the other hand, a directionally selective filter for diagonal features with positive gradient must have passbands only in quadrants 2 and 4 of the frequency plane, while a filter for diagonals with negative gradient must have passbands only in quadrants 1 and 3. The poor directional properties of real separable filters make it difficult to generate steerable or directionally selective algorithms, based on the separable real DWT.

5 Properties of complex wavelets

It is found that both of the above problems can be solved effectively by the *complex wavelet transform* (CWT). The structure of the CWT is the same as in figure 1a, except that the CWT filters have complex coefficients and generate complex output samples. This is shown in figure 6, in which each block is a complex filter and includes down-sampling by 2 (not shown) at its outputs. Since the output sampling rates are unchanged from the DWT, but each sample contains a real and imaginary part, a redundancy of 2:1 is introduced (we shall show later that this becomes 4:1 in two dimensions).

The complex filters may be designed such that the *magnitudes* of their step responses vary slowly with input shift; only the *phases* vary rapidly. This is shown in figure 5a, in which the real and imaginary parts of a typical complex wavelet step response are superimposed and the uppermost curve represents the magnitude of the response. Note that the real part is an odd function while the imaginary part is even.

This wavelet was derived from the following simple (4,4)-tap Lo and Hi filters:

$$\left.\begin{aligned} h_0 &= \tfrac{1}{1}0[1 - j, 4 - j, 4 + j, 1 + j], \\ h_1 &= \tfrac{1}{4}8[-3 - 8j, 15 + 8j, -15 + 8j, 3 - 8j]. \end{aligned}\right\} \tag{5.1}$$

The level 1 filters, Lo_p and Hi_p in figure 6, include an additional 2-tap prefilter, which has a zero at $z = -j$, in order to simulate the effect of a filter tree extending further levels to the left of level 1, as discussed by Magarey & Kingsbury (1998).

Figure 5b plots the energy at each level versus input step position for this CWT, and, in contrast to figure 4b, shows that it is approximately constant at all levels. Hence, the energy of each CWT band may be made approximately *shift invariant*. Simoncelli *et al.*(1992) have shown that shift invariance of energy distribution is equivalent to the property of interpolability.

The other key property of these complex wavelets is that their phases vary approximately linearly with input shift (as with Fourier coefficients). Thus, based on measurement of phase shifts, efficient displacement estimation is possible and interpolation between consecutive complex samples can be relatively simple and accurate. Further details of the design of these wavelets and of their application to motion estimation are given in Magarey & Kingsbury (1998).

5.1   Extension to multiple dimensions

Extension of complex wavelets to two dimensions is achieved by separable filtering along rows and then columns. However, if row and column filters both suppress negative frequencies, then only the first quadrant of the two-dimensional signal spectrum is retained. Two adjacent quadrants of the spectrum are required to represent fully a real two-dimensional signal, so an extra 2:1 factor of redundancy is required, producing 4:1 redundancy overall in the transformed two-dimensional signal. This is achieved by additional filtering with complex conjugates of either the row or column filters. If the signal exists in m-dimensional space ($m > 2$), then further conjugate pairs of filters are needed for each dimension leading to redundancy of 2^m:1.

The most computationally efficient way to achieve the pairs of conjugate filters is to maintain separate imaginary operators, j_1 and j_2, for the row and column processing, as shown in figure 7. This produces four-element 'complex' vectors: $\{a, b, c, d\} = a + bj_1 + cj_2 + dj_1j_2$. Note that these are not quaternions as they have different properties. Each 4-vector can be converted into a pair of conventional complex 2-vectors, by letting $j_1 = j_2 = j$ in one case and $j_1 = -j_2 = -j$ in the other case. This corresponds to sum and difference operations on the $\{a, d\}$ and $\{b, c\}$ pairs and produces two complex outputs, $(a-d)+(b+c)j$ and $(a+d)+(-b+c)j$, corresponding to first and second quadrant directional filters, respectively. The $\sum/\Delta$ blocks in figure 7 do this.

Complex filters in multiple dimensions provide true directional selectivity, despite being implemented separably, because they are still able to separate all parts of the m-dimensional frequency space. For example, a two-dimensional CWT

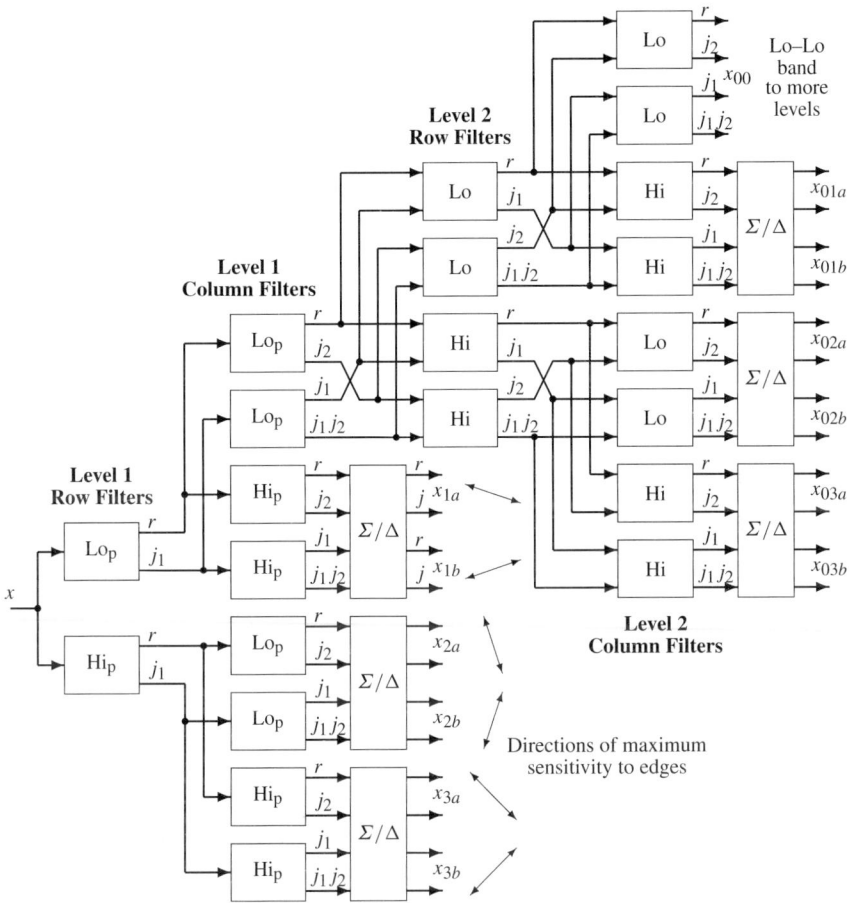

Fig. 7: Two levels of the complex wavelet tree for a real two-dimensional input image x, giving six directional bands at each level (the directions are shown for level 1). Components of four-element 'complex' vectors are labelled r, j_1, j_2, $j_1 j_2$.

produces six bandpass subimages of complex coefficients at each level, which are strongly orientated at angles of $\pm 15°$, $\pm 45°$ and $\pm 75°$, depicted by the double-headed arrows in figure 7 and by the two-dimensional impulse responses in figure 12.

6 The dual-tree complex wavelet transform

For many applications it is important that the transform be perfectly invertible. A few authors, such as Lawton (1993) and Belzer *et al.*(1995), have experimented with complex factorizations of the standard Daubechies polynomials and obtained

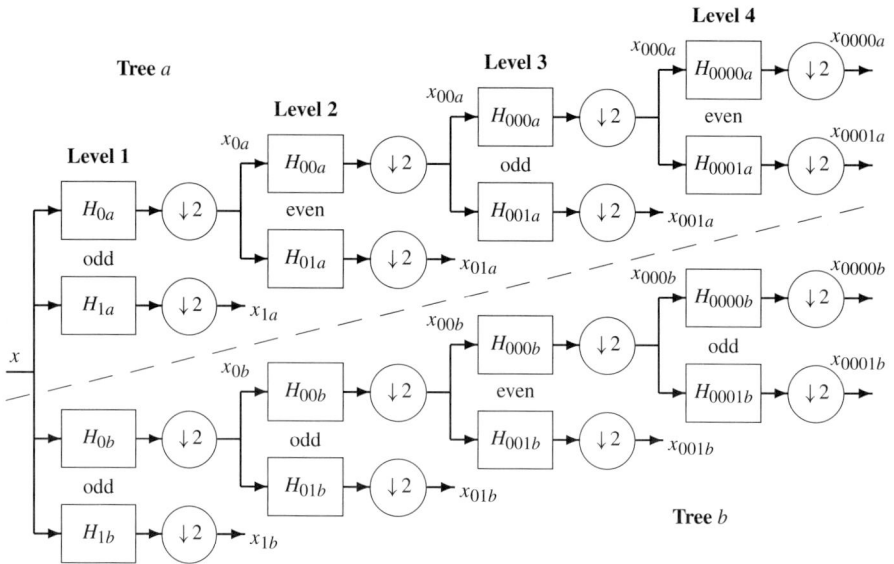

Fig. 8: Dual tree of real filters for the DT CWT, showing whether the filters have odd or even length. The two trees give the real and imaginary parts of the complex coefficients.

PR complex filters, but these do not give filters with good frequency-selectivity properties. To provide shift invariance and directional selectivity, all of the complex filters should emphasize positive frequencies and reject negative frequencies, or vice versa (see Section 5 and 7).

Unfortunately, it is very difficult to design an inverse transform, based on complex filters of the type defined at the start of Section 5, which has good frequency selectivity *and* PR at all levels of the transform. Although such filters can be designed to give PR quite easily at level 1 of the tree by applying the constraint that the reconstructed output signal must be real, a similar constraint cannot be applied at further levels where inputs *and* outputs are complex. For PR below level 1, the set of four filters in figure 1*b* must have a flat overall frequency response. However, this is not possible if all of the filters tend to reject negative frequencies. Hence, a different approach to generating a complex filter tree is needed.

In Kingsbury (1998*a*, *b*), we introduced the DT CWT, which added perfect reconstruction to the other attractive properties of complex wavelets: shift invariance; good directional selectivity; limited redundancy; and efficient order-N computation.

The dual-tree transform was developed by noting that approximate shift invariance can be achieved with a *real* DWT by doubling the sampling rate at each

Input samples ⟵——— Block of 16 input samples ———⟶

x: • • • • • • • • | • • • • • • • •

Level 1 samples

odd Lo x_{0a}:	a	a	a	a	a	a	a	a
odd Lo x_{0b}:	b	b	b	b	b	b	b	b
odd Hi x_{1a}:	a	a	a	a	a	a	a	a
odd Hi x_{1b}:	b	b	b	b	b	b	b	b

Level 2 samples

even Lo x_{00a}:	a		a		a		a
odd Lo x_{00b}:		b		b		b		b
Hi x_{01a}, x_{01b}:	$*$		$*$		$*$		$*$

Level 3 samples

odd Lo x_{000a}:	a		a
even Lo x_{000b}:		b		b
Hi x_{001a}, x_{001b}:		$*$		$*$

Level 4 samples

even Lo x_{0000a}:	a
odd Lo x_{0000b}:		b
Hi x_{0001a}, x_{0001b}:	$*$

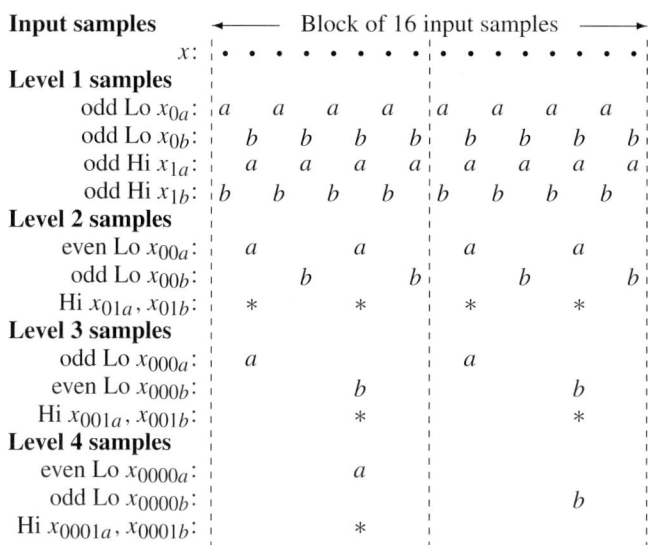

Fig. 9: Effective sampling instants of odd and even filters in figure 8 assuming zero phase responses. (a = tree a, b = tree b, $*$ = combined samples.)

level of the tree. For this to work, the samples must be evenly spaced. We can double all the sampling rates in figure 1a by eliminating the down-sampling by 2 after the level 1 filters. This is equivalent to having two parallel fully decimated trees, a and b in figure 8, provided that the delays of H_{0b} and H_{1b} are one sample offset from H_{0a} and H_{1a}. We then find that, to get uniform intervals between samples from the two trees *below* level 1, the filters in one tree must provide delays that are half a sample different (at each filter input rate) from those in the other tree. For linear phase, this requires *odd-length* filters in one tree and *even-length* filters in the other. Greater symmetry between the two trees occurs if each tree uses odd and even filters alternately from level to level, but this is not essential. In figure 9 we show the effective sampling instants of the filter output samples when the filters are odd and even, as in figure 8. For example, at level 2, the tree a filters are even length, so the a output samples (x_{00a}) occur midway between pairs of samples from the a lowpass filter at level 1 (x_{0a}), whereas the tree b filters are odd length, so the b output samples (x_{00b}) are aligned with samples from the b lowpass filter at level 1 (x_{0b}).

To invert the transform, the PR filters G are applied in the usual way to invert each tree separately, and, finally, the two results are averaged. Compared with the undecimated (à trous) wavelet tree, which eliminates down-sampling after *every* level of filtering, the dual tree effectively eliminates down-sampling *only* after the level 1 filters, and, hence, its redundancy is much less.

Thus far, the dual tree does not appear to be a complex transform at all. However,

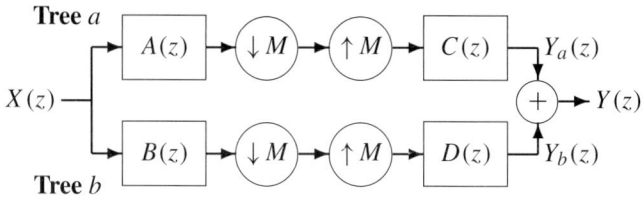

Fig. 10: Basic configuration of the dual tree if either wavelet or scaling-function coefficients from just level m are retained ($M = 2^m$).

when the outputs from the two trees in figure 8 are interpreted as the real and imaginary parts of complex wavelet coefficients, the transform effectively becomes complex. If the filters are from linear-phase PR biorthogonal sets, the odd-length highpass filters have even symmetry about their midpoint, while the even-length highpass filters have odd symmetry. The impulse responses of these then look very like the real and imaginary parts of the complex wavelets of the previous section. In fact, the block diagrams of figures 6 and 7 still apply to the dual-tree transform, although the operations within each filter block do change: the real and imaginary parts of each block's output are no longer calculated based on the usual rules of complex algebra, but, instead, each is based on filtering just one part of the block's input with either an odd- or even-length filter.

7   Shift-invariant filter design

In order to show the shift-invariant properties of the dual tree, we consider what happens when the signal is reconstructed using coefficients of just one type (wavelet or scaling function) from just one level of the dual tree. This models (in an extreme way) the virtual elimination of certain sub-bands that commonly occur in many algorithms. For example, we might choose to retain only the level-3 wavelet coefficients, x_{001a} and x_{001b}, from figure 8, and set all others to zero. If the signal y, reconstructed from just these coefficients, is free of aliasing, then the transform is defined to be shift invariant at that level.

Figure 10 shows the simplified analysis and reconstruction parts of the dual tree when coefficients of just one type and level are retained. All down(up)-sampling operations are moved to the output (input) of the analysis (reconstruction) filter banks and the cascaded filter transfer functions are combined. $M = 2^m$ is the total down-sampling factor. For example, if we retain only x_{001a} and x_{001b}, then $M = 8$, $A(z) = H_{0a}(z)H_{00a}(z^2)H_{001a}(z^4)$ and $B(z)$, $C(z)$, $D(z)$ are obtained similarly.

Letting $W = e^{j2\pi/M}$, multirate analysis of figure 10 gives:

$$Y(z) = \frac{1}{M} \sum_{k=0}^{M-1} X(W^k z)[A(W^k z)C(z) + B(W^k z)D(z)]. \qquad (7.1)$$

For shift invariance, the aliasing terms (for which $k \neq 0$) must be negligible. So we design $B(W^k z)D(z)$ to cancel out $A(W^k z)C(z)$ for all non-zero k that give overlap of the pass or transition bands of the filters $C(z)$ or $D(z)$ with those of the shifted filters $A(W^k z)$ or $B(W^k z)$. Separate strategies are needed depending on whether the filters are lowpass (for scaling functions) or bandpass (for wavelets).

For level m in the dual tree, the lowpass filters have passbands from $(-f_s/2M)$ to $(f_s/2M)$, where f_s is the input sampling frequency. The W^k terms in (7.1) shift the passbands in multiples, k, of f_s/M. If $A(z)$ and $C(z)$ have similar frequency responses (as required for near-orthogonal filter sets) and significant transition bands, it is not possible to make $A(Wz)C(z)$ small at all frequencies $z = e^{j\theta}$, because the frequency shift, f_s/M, of $A(z)$ due to W is too small (the A and C responses tend to overlap at their -3 dB points). However, it is quite easy to design $A(W^2z)C(z)$ to be small since the frequency shift of A is twice as great and the responses no longer overlap significantly. Hence, for the lowpass case, we design $B(W^k z)D(z)$ to cancel $A(W^k z)C(z)$ when k is odd by letting

$$B(z) = z^{\pm M/2}A(z) \text{ and } D(z) = z^{\mp M/2}C(z), \qquad (7.2)$$

so that $B(W^k z)D(z) = (-1)^k A(W^k z)C(z)$. In this way, the unwanted aliasing terms, mainly at $k = \pm 1$, are approximately cancelled out. This is equivalent to a single tree with a decimation by $\frac{1}{2}M$ rather than M.

Now consider the bandpass case. Here we find that the edges of the positive frequency passband of C or D, $(f_s/2M) \to (f_s/M)$, will tend to overlap with the edges of the negative frequency passband of A or B, which gets shifted either to $0 \to (f_s/2M)$ or to $(f_s/M) \to (3f_s/2M)$ when $k = 1$ or 2, respectively. Similarly for the opposite passbands when $k = -1$ or -2. We find that the main aliasing terms are always caused by the overlap of opposing frequency passbands (i.e. passbands that have opposite polarity of centre frequency in the unshifted filters). It happens that the solution here is to give B and D positive and negative passbands of opposite polarity while A and C have passbands of the same polarity (or vice versa).

Suppose we have prototype *complex* filters $P(z)$ and $Q(z)$, each with just a single passband $(f_s/2M) \to (f_s/M)$ and zero gain at all negative frequencies, then we let

$$\left. \begin{aligned} A(z) &= \text{Re}[2P(z)] = P(z) + P^*(z), \\ B(z) &= \text{Im}[2P(z)] = -j[P(z) - P^*(z)], \\ C(z) &= \text{Re}[2Q(z)] = Q(z) + Q^*(z), \\ D(z) &= \text{Im}[-2Q(z)] = j[Q(z) - Q^*(z)], \end{aligned} \right\} \qquad (7.3)$$

where conjugation is given by

$$P^*(z) = \sum_r p_r^* z^{-r},$$

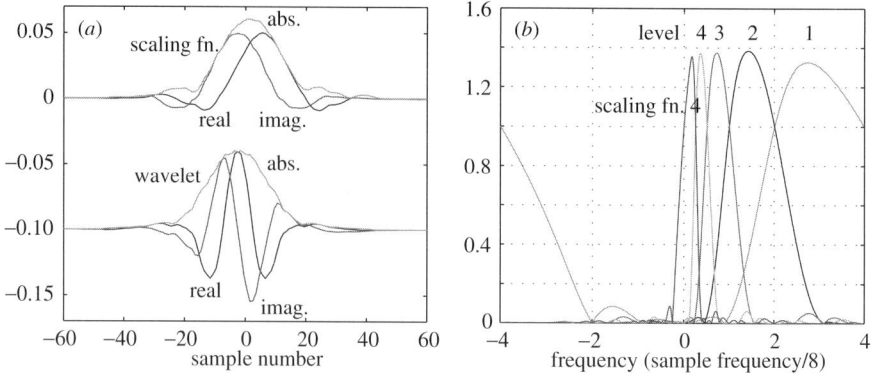

Fig. 11: (*a*) Impulse responses at level 4 of the DT CWT scaling function, and wavelet; (*b*) frequency responses of the wavelets at levels 1–4 and of the level 4 scaling function.

and produces negative frequency passbands. The overlap terms are of the form $Q(z)P^*(W^k z)$ for $k = 1, 2$, and $Q^*(z)P(W^k z)$ for $k = -1, -2$, which all cancel when $B(W^k z)D(z)$ is added to $A(W^k z)C(z)$ in (7.1) to give

$$A(W^k z)C(z) + B(W^k z)D(z) = 2P(W^k z)Q(z) + 2P^*(W^k z)Q^*(z). \qquad (7.4)$$

Hence, we now need only design the filters such that the positive frequency complex filter $Q(z)$ does not overlap with shifted versions of the similar filter $P(z)$. This is quite easy since the complex filter bandwidths are only $f_s/2M$, while the shifts are in multiples of f_s/M. The formulations in equation (7.3) show that the highpass filter outputs from trees a and b should be regarded as the *real and imaginary parts of complex processes*.

For the lowpass filters, equation (7.2) implies that the tree b samples should interpolate midway between the tree a samples, effectively doubling the sampling rate, as shown by the interleaved a and b samples at each level in figure 9. This may be achieved by two identical lowpass filters (either odd or even) at level 1, offset by 1 sample delay, and then by pairs of odd and even length filters at further levels to achieve the extra delay difference of $\frac{1}{4}M$ samples, to make the total difference $\frac{1}{2}M$ at each level.

The responses of $A(z)$ and $B(z)$ also need to match, which can be achieved exactly at level 1, but only approximately beyond this. We do this by designing the even-length $H_{00a}(z)$ to give minimum mean squared error in the approximation

$$z^{\pm 2}H_{0a}(z)H_{00a}(z^2) \approx H_{0b}(z)H_{00b}(z^2).$$

Note that $H_{00b}(z) = H_{0b}(z) = zH_{0a}(z)$, so this is just a matrix pseudo-inverse problem. Then the H_{01a} can be designed to form a perfect reconstruction set with

Table 1 Table of coefficients for the DT CWT analysis filters (The reconstruction filters are obtained by negating alternate coefficients and swapping bands.)

odd $H_{...0}$ 13-tap	odd $H_{...1}$ 19-tap	even $H_{...0}$ 12-tap	even $H_{...1}$ 16-tap
	−0.000 070 6		
	0		−0.000 464 5
−0.001 758 1	0.001 341 9		0.001 334 9
0	−0.001 883 4	−0.005 810 9	0.002 200 6
0.022 265 6	−0.007 156 8	0.016 697 7	−0.013 012 7
−0.046 875 0	0.023 856 0	−0.000 064 1	0.001 536 0
−0.048 242 2	0.055 643 1	−0.083 491 4	0.086 900 8
0.296 875 0	−0.051 688 1	0.091 953 7	0.083 355 2
0.555 468 8	−0.299 757 6	0.480 715 1	−0.488 595 7
0.296 875 0	0.559 430 8	0.480 715 1	0.488 595 7
−0.048 242 2	−0.299 757 6	0.091 953 7	−0.083 355 2
⋮	⋮	⋮	⋮

H_{00a}, such that the reconstruction filters, G_{00a} and G_{00b}, also match each other closely.

Finally, the symmetry of the odd-length highpass filters and the anti-symmetry of the even-length highpass filters produce the required phase relationships between the positive and negative frequency passbands, and equation (7.3) is approximately satisfied too.

These filters can then be used for all subsequent levels of the transform. Good shift invariance (and wavelet smoothness) requires that frequency response sidelobes of the cascaded multi-rate filters should be small. This is achieved if each lowpass filter has a stopband covering $\frac{1}{3}$ to $\frac{2}{3}$ of its sample rate, so as to reject the image frequencies due to subsampling in the next lowpass stage. If the highpass filters then mirror this characteristic, the conditions for no overlap of the shifted bandpass responses in (7.4) are automatically satisfied.

As an example, we selected two linear-phase PR biorthogonal filter sets that meet the above conditions quite well and are also nearly orthogonal. For the odd-length set, we took the (13,19)-tap filters, designed using the transformation of variables method given at the end of Section 3, and then designed a (12,16)-tap even-length set to match (as above). Figure 11 shows impulse responses and frequency responses of the reconstruction filter bank; and the analysis filters are very similar. The coefficients are listed in table 1 and the two-dimensional versions of the level 4 impulse responses are shown in figure 12 (note their strong directionality).

Figure 13a demonstrates the shift invariance in one dimension of the DT CWT

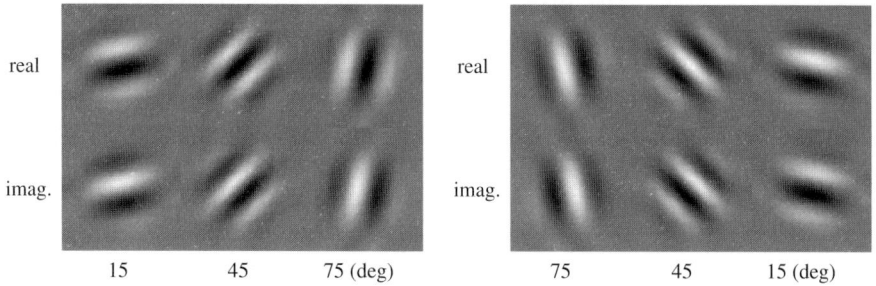

Fig. 12: Real and imaginary parts of two-dimensional impulse responses for the six bandpass bands at level 4. (See table 1 for the coefficients for the DT CWT analysis filters.)

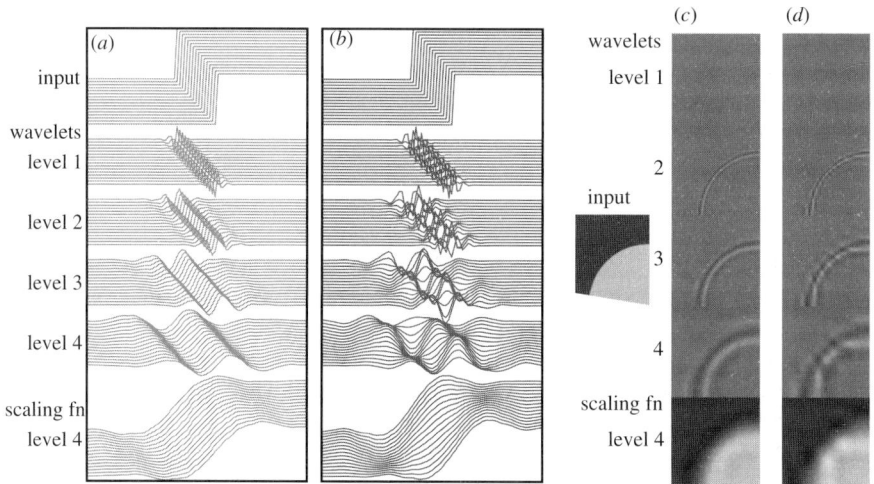

Fig. 13: Wavelet and scaling function components at levels 1–4 of 16 shifted step responses using the DT CWT (*a*) and real DWT (*b*); and of one quadrant of a 'disc' image using the DT CWT (*c*) and real DWT (*d*).

with these filters. The input of the transform (at the top) is a unit step, shifted to 16 adjacent sampling instants in turn (each shift is displaced down a little to give a waterfall style of display). Below this are shown the components of the output of the *inverse* transform, reconstructed in turn from the wavelet coefficients at each of levels 1–4, and from the scaling function coefficients at level 4. This follows our definition of shift invariance given at the start of this section. Note that summing these components reconstructs the input steps perfectly. Good shift invariance is seen from the fact that each of the reconstructed output components in figure 13*a* has a shape and amplitude that hardly varies as the input is shifted. This shows that the DT CWT has decomposed the input step into five separate components, which are virtually independent of the location of the step relative to the sub-sampling points of the transform.

For comparison, figure 13*b* shows the equivalent components if the real DWT is used. The DWT components are much less consistent with shift. The energies of the DT CWT coefficients at any given level vary over the 16 shifts by no more than 1.025:1, whereas the DWT coefficient energies vary by up to 5.45:1: a big difference! Although the five components of the DWT still sum to give perfectly reconstructed steps at each shift position, the decomposition varies considerably with shift. If we are trying to detect features of a signal from its wavelet decomposition, then it is highly confusing to have the sort of unwanted variation with shift shown in figure 13*b*.

Figure 13*c*, *d* shows the equivalent comparison in two dimensions using an image of a circular disc as input. The gradual shift and rotation of the edge of the disc with respect to the rectangular sub-sampling grids of the transform (not shown) form a good test of shift and rotational dependencies. In these images, all bandpass coefficients at a given wavelet level are retained. In figure 13*c* we see the near-perfect circular arcs, generated by the components at each level for the DT CWT, which show good shift and rotational invariance. Contrast these with the badly distorted arcs for the DWT in figure 13*d*, caused by aliasing.

8   Applications of the complex wavelet transform

The shift invariance and directionality of the CWT may be applied to advantage in many areas of image processing, for example: denoising, restoration, texture modelling, steerable filtering, registration/motion processing, object segmentation, and image classification. We have space for only one example here.

In figure 14 we show an example of denoising. Image (*d*) is the result of denoising image (*a*) using the DT CWT and a simple soft thresholding method that suppresses all complex wavelet coefficients x of low amplitude with a raised cosine gain law: $g(x) = \frac{1}{2}(1 - \cos\{\pi |x|/T\})$ for $|x| < T$, and $g(x) = 1$ elsewhere. For comparison, we show images (*b*) and (*c*), which were obtained using the same soft thresholding method with the real DWT in its decimated and undecimated forms, respectively. (*b*) shows significantly worse artefacts than (*d*), while (*c*) is very similar to (*d*) but requires about five times as much computation. In all cases, the thresholds T were selected so as to get minimum mean-squared error from the original (clean) image. In practice, more complicated thresholding methods may be used, such as in Malfait & Roose (1997), which uses Markov random fields in conjunction with an undecimated WT. It is likely that, by replacing the undecimated WT with the CWT, the effectiveness of the MRFs at coarser wavelet levels can be improved, owing to the more appropriate sampling rates of the CWT.

In conclusion, we are now investigating a number of applications of the CWT and are finding that it may have many uses as a multiresolution front end for processing images and other multidimensional signals.

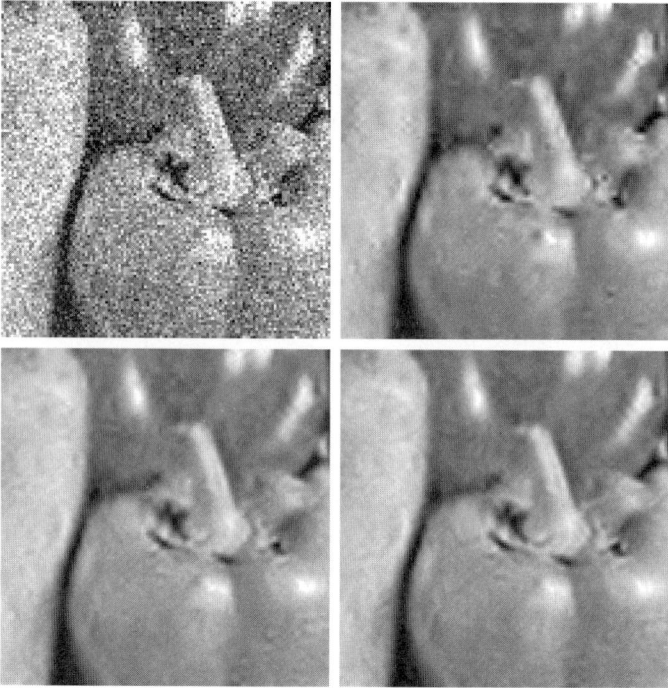

Fig. 14: 128×128 pel portions of the pepper's image: (*a*) with white Gaussian noise added to give SNR = 3.0 dB; (*b*) denoised with real DWT, SNR = 12.24 dB; (*c*) denoised with undecimated WT, SNR = 13.45 dB; (*d*) denoised with dual-tree CWT, SNR = 13.51 dB.

9 References

Belzer, B., Lina, J.-M. & Villasenor, J. 1995 Complex, linear-phase filters for efficient image coding. *IEEE Trans. Signal Proc.* **43**, 2425–2427.

Daubechies, I. 1990 The wavelet transform, time-frequency localisation and signal analysis. *IEEE Trans. Informat. Theory* **36**, 961–1005.

Kingsbury, N. G. 1998*a* The dual-tree complex wavelet transform: a new technique for shift invariance and directional filters. In *Proc. 8th IEEE DSP Workshop, Bryce Canyon, UT, USA*, paper no. 86.

Kingsbury, N. G. 1998*b* The dual-tree complex wavelet transform: a new efficient tool for image restoration and enhancement. In *Proc. EUSIPCO 98, Rhodes, Greece*, pp. 319–322.

Kovacevic, J. & Daubechies, I. (eds) 1996 Special issue on wavelets. *Proc. IEEE* **84**, 507–685.

Lawton, W. 1993 Applications of complex valued wavelet transforms to sub-band decomposition. *IEEE Trans. Signal Proc.* **41**, 3566–3568.

Magarey, J. F. A. & Kingsbury, N. G. 1998 Motion estimation using a complex-valued wavelet transform. *IEEE Trans. Signal Proc.* **46**, 1069–1084.

Malfait, M. & Roose, D. 1997 Wavelet-based image denoising using a Markov random field a priori model. *IEEE Trans. Image Proc.* **6**, 549–565.

Mallat, S. G. 1989 A theory for multiresolution signal decomposition: the wavelet representation. *IEEE Trans. PAMI* **11**, 674–693.

Rioul, O. & Vetterli, M. 1991 Wavelets and signal processing. *IEEE Signal Processing Mag.* **8**, 14–38.

Simoncelli, E. P., Freeman, W. T., Adelson, E. H. & Heeger, D. J. 1992 Shiftable multiscale transforms. *IEEE Trans. Informat. Theory* **38**, 587–607.

Tay, D. B. H. & Kingsbury, N. G. 1993 Flexible design of multidimensional perfect reconstruction FIR 2-band filters using transformations of variables. *IEEE Trans. Image Proc.* **2**, 466–480.

Vetterli, M. & Kovacevic, J. 1995 *Wavelets and sub-band coding.* Englewood Cliffs, NJ: Prentice-Hall.

10
Application of wavelets to filtering of noisy data

Ue-Li Pen

Canadian Institute for Theoretical Astrophysics, University of Toronto,
60 St George Street, Toronto, Canada M5S 3H8

Abstract

I discuss approaches to optimally remove noise from images. A generalization of
Wiener filtering to non-Gaussian distributions and wavelets is described, as well as
an approach to measure the errors in the reconstructed images. We argue that the
wavelet basis is highly advantageous over either Fourier or real-space analysis if
the data are intermittent in nature, i.e. if the filling factor of objects is small.

Keywords: wavelets; optimal filtering; Wiener filtering

1 Introduction

In astronomy, the collection of data is often limited by the presence of background
noise. Various methods are used to filter the noise while retaining as much 'useful'
information as possible. In recent years, wavelets have played an increasing role
in astrophysical data analysis. It provides for a general parameter-free procedure
to look for objects of varying size scales. In the case of the cosmic microwave
background (CMB), one is interested in the non-Gaussian component in the
presence of Gaussian noise and signal. An application of wavelets is presented by
Tenorio *et al.*(1999). This paper generalizes their analysis beyond the thresholding
approximation. X-ray images are also frequently noise dominated, caused by
instrumental and cosmic background. Successful wavelet reconstructions were
achieved by Damiani *et al.*(1997*a, b*).

At times, generic tests for non-Gaussianity are desired. Inflationary theories
predict, for example, that the intrinsic fluctuations in the CMB are Gaussian, while
topological defect theories predict non-Gaussianity. A full test for non-Gaussianity
requires measuring all N-point distributions, which is computationally not tractable
for realistic CMB maps. Hobson *et al.*(1999) have shown that wavelets are a
more sensitive discriminant between cosmic string and inflationary theories if one
examines only the one-point distribution function of basis coefficients.

For Gaussian random processes, Fourier modes are statistically independent.
Current theories of structure formation start from an initially linear Gaussian random
field which grows nonlinear through gravitational instability. Nonlinearity occurs

through processes local in real space. Wavelets provide a natural basis that compromises between locality in real and Fourier space. Pando & Fang (1996) have applied the wavelet decomposition in this spirit to the high redshift L_α systems which are in the transition from linear to nonlinear regimes, and are thus well analysed by the wavelet decomposition.

We will concentrate in this paper on the specific case of data laid out on a two-dimensional grid, where each grid-point is called a *pixel*. Such images are typically obtained through various imaging instruments, including CCD arrays on optical telescopes, photomultiplier arrays on X-ray telescopes, differential radiometry measurements using bolometers in the radio band, etc. In many instances, the images are dominated by noise. In the optical, the sky noise from atmospheric scatter, zodiacal light and extragalactic backgrounds sets a constant flux background to any observation. CCD detectors essentially count photons, and are limited by the Poissonian discreteness of their arrival. A deep exposure is dominated by sky background, which is subtracted from the image to obtain the features and objects of interest. Since the intensity of the sky noise is constant, it has a Poissonian error with standard deviation $e \propto n^{1/2}$, where n is the expected photon count per pixel. After subtracting the sky average, this fluctuating component remains as white noise in the image. For large modern telescopes, images are exposed to near the CCD saturation limit, with typical values of $n \sim 10^4$. The Poisson noise is well described by Gaussian statistics in this limit.

We would like to pose the problem of filtering out as much of the noise as possible, while maximally retaining the data. In certain instances, optimal methods are possible. If we know the data to consist of astronomical point objects, which have a shape on the grid given by the atmospheric spreading or telescope optics, we can test the likelihood at each pixel that a point source was centred there. The iterative application of this procedure is implemented in the routine CLEAN of the Astronomical Image Processing Software (AIPS) (Cornwell & Braun 1989).

If the sources are not point-like, or the atmospheric point-spread function varies significantly across the field, the approach is no longer optimal. In this paper we examine an approach to implement a generic noise filter using a wavelet basis. In Section 2 we first review two popular filtering techniques, thresholding and Wiener. In Section 3 we generalize Wiener filtering to inherit the advantages of thresholding. A Bayesian approach to image reconstruction (Vidakovic 1998) is used, where we use the data itself to estimate the prior distribution of wavelet coefficients. We recover Wiener filtering for Gaussian data. Some concrete examples are shown in Section 4.

2   Classical filters

2.1   Thresholding

A common approach to suppressing noise is known as thresholding. If the amplitude of the noise is known, one picks a specific threshold value, for example $\tau = 3\sigma_{\text{noise}}$

to set a cut-off at three times the standard deviation of the noise. All pixels less than this threshold are set to zero. This approach is useful if we wish to minimize false detections, and if all sources of signal occupy only a single pixel. It is clearly not optimal for extended sources, but often used due to its simplicity. The basic shortcoming is its neglect of correlated signals which covers many pixels. The choice of threshold also needs to be determined heuristically. We will attempt to quantify this procedure.

2.2 Wiener filtering

In the specific case that both the signal and the noise are Gaussian random fields, an optimal filter can be constructed which minimizes the impact of the noise. If the noise and signal are stationary Gaussian processes, Fourier space is the optimal basis where all modes are uncorrelated. In other geometries, one needs to expand in signal-to-noise eigenmodes (see, for example, Vogeley & Szalay 1996). One needs to know both the power spectrum of the data, and the power spectrum of the noise. We use the least-square norm as a measure of goodness of reconstruction. Let E be the reconstructed image, U the original image and N the noise. The noisy image is called $D = U + N$. We want to minimize the error $e = \langle (E - U)^2 \rangle$. For a linear process, $E = \alpha D$. For our stationary Gaussian random field, different Fourier modes are independent, and the optimal solution is $\alpha = \langle U^2 \rangle / \langle D^2 \rangle$. U^2 is the intrinsic power spectrum. Usually, D^2 can be estimated from the data, and if the noise power spectrum is known, the difference can be estimated subject to measurement scatter as shown in figure 1. Often, the power spectrum decays with increasing wavenumber (decreasing length scale): $\langle U^2 \rangle = ck^{-n}$. For white noise with unit variance, we then obtain $\alpha = c/(k^n + c)$, which tends to unity for small k and zero for large k. We really only need to know the parameters c, n in the crossover region $E^2 \sim 1$. In Section 4 we will illustrate a worked example.

Wiener filtering is very different from thresholding, since modes are scaled by a constant factor independent of the actual amplitude of the mode. If a particular mode is an outlier far above the noise, the algorithm would still force it to be scaled back. This can clearly be disadvantageous for highly non-Gaussian distributions. If the data are localized in space, but sparse, the Fourier modes dilute the signal into the noise, thus reducing the signal significantly as is seen in the examples in Section 4. Furthermore, choosing α independent of D is only optimal for Gaussian distributions. One can generalize as follows.

3 Non-Gaussian filtering

We can extend Wiener filtering to non-Gaussian probability density functions (PDFs) if the PDF is known and the modes are still statistically independent. We will denote the PDF for a given mode as $\Theta(u)$, which describes a random variable U. When Gaussian white noise with unit variance $\mathcal{N}(0, 1)$ is added, we obtain a new random

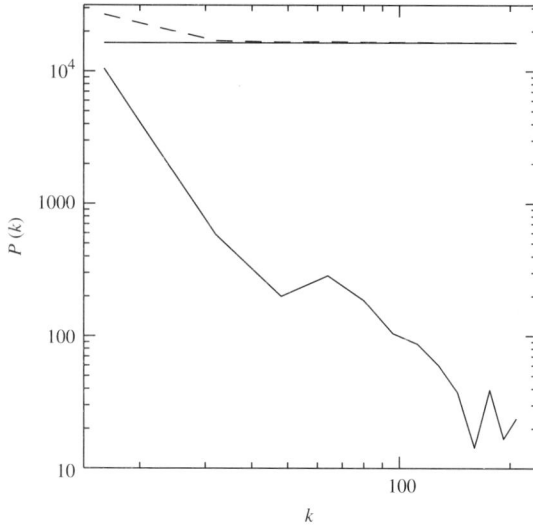

Fig. 1: The power spectrum of figure 3. The dashed line is the power spectrum measured from the noisy data. The horizontal line is the noise. The lower solid line is the difference between the measured spectrum and the noise. We see that the measurement of the difference becomes noise limited at large k.

variable $D = U + \mathcal{N}(0, 1)$ with PDF

$$f(d) = (2\pi)^{-1/2} \int \Theta(u) \exp(-\tfrac{1}{2}(u - d)^2)\, du.$$

We can calculate the conditional probability $P(U \mid D) = P(D \mid U)P(U)/P(D)$ using Bayes's theorem. For the posterior conditional expectation value we obtain

$$\langle U \mid D = d \rangle = \frac{1}{\sqrt{2\pi} f(d)} \int \exp[-\tfrac{1}{2}(u - d)^2]\Theta(u)u\, du$$

$$= D + \frac{1}{\sqrt{2\pi} f(d)} \partial_d \int \exp[-\tfrac{1}{2}(u - d)^2]\Theta(u)\, du$$

$$= D + (\ln f)'(d). \tag{3.1}$$

Similarly, we can calculate the posterior variance

$$\langle (U - \bar{U})^2 \mid D = d \rangle = 1 + (\ln f)''(d). \tag{3.2}$$

For a Gaussian prior with variance σ, equation (3.1) reduces to Wiener filtering. We have a generalized form for $\alpha = 1 + (\ln f)'/D$. For distributions with long tails, $(\ln f)' \sim 0$, $\alpha \sim 1$, and we leave the outliers alone, just as thresholding would suggest.

For real data, we have two challenges: (1) estimating the prior distribution Θ; and (2) finding a basis in which Θ is most non-Gaussian.

3.1 Estimating prior Θ

The general non-Gaussian PDF on a grid is a function of N variables, where N is the number of pixels. It is generally not possible to obtain a complete description of this large dimensional space (Field, this issue). It is often possible, however, to make simplifying assumptions. We consider two descriptions: Fourier space and wavelet space. We will assume that the one-point distributions of modes are non-Gaussian, but that they are still statistically independent. In that case, one needs only to specify the PDF for each mode. In a hierarchical basis, where different basis functions sample characteristic length scales, we further assume a scaling form of the prior PDF $\Theta_l(u) = l^{-\beta}\Theta(u/l^{\beta})$. Here $l \sim 1/k$ is the characteristic length scale, for example the inverse wavenumber in the case of Fourier modes. For images we often have $\beta \sim 1$.

Wavelets still have a characteristic scale, and we can similarly assume scaling of the PDF. In analogy with Wiener filtering, we first determine the scale dependence. For computational simplicity, we use Cartesian product wavelets (Meyer 1992). Each basis function has two scales; call them $2^i, 2^j$. The real-space support of each wavelet has area $A \propto 2^{i+j}$, and we find empirically that the variance depends strongly on that area. The scaling relation does not directly apply for $i \neq j$, and we introduce a lowest-order correction using $\ln(\sigma) = c_1(i + j) + c_2(i - j)^2 + c_3$. We then determine the best-fit parameters c_i from the data. The actual PDF may depend on the length scale $i + j$ and the elongation $i - j$ of the wavelet basis. One could parameterize the PDF, and solve for this dependence (Vidakovic 1998), or bin all scales together to measure a non-parametric scale-averaged PDF. We will pursue the latter.

The observed variance is the intrinsic variance of Θ plus the noise variance of $\mathcal{N}$, so the variance $\sigma^2_{\text{intrinsic}} = \sigma^2_{\text{obs}} - \sigma^2_{\text{noise}}$ has error $\propto \sigma_{\text{obs}^2}/n$ where n is the number of coefficients at the same length scale. We weigh the data accordingly. Because most wavelet modes are at short scales, most of the weight will come near the noise threshold, which is what we desire. We now proceed to estimate $f(d)$. Our ansatz now assumes $\Theta_{ij}(u) \propto \Theta(u/\exp[c_1(i + j) + c_2(i - j)^2 + c_3])$, where $\Theta(u)$ has unit variance. We can only directly measure f_{ij}. We sort these in descending order of variance, f_m. Again, typically the largest scale modes will have the largest variance. In the images explored here, we find typical values of c_1 between 1 and 2, while $c_2 \sim -0.2$. For the largest variance modes, noise is least important. From the data, we directly estimate a binned PDF for the largest scale modes. By hypothesis, $D = U/l^{\beta} + \mathcal{N}(0, \sigma_{l,\text{noise}})$. We reduce the larger scale PDF by convolving it with the difference of noise levels to obtain an initial guess for the smaller scale PDF:

$$f'_{l'}(d) = \frac{(l'/l)^{\beta}}{\sqrt{\pi}} \int_{-\infty}^{\infty} f_l[u(l/l')^{\beta}] \exp\left[-\frac{(u - d)^2}{2(\sigma^2_{l',\text{noise}} - \sigma^2_{l,\text{noise}})}\right] du. \tag{3.3}$$

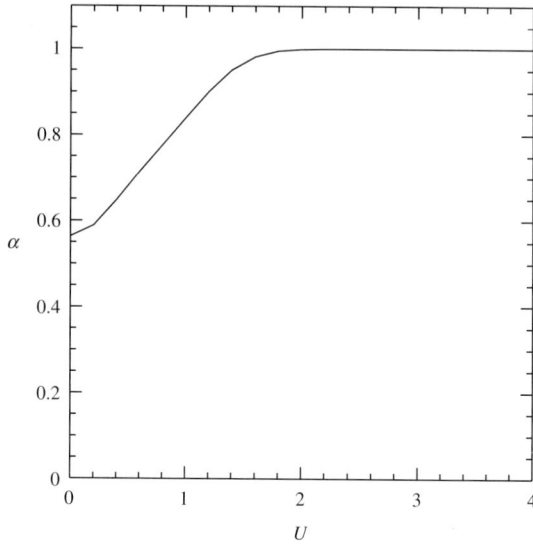

Fig. 2: The optimal filter function α for the non-Gaussian wavelet model at $\sigma_{\text{noise}} = \sigma_{\text{data}}$. U is given in units of the total standard deviation $\sigma_{\text{abs}}^2 = \sigma_{\text{noise}} + \sigma_{\text{data}}$. At small amplitudes, it is similar to a Wiener filter $\alpha = \frac{1}{2}$, but limits to unity for large outliers.

To this we add the actual histogram of wavelet coefficients at the smaller scale. We continue this hierarchy to obtain an increasingly better estimate of the PDF, having used the information from each scale. In figure 2 we show the optimal weighting function obtained for the examples in Section 4.

On the largest scales, the PDF will be poorly defined because relatively few wavelets lie in that regime. The current implementation performs no filtering, i.e. sets $\alpha = 1$ for the largest scales. A potential improvement could be implemented: within the scaling hypothesis, we can deconvolve the noisy $f(D)$ obtained from small scales to estimate the PDF on large scales. The errors in the PDF estimation are themselves Poissonian, and in the limit that we have many points per PDF bin, we can treat those as Gaussian. The deconvolution can then be optimally filtered to maximize the use of the large number of small-scale wavelets to infer the PDF of large-scale wavelets. Of course, the non-Gaussian wavelet analysis could then be recursively applied to estimate the PDF. Instead of the Bayesian prior PDF, we would then specify a prior for the prior. This possibility will be explored in future work.

3.2   Maximizing non-Gaussianity using wavelets

Errors are smallest if a large number of coefficients are near zero and when the modes are close to being statistically independent. Let us consider several extreme cases and their optimal strategies. Imagine that we have an image consisting of true uncorrelated point sources, and each point source only occupies one pixel. Further assume that only a very small fraction ϵ of possible pixels are occupied, but when a point source is present, it has a constant luminosity L, and then add a uniform white-noise background with unit variance. In Fourier space, each mode has unit variance from the noise, and variance $L^2\epsilon$ from the point sources. We easily see that it will be impossible to distinguish signal from noise if $L^2\epsilon < 1$. In real space, white noise is also uncorrelated, so we are justified to treat each pixel separately. Now we can easily distinguish signal from noise if $L > \sqrt{-\ln(\epsilon)}$. If $L = 10$ and $\epsilon = 0.001$, we have a situation where the signal is easy to detect in real space and difficult in Fourier space, and in fact the optimal filter (3.1) is optimal in real space where the points are statistically independent. In Fourier space, even though the covariance between modes is zero, modes are not independent.

Now consider the more realistic case that objects occupy more than one pixel, but are still localized in space, and only have a small covering fraction. This is the case of intermittent information. The optimal basis will depend on the actual shape of the objects, but it is clear that we want basis functions which are localized. Wavelets are a very general basis to achieve this, which sample objects of any size scale, and are able to effectively excise large empty regions. We expect PDFs to be more strongly non-Gaussian in wavelet space than either real or Fourier space.

In this formulation, we obtain not only a filtered image, but also an estimate of the residual noise, and a noise map. For each wavelet coefficient we find its posterior variance using (3.2). The inverse-wavelet transform then constructs a noise-variance map on the image grid.

4 Examples

In order to be able to compare the performance of the filtering algorithm, we use as an example an image to which the noise is added by hand. The 'denoised' result can then be compared to the 'truth'. We have taken a random image from the Hubble space telescope (HST), in this case the 100 000th image (PI: C. Steidel). The original picture is shown in figure 3. The grey scale is from 0 to 255. At the top are two bright stars with the telescope support structure diffraction spikes clearly showing. The extended objects are galaxies. We then add noise with variance 128, which is shown in figure 4. The mean signal-to-noise ratio of the image is $\frac{1}{4}$. We can tell by eye that a small number of regions still protrude from the noise.

The power spectrum of the noisy image is shown in figure 1. We use the known expectation value of the noise variance. The subtraction of the noise can be performed even when the noise substantially dominates over the signal, as can be seen in the image. In most astronomical applications, noise is instrumentally induced and the distribution of the noise is very well documented. Blank field exposures, for example, often provide an empirical measurement.

We first apply a Wiener filter, with the result shown in figure 5. We notice

Fig. 3: The original image, taken from the HST Web page www.stsci.edu. It is the 100 000th image taken with HST for C. Steidel of Caltech.

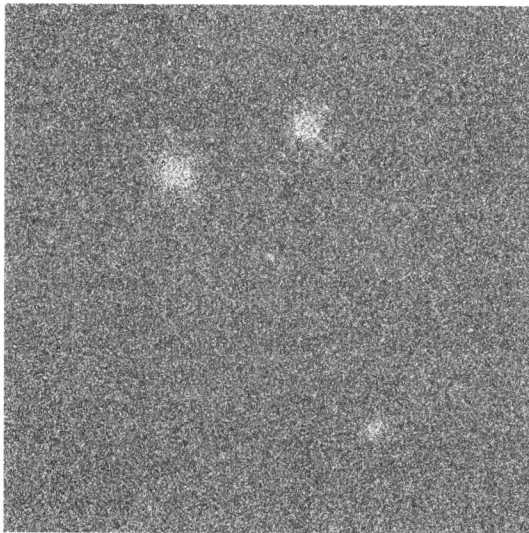

Fig. 4: Figure 3 with substantial noise added.

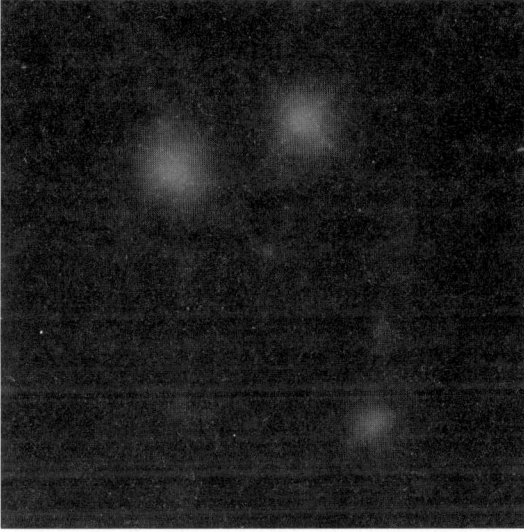

Fig. 5: Wiener filtered. We see that the linear scaling function substantially reduces the flux in the bright stars.

immediately the key feature: all amplitudes are scaled by the noise, so the bright stars have been downscaled significantly. The noise on the star was less than unity, but each Fourier mode gets contributions from the star as well as the global noise of the image. The situation worsens if the filling factor of the signal regions is small. The mean intensity of the image is stored in the $k = 0$ mode, which is not significantly affected by noise. While total flux is approximately conserved, the flux on each of the objects is non-locally scattered over the whole image by the Wiener filter process.

The optimal Bayesian wavelet filter is shown in figure 6. A Daubechies 12 wavelet was used, and the prior PDF reconstructed using the scaling assumption described in Section 3. We see immediately that the amplitudes on the bright objects are much more accurate. We see also that the faint vertical edge-on spiral on the lower right just above the bright elliptical is clearly visible in this image, while it had almost disappeared in the Wiener filter.

The Bayesian approach allows us to estimate the error in the reconstruction using equation (3.2). We show the result in figure 7. We can immediately see that some features in the reconstructed map, for example the second faint dot above the bright star on the upper left, have large errors associated with them, and are indeed artefacts of reconstruction. Additionally, certain wavelets experience large random errors. These appear as checkered 'wavelet' patterns on both the reconstructed image and the error map.

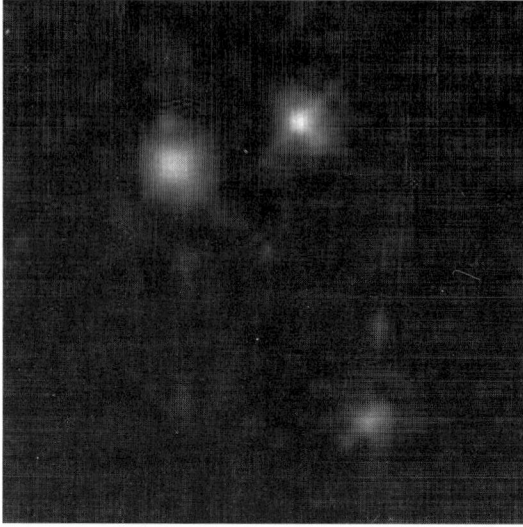

Fig. 6: Non-Gaussian wavelet filtered. Several of the features that had been lost in the Wiener filtering process are recovered here.

Fig. 7: Error map. Plotted is the posterior Bayesian variance. We see that some features, for example the small dot in the upper left, have large errors associated with them, and are therefore artefacts.

5 Discussion

Fourier space has the advantage that for translation-invariant processes, different modes are pairwise uncorrelated. If modes were truly independent, the optimal filter for each k mode would also be globally optimal. As we have seen from the example in Section 3 b, processes which are local in real space are not optimally processed in Fourier space, since different Fourier modes are not independent; wavelet modes are not independent, either. For typical data the correlations are relatively sparse. In the astronomical images under consideration, the stars and galaxies are relatively uncorrelated with each other. Wavelets with compact support sample a limited region in space, and wavelets which do not overlap on the same objects on the grid will be close to independent. Even for Gaussian random fields, wavelets are close to optimal since they are relatively local in Fourier space. Their overlap in Fourier space leads to residual correlations which are neglected. We see that wavelets are typically close to optimal, even though they are never truly optimal, but in the absence of a full prior, they allow us to work with generic datasets and usually outperform Wiener filtering.

In our analysis, we have used Cartesian-product Daubechies wavelets. These are preferentially aligned along the grid axes. In the wavelet filtered map (figure 6) we see residuals aligned with the coordinate axes. Recent work by Kingsbury (this issue) using complex wavelets would probably alleviate this problem. The complex wavelets have a factor of two redundancy, which is used in part to sample spatial translations and rotational directions more homogeneously and isotropically.

6 Conclusions

We have presented a generalized noise-filtering algorithm. Using the ansatz that the PDF of mode or pixel coefficients is scale invariant, we can use the observed dataset to estimate the PDF. By application of Bayes's theorem, we reconstruct the filter map and noise map. The noise map gives us an estimate of the error, which tells us the performance of the particular basis used and the confidence level of each reconstructed feature. Based on comparison with controlled data, we find that the error estimates typically overestimate the true error by about a factor of two.

We argued that wavelet bases are advantageous for data with a small duty cycle that is localized in real space. This covers a large class of astronomical images, and images where the salient information is intermittently present.

Acknowledgements

I thank Iain Johnstone, David Donoho and Robert Crittenden for helpful discussions. I am most grateful to Bernard Silverman and the Royal Society for organizing this Discussion Meeting.

7 References

Cornwell, T. & Braun, R. 1989 Deconvolution. In *Synthesis imaging in radio astronomy* (ed. R. A. Perley, F. R Schwab & A. H. Bridle), pp. 167–184. Astronomical Society of the Pacific.

Damiani, F., Maggio, A., Micela, G. & Sciortino, S. 1997*a* A method based on wavelet transforms for source detection in photon-counting detector images. I. Theory and general properties. *Astrophys. J.* **483**, 350–369.

Damiani, F., Maggio, A., Micela, G. & Sciortino, S. 1997*b* A method based on wavelet transforms for source detection in photon-counting detector images. II. Application to ROSAT PSPC images. *Astrophys. J.* **483**, 370–389.

Hobson, M. P., Joens, A. W. & Lasenby, A. N. 1999 Wavelet analysis and the detection of non-Gaussianity in the CMB. *Mon. Not. R. Astr. Soc.* **310**, 565–570

Meyer, Y. 1992 *Wavelets and operators*, ch. 3.3, p. 81. Cambridge University Press.

Pando, J. & Fang, L.-Z. 1996 A wavelet space-scale decomposition analysis of structures and evolution of QSO Lyα absorption lines. *Astrophys. J.* **459**, 1–11.

Tenorio, L., Jaffe, A. H., Hanany, S. & Lineweaver, C. H. 1999 Application of wavelets to the analysis of cosmic microwave background maps. *Mon. Not. R. Astr. Soc.* **310**, 823–834.

Vidakovic, B. 1998 Wavelet-based nonparametric bayes method. In *Practical nonparametric and semiparametric Bayesian statistics* (ed. D. D. Dey, P. Müller & D. Sinha), pp. 133–155. Springer.

Vogeley, M. S. & Szalay, A. S. 1996 Eigenmode analysis of galaxy redshift surveys. I. Theory and methods. *Astrophys. J.* **465**, 34–53.

11

Approximation and compression of piecewise smooth functions

Paolo Prandoni

Laboratory for Audio-Visual Communications, Swiss Federal Institute of Technology, CH-1015 Lausanne, Switzerland

Martin Vetterli

Department of Electrical Engineering and Computer Sciences, University of California, Berkeley, CA 94720, USA

Abstract

Wavelet or sub-band coding has been quite successful in compression applications, and this success can be attributed in part to the good approximation properties of wavelets. In this paper, we revisit rate–distortion (RD) bounds for the wavelet approximation of piecewise smooth functions, and piecewise polynomial functions in particular. We contrast these results with RD bounds achievable using an oracle-based method. We then introduce a practical dynamic programming algorithm, which achieves performance similar to the oracle method, and present experimental results.

Keywords: data compression; data modelling; wavelet approximation; rate–distortion optimality; dynamic allocation; joint segmentation quantization

1 Introduction

Wavelets have had an important impact on signal processing theory and practice and, in particular, wavelets play a key role in compression, image compression being a prime example (Vetterli & Kovačević 1995). This success is linked to the ability of wavelets to capture efficiently both stationary and transient behaviours. In signal processing parlance, wavelets avoid the problem of the window size (as in the short-time Fourier transform, for example), since they work with many windows due to the scaling property.

An important class of processes encountered in signal processing practice can be thought of as 'piecewise stationary'. As an example, speech is often analysed using such a model, for example in local linear predictive modelling as used in speech compression; such processes can be generated by switching between various stationary processes. Wavelet methods are possible models as well, being able both to fit the stationary part and capture the breakpoints. In the deterministic case, 'piecewise smooth functions' are a class of particular interest. For an example consider piecewise polynomial functions. Again, wavelets are good approximants if simple nonlinear approximation schemes are used. The performance of wavelets in such a context is again linked to their ability to fit polynomials by scaling functions (up to the appropriate approximation order) while capturing the breakpoints efficiently by a small number of wavelet coefficients.

When one is interested in compression applications, a key question is not just the approximation behaviour, but the effective rate-distortion (RD) characteristic of schemes where wavelets and scaling functions are used as elementary approximation atoms. Here, we are moving within an *operational* framework in which a well-defined implementable algorithmic procedure is associated to each coding scheme. In this operational scenario, a fundamental measure of performance is provided by the RD behaviour specific to the algorithm: given a certain rate of, say, r bits, the RD curve yields the overall amount of distortion (measured by some suitably chosen metric) between the original data and the data which are reconstructed at the output of an encoding–decoding scheme when the amount of bits exchanged between the coder and the decoder is exactly r. In this sense, the rate is always finite (we deal with implementable algorithms) and the performance measure becomes dependent on much more than the choice of basis functions. Pioneering work by Shannon (the RD concept) and Kolmogorov (the epsilon-entropy concept) spurred much interesting theoretical work on these types of ultimate lower bound on coding performance (Donoho *et al*.1998). Most of the results, however, albeit very intriguing, are of limited practical significance and in many cases only (hopefully tight) upper bounds on the performance of a *particular* coding algorithm can be obtained with relative ease; this is the case with our contribution. It would be beyond the scope of the paper to provide a more general introduction to algorithmic RD analysis; a very exhaustive overview of the topic, with a particular attention to transform coding methods, can be found in a recent article by Donoho *et al*.(1998).

The rest of the paper is organized as follows: first we will review some recent work in wavelets and sub-band coding, recalling some now classic results on approximating piecewise smooth functions or piecewise stationary processes[1]. Then, piecewise polynomial functions will be specifically analysed, and the differences between a wavelet-based and an oracle-based method will be shown. An oracle method is a coding scheme in which we assume we can obtain all function parameters with arbitrary accuracy from an omniscient 'oracle'; it is a commonly

---

[1] A word of caution on the term 'stationary' is perhaps necessary here. The standard definition of a stationary process involves statistics of all orders across an ensemble of realizations. When the stationarity conditions are met, a fundamental result of signal processing states that the spectral properties of linear function operators (e.g. filters) defined in a deterministic setting retain their meaningfulness in the stochastic domain; in terms of signal modelling, this is equivalent to saying that a single model is able to capture the entirety of the process. In practice, signal processing applications almost always deal with a single *finite-length* realization of an unknown process whose stochastic description is very sketchy. In this case, it is often advantageous to split the signal into pieces and use a different linear model for each piece. In this sense, piecewise stationarity is operatively defined by the ability to use a single model over a given piece of data. Alternatively, one could look at a piecewise stationary signal as the output of a system in which an unknown switching process selects between many possible processes, all of which are assumed stationary. This liberal use of the term 'stationary' is clearly an abuse of language, but its connotations are usually univocally clear in a signal processing scenario.

used theoretical device to infer an ideal limit on the performance of practical coding schemes. Next, we will present a practical method based on dynamic programming whose performance approaches that of the oracle method, and conclude with some experimental results.

2 The compression problem

Compression is the trade-off between description complexity and approximation quality. Given an object of interest, or a class of objects, one studies this trade-off by choosing a representation (e.g. an orthonormal basis) and then deciding how to describe the object parsimoniously in the representation. Such a parsimonious representation typically involves approximation.

For example, for a function described with respect to an orthonormal basis, only a subset of basis vectors might be used (subspace approximation) and the coefficients used in the expansion are approximated (quantization of the coefficients). Thus, both the subspace approximation and the coefficient quantization contribute to the approximation error. More formally, for a function f in $L_2(R)$ for which $\{\varphi_n\}$ is an orthonormal basis, we have the approximate representation

$$\hat{f} = \sum_{n \in I} \hat{\alpha} \varphi_n, \quad \hat{\alpha}_n = Q[\langle \varphi_n, f \rangle], \tag{2.1}$$

where I is an index subset and $Q[\cdot]$ is a quantization function, like, for example, the rounding to the nearest multiple of a quantization step $\triangle$.

Typically, the approximation error is measured by the L_2 norm, or (mean) squared error

$$\epsilon = \|f - \hat{f}\|_2^2. \tag{2.2}$$

The description complexity corresponds to describing the index set I, as well as describing the quantized coefficients $\hat{\alpha}_n$. The description complexity is usually called the rate R, corresponding to the number of binary digits (or bits) used. Therefore the approximation $\hat{f}$ of f leads to a RD pair (R, ϵ), indicating one possible trade-off between description complexity and approximation error. The example just given, despite its simplicity, is quite powerful and is actually used in practical compression standards. It also raises the following questions.

(Q1) What are classes of objects of interest and for which the RD trade-off can be well understood?

(Q2) If approximations are done in bases, what are good bases to use?

(Q3) How do we choose the index set and the quantization?

(Q4) Are there objects for which approximation in bases is suboptimal?

Historically, (Q1) has been addressed by the information theory community in the context of RD theory. Shannon (1948) posed the problem in his landmark paper and proved RD results in Shannon (1959). The classic book by Berger (1971) is still a reference on the topic. Yet, RD theory has been mostly concerned with exact results within an asymptotic framework (the so-called large-block-size assumption together with random coding arguments). Thus, only particular processes (e.g. jointly Gaussian processes) are amenable to this exact analysis, but the framework has been used extensively, in particular in its operational version (when practical schemes are involved) (see, for example, the review by Ortega & Ramchandran (1998)). It is to be noted that RD analysis covers all cases (e.g. small rates with large distortions) and that the case of very fine approximation (or very large rates) is usually easier but less useful in practice.

In the stationary jointly Gaussian case (Q2) has a simple answer, based on RD theory. For any process, the Karhunen–Loève basis leads to the best linear approximation, due to its decorrelating properties. In the jointly Gaussian case the best possible approximation indeed happens to be a linear approximation, since decorrelation implies statistical independence. Yet, not all things in life are jointly Gaussian, and more powerful techniques than linear approximation can achieve a better RD trade-off when the rate is constrained; that is where wavelets come into play, in conjunction with more general nonlinear approximation strategies[2]. For processes which are piecewise smooth (e.g. images), the abrupt changes are well captured by wavelets, and the smooth or stationary parts are efficiently represented by coarse approximations using scaling functions. Both practical algorithms (e.g. the EZW algorithm of Shapiro (1993)) and theoretical analyses (Cohen *et al.* 1997; Mallat & Falzon 1998) have shown the power of approximation within a wavelet basis. An alternative is to search large libraries of orthonormal bases, based for example on binary sub-band coding trees. This leads to wavelet packets (Coifman & Wickerhauser 1992) and RD optimal solutions (Ramchandran & Vetterli 1993).

(Q3) is more complex than it looks at first sight. If there was no cost associated with describing the index set, then clearly I should be the set $\{n\}$ such that

$$|\langle \varphi_n, f \rangle_{n \in I}| \geq |\langle \varphi_m, f \rangle|_{m \notin I}. \tag{2.3}$$

However, when the rate for I is accounted for, it might be more efficient to use a fixed set I for a class of objects. For example, in the jointly Gaussian case, the

---

[2]The notion of 'best' basis becomes slightly tricky if we allow for signal-dependent bases (to which the KLT also belongs). Indeed, suppose we want to code the Mona Lisa image; then the best basis is clearly that in which the first vector is the Mona Lisa image itself: with this choice we need only one bit to code the data. Yet the coding gain is entirely offset by the cost of informing the decoder of the basis 'structure'. In fact, the choice of transform must be a compromise between sufficient generality and power of representation within a class of signals. Piecewise smooth functions are well approximated by wavelets with enough vanishing moments and by local polynomial expansions. These models are sufficiently general to apply to a wide variety of signals, even departing in some degree from the piecewise polynomial template. Other issues in the choice of transform include computational efficiency and RD behaviour in the case of quantized truncated representations. The comparison in the paper addresses these last two issues specifically.

optimal procedure chooses a fixed set of Karhunen–Loève basis vectors (namely those corresponding to the largest eigenvalues) and spends all the rate to describe the coefficients with respect to these vectors. Note that a fixed subset corresponds to a linear approximation procedure (before quantization, which is itself nonlinear), while choosing a subset as in (2.3) is a nonlinear approximation method.

It is easy to come up with examples of objects for which nonlinear approximation is far superior to linear approximation. Consider a step function on [0, 1], where the step location is uniformly distributed on [0, 1]. Take the Haar wavelet basis as an orthonormal basis for [0, 1]. It can be verified that the approximation error using M terms is of the order of $\epsilon_L \sim 1/M$ for the linear case, and of $\epsilon_{NL} \sim 2^{-M}$ for a nonlinear approximation using the M largest terms. However, this is only the first part of the RD story, since we still have to describe the M chosen terms. This RD analysis takes into account that a certain number of scales J have to be represented, and at each scale, the coefficients require a certain number of bits. This split leads to a number of scales $J \sim \sqrt{R}$. The error is the sum of errors of each scale, each of which is on the order of $2^{-R/J}$. Together, we get

$$D_{NL}(R) \sim \sqrt{R}2^{-\sqrt{R}} \tag{2.4}$$

The quantization question is relatively simple if each coefficient α_n is quantized by itself (so-called scalar quantization). Quantizing several coefficients together (or vector quantization) improves the performance, but increases complexity. Usually, if a 'good' basis is used and complexity is an issue, scalar quantization is the preferred method.

The fourth question is a critical one. While approximation in orthonormal bases is very popular, it cannot be the end of the story. Just as not every stochastic process is Gaussian, not all objects will be well represented in an orthonormal basis. In other words, fitting a linear subspace to arbitrary objects is not always a good approximation, but even for objects where basis approximation does well, some other approximation method might do much better. In our step-function example studied earlier, a simple-minded coding of the step location and the step value leads to a RD behaviour

$$D'(R) \sim 2^{-R/2}. \tag{2.5}$$

In the remainder of this paper we are going to study in more detail the difference between wavelet and direct approximation of piecewise polynomial signals.

3   RD upper bounds for a piecewise polynomial function

Consider a continuous-time signal $s(t)$, $t \in [a, b]$, composed of M polynomial pieces; assume that the maximum degree of any polynomial piece is less than or equal to N and that each piece (and therefore the entire signal) is bounded in magnitude by some constant A. The signal is uniquely determined by the M

polynomials and by $M-1$ internal breakpoints; by augmenting the set of breakpoints with the interval extremes a and b, we can write

$$s(t) = \sum_{n=0}^{N} p_n^{(i)} t^n = p_i(t) \quad \text{for } t_i \le t < t_{i+1}, \tag{3.1}$$

where $a = t_0 < t_1 < \cdots < t_{M-1} < t_M = b$ are the breakpoints and the $p_n^{(i)}$ are the ith polynomial coefficients (with $p_n^{(i)} = 0$ for n larger than the polynomial degree); let $T = (b-a)$.

3.1   Polynomial approximation

In this section we will derive an upper bound on the RD characteristic for a quantized piecewise polynomial approximation of $s(t)$. For the time being, assume that the values for M, for the degrees of the polynomial pieces and for the internal breakpoints, are provided with arbitrary accuracy by an oracle. The derivation of the operational RD bound will be carried out in three steps: first we will determine a general RD upper bound for the single polynomial pieces; secondly, we will determine an RD upper bound for encoding the breakpoint values; finally, we will determine the jointly optimal bit allocation for the whole signal.

3.1.1   Encoding of one polynomial piece

Consider the ith polynomial of degree N_i, defined over the support $I_i = [t_i, t_{i+1}]$ of width S_i. Using a local Legendre expansion (see Appendix A) we can write (the subscript i is dropped for clarity throughout this section):

$$p(t) = \sum_{n=0}^{N} p_n t^n = \sum_{n=0}^{N} \frac{2n+1}{S} l_n L_I(n; t), \tag{3.2}$$

where $L_I(n; t)$ is the nth-degree Legendre polynomial over I; due to the properties of the expansion it may be shown that

$$|l_n| \le AS \tag{3.3}$$

for all n. The squared error after quantizing the coefficients can be expressed as

$$e^2 = \sum_{n=0}^{N} \left(\frac{2n+1}{S}\right)^2 (l_n - \hat{l}_n)^2 \int_{t_i}^{t_{i+1}} L_I^2(n; t)\, dt$$

$$= S^{-1} \sum_{n=0}^{N} (2n+1)(l_n - \hat{l}_n)^2, \tag{3.4}$$

where $\hat{l}_n$ are the quantized values. Assume using for each coefficient a different b_n-bit uniform quantizer over the range specified by (3.3) for a step size of $2AS2^{-b_n}$;

the total squared error can be upper bounded as

$$e^2 \le D_p = A^2 S \sum_{n=0}^{N} (2n + 1) 2^{-2b_n}. \tag{3.5}$$

For a global bit budget of R_p bits, the optimal allocation is found by solving the following reverse water-filling problem:

$$\frac{\partial D_p}{\partial b_n} = \text{const}, \qquad \sum b_n = R_p, \tag{3.6}$$

which yields

$$b_n = \frac{R_p}{N + 1} + \log_2 \sqrt{\frac{2n + 1}{\bar{C}}}, \tag{3.7}$$

with

$$\bar{C} = \left[\prod_{n=0}^{N} (2n + 1) \right]^{1/(N+1)}; \tag{3.8}$$

since the geometric mean is always less than or equal to the arithmetic mean we have $\bar{C} \le (N+1)$, and we finally obtain the following upper bound for the ith polynomial piece:

$$D_p(R_p) \le A^2 S(N + 1)^2 2^{-(2/N+1)R_p}. \tag{3.9}$$

3.1.2 Encoding of switchpoints

Assume that the $M + 1$ switchpoints t_i, as provided by the oracle, are quantized with a uniform quantizer over the entire support of the signal. In terms of the overall mean-squared error (MSE), the error relative to each quantized switchpoint can be upper bounded by (see figure 1a):

$$e_{t_i}^2 \le 4A^2 |t_i - \hat{t}_i|. \tag{3.10}$$

Again, the magnitude of the error is at most one half of the quantizer's step size, so that for a given switchpoint we have

$$D_t(R_t) \le 2A^2 T 2^{-R_t}, \tag{3.11}$$

where R_t is the quantizer's rate.

3.1.3 Composite RD bound

The global distortion bound for $s(t)$ is obtained additively as

$$D \le \sum_{i=1}^{M} D_{p_i}(R_{p_i}) + \sum_{i=0}^{M+1} D_{t_i}(R_{t_i}), \tag{3.12}$$

(a)   (b)

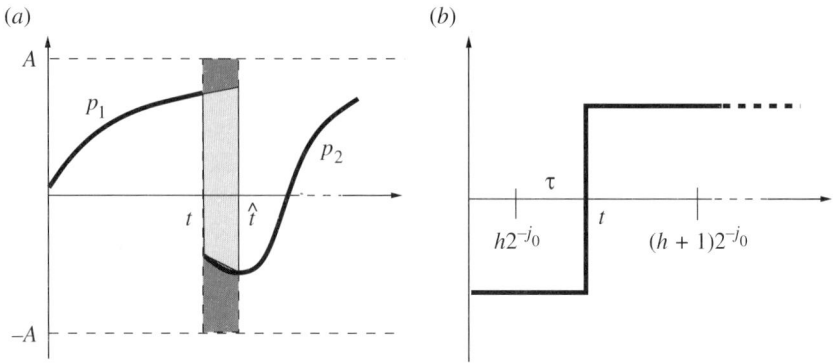

Fig. 1: Encoding of switchpoints: (*a*) true error (light area) and general upper bound (dark area); (*b*) location of the jump at a given wavelet scale.

where $D_{p_i}(R_{p_i})$ and $D_{t_i}(R_{t_i})$ are the bounds in (3.9) and (3.11), respectively, and where the subscript denotes the index of the polynomial pieces.

In order to obtain the optimal bit allocation for the composite polynomial function given an overall rate, it would be necessary to find the constant-slope operating points for all the summation terms in (3.12), as shown in (3.6); the resulting formulae, however, would be entirely impractical due to their dependence on all the polynomial parameters across the whole function. Instead, we choose to derive a coarser but general upper bound by introducing the following simplifications: (1) all polynomial pieces are assumed of maximum degree N; this implies that, for polynomials of lower degree, bits are also allocated to the zero coefficients; (2) the support of each polynomial piece S_i is 'approximated' by T, the entire function's support; together with the previous assumption, this means that the water-filling algorithm will assign the same number of bits R_p to each polynomial piece; (3) the origin of the function support (a) is either known or irrelevant; this reduces the number of encoded switchpoints to M; and (4) all switchpoints are encoded at the same rate R_t. With these simplifications the RD bound becomes

$$D(R) \le A^2 T M (2^{-R_t+1} + (N+1)^2 2^{-(2/N+1)R_p}), \qquad (3.13)$$

where the total bit rate is $R = M(R_t + R_p)$. By the usual reverse water-filling argument we obtain the optimal allocation

$$R_p = \frac{N+1}{N+3} \frac{R}{M} + \log_2 K, \qquad (3.14)$$

$$R_t = \frac{2}{N+3} \frac{R}{M} - \log_2 K, \qquad (3.15)$$

with $K = (2N+2)^{(N+1)/(N+3)}$. Using the relation (for $N > 0$)

$$2K + (N+1)^2 K^{-2/N+1} \le 2(N+1)^2, \qquad (3.16)$$

a simplified global upper bound is finally

$$D_P(R) \leq 2A^2 T M(N+1)^2 2^{-(2/N+3)(R/M)}. \tag{3.17}$$

3.2   Wavelet-based approximation

In this section we will obtain an upper bound for the case of a quantized nonlinear approximation of $s(t)$ using a wavelet basis over $[a, b]$. The derivation follows the lines in the manuscript by Cohen *et al.*(1997) and assumes the use of compact support wavelets with at least $N+1$ vanishing moments (Cohen *et al.*1993).

3.2.1   Distortion

If the wavelet has $N+1$ vanishing moments, then the only non-zero coefficients in the expansion correspond to wavelets straddling one or more switchpoints; since the wavelet also has a compact support, each switchpoint affects only a finite number of wavelets at each scale, which is equal to the length of the support itself. For $N+1$ vanishing moments, the wavelet support L satisfies $L \geq 2N+1$ and therefore, at each scale j in the decomposition, the number of non-zero coefficients C_j is bounded as

$$L \leq C_j \leq ML. \tag{3.18}$$

For a decomposition over a total of J levels, if we neglect the overlaps at each scale corresponding to wavelets straddling more than a single switchpoint we can upper bound the total number of non-zero coefficients C as

$$C \leq MLJ. \tag{3.19}$$

It can be shown (see, for example, Mallat 1997) that the non-zero coefficients decay with increasing scale as

$$|c_{j,k}| \leq ATW2^{-j/2}, \tag{3.20}$$

where W is the maximum of the wavelet's modulus. Using the same high-resolution b-bit uniform quantizer for all the coefficients with a step-size of $2ATW2^{-b}$ we obtain the largest scale before all successive coefficients are quantized to zero:

$$J = 2b - 2. \tag{3.21}$$

With this allocation choice the total distortion bound is $D = D_q + D_t$ where

$$D_q = \sum_k \sum_{j=0}^{J} (c_{j,k} - \hat{c}_{j,k})^2 \tag{3.22}$$

is the quantization error for the coded non-zero coefficients and where

$$D_t = \sum_k \sum_{j=J+1}^{+\infty} c_{j,k}^2 \tag{3.23}$$

is the error due to the wavelet series truncation after scale J (in both summations the index k runs over the non-zero coefficients in each scale). Upper bounding the quantization error in the usual way and using the bound in (3.20) for each discarded coefficient, we obtain

$$D \le C(ATW)^2 2^{-2b} + ML(ATW)^2 2^{-J} = ML(ATW)^2(1 + \tfrac{1}{4}J)2^{-J}. \quad (3.24)$$

3.2.2   Rate

Along with the quantized non-zero coefficients, we must supply a significance map indicating their position; due to the structure of $s(t)$, two bits per coefficients suffice to indicate which of the next-scale wavelet siblings (left, right, both or none) are non-zero. The total rate therefore is

$$R = C(b + 2) \le MLJ(\tfrac{1}{2}J + 3), \quad (3.25)$$

where we have used (3.19) and (3.21). In our high-resolution hypothesis it is surely going to be $b \ge 4$ and therefore we can approximate (3.25) as

$$R \le MLJ^2. \quad (3.26)$$

3.2.3   Global upper bound

Equation (3.24) provides a distortion bound as a function of J; in turn, J is a function of the overall rate as in (3.26). Combining the results we obtain the overall RD bound

$$D_W(R) \le (ATW)^2 M(2N + 1)\left(1 + \frac{1}{4}\sqrt{\frac{1}{2N+1}\frac{R}{M}}\right)2^{-\sqrt{(1/2N+1)(R/M)}}, \quad (3.27)$$

where we have assumed a minimum support wavelet, for which $L = 2N + 1$.

3.3   Commentary

To recapitulate, the two upper bounds obtained in the previous sections are of the form:

$$\text{polynomial approximation} \quad D_P(R) = C'_p 2^{-C_p R},$$

$$\text{wavelet approximation} \quad D_W(R) = C'_w(1 + \alpha\sqrt{C_w R})2^{-\sqrt{C_w R}}.$$

Since these are upper bounds, we are especially concerned with their tightness. Unfortunately, as we have seen, many simplifications have been introduced in the derivation, some of which are definitely rather crude; we will therefore concentrate on the rate of decay of the RD function rather than on the exact values of the constants. In order to gauge the applicability of the theoretical bounds, we can try to compare them to the actual performance of practical coding systems. A word of caution is, however, necessary: in order to implement the coding schemes described above, which are derived for continuous-time functions, a

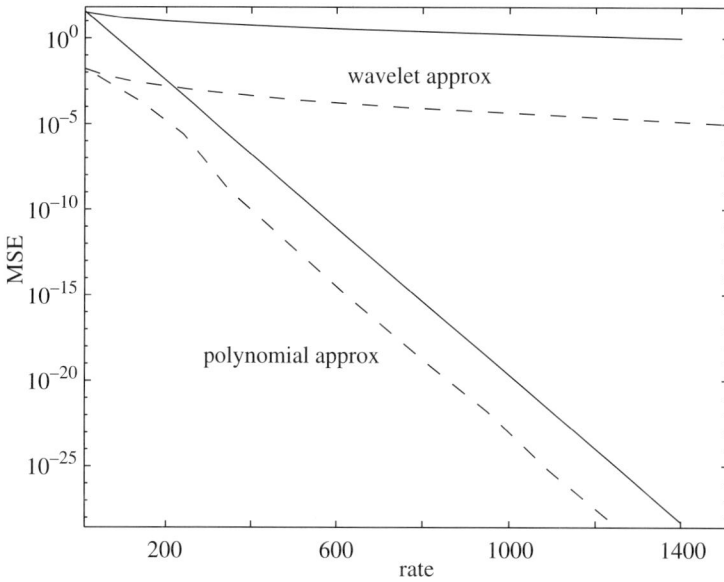

Fig. 2: Theoretical (solid) and experimental (dashed) RD curves.

discretization of the test data is necessary; as a consequence, an implicit granularity of the time axis is introduced, which limits the allowable range for both the breakpoint quantization rate and for the number of decomposition levels in the wavelet transformation. Unfortunately, computational requirements soon limit the resolution of the discretization: in our experiments we have used 2^{16} points. The two approximation techniques have been applied to randomly generated piecewise polynomial functions with parameters $A = 1$, $T = 1$, $N = 4$ and $M = 4$; Daubechies wavelets with five vanishing moments on the $[0, 1]$ interval have been used for the decomposition. The results are shown in figure 2: the solid lines and the dashed lines display the RD bound and the operational RD curve, respectively, for the polynomial and wavelet approximation strategies averaged over 50 function realizations.

A closer inspection of the RD curves shows that, especially for the wavelet case, there appears to be a large numerical offset between theoretical and practical values even though the rate of decay is correct. This simply indicates that the bounds for the constants in (3.27) are exceedingly large and the question is whether we can arrive at tighter estimates. In the absence of a detailed statistical description for the characteristic parameters of $s(t)$, the answer remains rather elusive in the general case; we can, however, try to develop our intuition by studying in more detail a toy problem involving minimal-complexity elements and a simple statistical model for the approximated function. The availability of a particular statistical model for the generating process allows us to derive an RD result in expectation, which is hopefully tighter. Consider the simple case in which $N = 0$, $M = 2$, $T = 1$ and $A = \frac{1}{2}$: the resulting $s(t)$ is simply a step function over, say, $[0, 1)$; we will assume that the location of the step transition t_0 is uniformly distributed over the support and that the values of the function left and right of the discontinuity are uniformly distributed over $[-\frac{1}{2}, \frac{1}{2}]$. Having a piecewise constant function allows us to use a Haar wavelet decomposition over the $[0, 1]$ interval; since the Haar family possesses a single vanishing moment and since there is no overlap between wavelets within a scale, the following facts hold:

(1) because of the absence of overlap, at each scale we have exactly one non-zero coefficient; the relation in (3.19) becomes exact:

$$C = J; \tag{3.28}$$

(2) under the high resolution hypothesis for a b-bit quantizer, the quantization error becomes a uniform random variable over an interval half a step wide; the expected error for each quantized coefficient is therefore $\frac{1}{1}2 2^{-2b}$;

(3) the series truncation error (3.23) is, in expectation (see Appendix B),

$$E[D_t] = (\tfrac{1}{3}6)2^{-(J+1)}; \tag{3.29}$$

(4) again, due to the non-overlapping properties of the Haar wavelet, we can rewrite (3.26) simply as $R \leq J^2$.

With these values, the RD curve, *in expectation*, becomes

$$D_W(R) \leq \tfrac{1}{7}2(1 + \tfrac{3}{4}\sqrt{R})2^{-\sqrt{R}}. \tag{3.30}$$

Figure 3a displays this curve along with the experimental results (dashed line); we can now see that the numerical values agree to within the same order of magnitude.

For completeness, the expected RD behaviour for the polynomial approximation of the above step function (obtained with a similar simplified analysis) turns out to be $D_P(R) = 1/(6\sqrt{2})\,2^{-R/2}$ and the curve, together with its experimental counterpart, is displayed in figure 3b.

4   RD optimal approximation

We have seen that the direct polynomial approximation displays a far better RD asymptotic behaviour than the standard nonlinear wavelet approximation. However, the polynomial bound was derived under two special hypotheses which are not generally met in practice: the availability of an 'oracle' and the use of high-resolution quantizers. Since the goal of many approximation techniques is a parsimonious representation of the data for compression purposes, the question

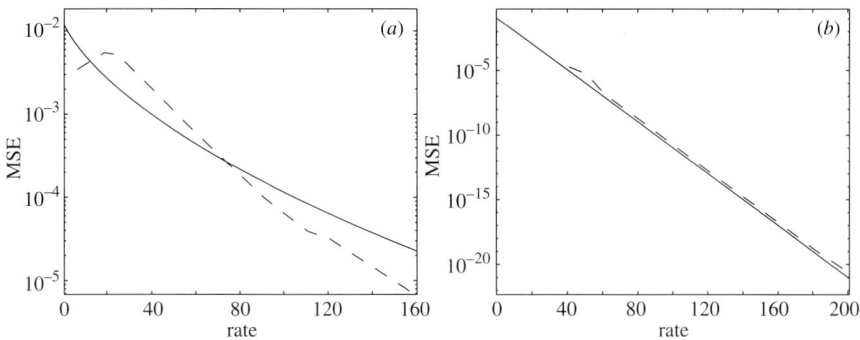

Fig. 3: Theoretical (solid) and experimental (dashed) RD curves for the step function approximation: (*a*) Haar wavelet approximation; (*b*) polynomial approximation.

arises naturally: what is the best coding strategy in a practical setting where the polynomial parameters are initially unknown and the bit rate is severely constrained? The problem can be cast in an *operational* RD framework where the objective is to 'fit' different polynomial pieces to the function subject to a constraint on the amount of resources used to describe the modelling and where the goodness of fit is measured by a global cost functional such as the MSE. In this constrained allocation approach, both breakpoints and local models must be determined jointly, since the optimal segmentation is a function of the family of approximants we allow for and of the global bit rate. In particular, for low rates, the available resources might not allow for a faithful encoding of all pieces and a globally optimal compromise solution must be sought for, possibly by lumping several contiguous pieces into one or by approximating the pieces by low-degree polynomials which have lighter description complexity.

In the following, we will illustrate a 'practical' algorithm which addresses and solves these issues, and whose performance matches and extends to the low-bit-rate case the oracle-based polynomial modelling. Note that now we are entering an algorithmic scenario where we perforce deal with discrete-time data vectors rather than continuous-time functions; similarly to the experimental results of the previous section, granularity of the involved quantities and computational requirements are now important factors.

4.1   Joint segmentation and allocation

Consider a K-point data vector $x = x_1^K$, which is a sampled version of a piecewise polynomial function $s(t)$ over a given support interval. The goal is to fit the data with local polynomial models so as to minimize the global MSE of the approximation under a given rate constraint. This defines an *operational* RD curve which is tied to the family of approximation models we choose to use. This initial choice is the crucial 'engineering' decision of the problem and is ruled by *a priori* knowledge on the input data (polynomial pieces of maximum degree N) and by economical considerations in terms of computational requirements. In particular, we choose a fixed limited set of possible rates associated to a polynomial model of given degree, with quantization of the individual coefficients following the line of (3.7). The validity of such design parameters can only be assessed via the performance measure yielded by the operational RD curve.

In the following we will assume a family of Q polynomial models, which is the aggregate set of polynomial prototypes from degree 0 to N with different quantization schemes for the parameters (more details later). For the data vector x define a *segmentation* t as a collection of $n + 1$ time indices:

$$t = \{t_0 = 1 < t_1 < t_2 < \cdots < t_{n-1} < t_n = K + 1\}.$$

The number of segments defined by t is $\sigma(t)$, $1 \le \sigma(t) \le K$, with the ith segment being $x_{t_i}^{t_{i+1}-1}$; segments are strictly disjoint. Let $T_{[1,K]}$ be the set of all possible segmentations for x, which we will simply write as T when the signal range is self-evident; it is clear that $|T_{[1,K]}| = 2^{K-1}$. Parallel to a segmentation t, define an *allocation* $w(t)$ as a collection of $\sigma(t)$ model indices w_i, $1 \le w_i \le Q$; let $W(t)$ be the set of all possible allocations for t, with $|W(t)| = Q^{\sigma(t)}$. Again, when the dependence on the underlying segmentation is clear, we will simply write w instead of $w(t)$.

For a given segmentation t and a related allocation w, define $R(t, w)$ as the cost, in bits, associated to the sequence of $\sigma(t)$ polynomial models and define $D(t, w)$ as the cumulative squared error of the approximation. Since there is no overlap between segments and the polynomial models are applied independently, we can write

$$D(t, p) = \sum_{i=1}^{\sigma(t)} \| \hat{p}(w_i) V_{(t_{i+1}-t_i)} - x_{t_i}^{t_{i+1}-1} \|^2 = \sum_{i=1}^{\sigma(t)} d^2(\hat{p}(w_i); t_i, t_{i+1}), \qquad (4.1)$$

where V is a Vandermonde matrix of size $N \times (t_{i+1} - t_i)$ and where $\hat{p}(w_i)$ is an $(N + 1)$-element vector containing the estimated polynomial coefficients for the ith segment quantized according to model w_i (the high-order coefficients being zero for model orders less than N). We will assume that the polynomial coefficients are coded independently and that their cost in bits is a function $b(\cdot)$ of the model's index only. An important remark at this point is that, by allowing for a data-dependent segmentation, *information about the segmentation and the allocation themselves must be provided along with the polynomial coefficients*. This takes the form of side information, which uses up part of the global bit budget of the RD optimization and must therefore be included in the expression for the overall rate. We will then write

$$R(t, w) = \sum_{k=1}^{\sigma(t)} (c + b(w_k)) = \sum_{k=1}^{\sigma(t)} r(p_k), \qquad (4.2)$$

where c is the side information associated to a new segment specifying its length and the relative polynomial order and quantization choice.

Our goal is to arrive at a minimization of the global squared error with respect to the local polynomials and to the data segmentation using the global rate as a parameter controlling the number of segments and the distribution of bits amongst the segments. Formally, this amounts to solving the following constrained problem:

$$\min_{t\in T}\ \min_{w\in W(t)} \{D(t, w)\}, \quad R(t, w) \le R_C. \tag{4.3}$$

While at first the task of minimizing (4.3) seems daunting, requiring $O(Q^N)$ explicit comparisons, we will show how it can be solved in polynomial time for almost all rates using standard optimization techniques.

4.2 Efficient solution

The problem of optimal resource allocation has been thoroughly studied in the context of quantization and coding for discrete datasets (Gersho & Gray 1992) and has been successfully applied to the context of signal compression and analysis (Ramchandran & Vetterli 1993; Xiong *et al.*1994; Prandoni *et al.*1997). In the following we will rely extensively on the results by Shoham & Gersho (1988), to which the reader is referred for details and proofs.

For the time being assume that a segmentation t_0 is given (a fixed-window segmentation, for instance) and that the only problem is to find the optimal allocation of polynomial pieces; each allocation defines an operational point in the RD plane as in figure 4*a* and the inner minimization in (4.3) requires us to find the allocation yielding the minimum distortion amongst all the allocations with the same given rate. However, if we restrict our search to the convex hull of the entire set of RD points, the minimization can be reformulated using Lagrange multipliers: define a functional $J(\lambda) = D(t, w) + \lambda R(t, w)$; if, for a given λ,

$$w^* = \arg \min_{w\in W(t_0)} \{J(\lambda)\}, \tag{4.4}$$

then w^* (star superscripts denote optimality) defines a point on the convex hull which solves the problem

$$\min_{w\in W(t_0)} \{D(t_0, w)\}, \quad R(t, w) \le R(t_0, w^*). \tag{4.5}$$

If we now let the segmentation vary, we simply obtain a larger population of operational RD points which are indexed by segmentation-allocation pairs as in figure 4*b*. Again, if we choose to restrict the minimization to the convex hull of the composite set of points, we can solve the associated Lagrangian problem as a double minimization:

$$J^*(\lambda) = \min_{t\in T}\ \min_{w\in W(t)} \{J(\lambda)\}; \tag{4.6}$$

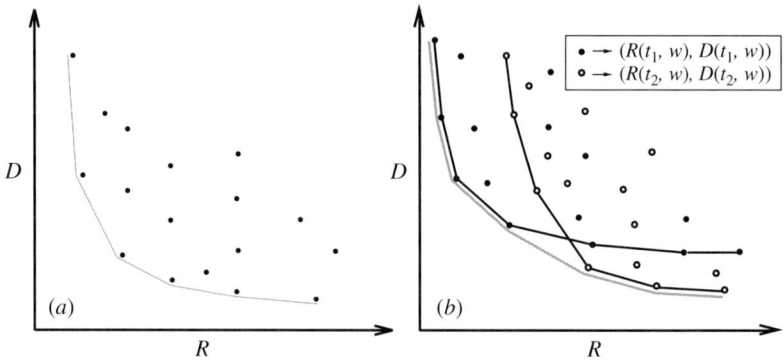

Fig. 4:   RD convex hulls: (*a*) convex hull for a single segmentation; (*b*) composite convex hull for two segmentations.

it should be noted that the restriction to the convex hull is of little practical limitation when the set of RD points is sufficiently dense; this is indeed the case given the cardinalities of T and W.

Even in the form of (4.6) the double minimization would still require an exhaustive search over all RD points, in addition to a search for the optimal λ. By taking the structure of rate and error into account, we can, however, rewrite (4.6) as

$$J^*(\lambda) = \min_{t \in T} \min_{w \in W(t)} \left\{ \sum_{k=1}^{\sigma(t)} (d^2(\hat{p}(w_k); t_k, t_{k+1}) + \lambda r(w_k)) \right\}. \qquad (4.7)$$

Since all quantities are non-negative and the segments are non-overlapping, the inner minimization over $W(t)$ can be carried out independently term-by-term, reducing the number of comparisons to $Q\sigma(t)$ per segmentation. Now the key observation is that, whatever the segmentation, all segments are coded with the same RD trade-off as determined by λ; therefore, for a given λ, we can determine the optimal t (in the sense of (4.3)) using dynamic programming (Bellman 1957). Indeed, suppose a breakpoint t belongs to t^*, the optimal segmentation; then it is easy to see that

$$J^*_{[1,N]}(\lambda) = J^*_{[1,t]}(\lambda) + \min_{t \in T_{[t,N]}} \min_{w \in W(t)} \{J(\lambda)\} \qquad (4.8)$$

(where subscripts indicate the signal range for the minimization). In other words, if t is an optimal breakpoint, the optimal cost functional for x_1^{t-1} is independent of subsequent data. This defines an incremental way to jointly determine the optimal segmentation and allocation as a recursive optimality hypothesis for all data points: for $0 \le t \le N$,

$$J^*_{[1,t]}(\lambda) = \min_{1 \le \tau \le t-1} \{J^*_{[1,\tau]}(\lambda) + \min_{1 \le w \le Q} \{d^2(\hat{p}(w); \tau, t) + \lambda r(w)\}\} \qquad (4.9)$$

(where $J_{[1,0]}^*(\lambda) = 0$). At each step t, only the new $J_{[1,t]}^*(\lambda)$ and the minimizing τ need be stored. The total number of comparisons for the double minimization is therefore $O(K^2)$. The latter must be iterated over λ until the rate constraint in (4.3) is met; luckily, the overall rate is a monotonically non-increasing function of λ (for a proof, see again Shoham & Gersho (1988)) so that the optimal value can be found with a fast bisection search (see Ramchandran & Vetterli 1993).

4.3 Implementation and results

In the implementation of the dynamic segmentation algorithm we have chosen a simplified set of quantization schemes. At low bit rates, equation (3.7) states that the optimal bit distribution for a set of polynomial coefficients is basically uniform. We choose four possible allocations of 4, 8, 12 and 16 bits for the single coefficient, with the total bit rate of a polynomial piece linearly dependent on its degree. A least-squares problem is solved for all orders from zero to N for each possible segment in an incremental fashion paralleling (4.9); this involves extending the QR factors of an order-N Vandermonde matrix by a new point at each step, which can be performed efficiently by means of Givens rotations. Side information for each segment is composed of two bits to signal the quantization scheme, $\lceil \log_2 N \rceil$ bits for the order of the polynomial model and $\lceil \log_2 K \rceil$ bits for the length of the segment. Finally, the computation in (4.9) can be efficiently organized on a trellis, where intermediate data are stored prior to the iteration over λ; further algorithmic details, omitted here, can be found in Prandoni (1999). The final computational requirements for the global minimization are on the order of $O(K^3)$, with storage on the order of $O(K^2)$.

We can now compare the experimental results of the optimal allocation algorithm with the polynomial approximation RD bound obtained using an oracle; however, since here the interest also lies in very low bit rates, we need to somehow refine the bound in (3.17). In fact, under severe rate constraints, there might not be enough bits to encode the exact structure of the function, and the dynamic algorithm will be forced to use a coarse segmentation in which several contiguous polynomial pieces are approximated by just one model; in the limit, when the rate goes to zero, the approximation error approaches the integral of $s^2(t)$ over the entire support of the function. This is not reflected by equation (3.17), where the expression for the error always assumes M distinct pieces. By approximating the maximum error by $4A^2T$, we can define a more appropriate RD bound for the polynomial case as

$$D_P'(R) = \min\{4A^2T, D_P(R)\}. \tag{4.10}$$

Figure 5 shows the numerical results obtained for a set of piecewise polynomial functions as in the previous experiment; the underlying sampling is, however, coarser here ($K = 2^8$), due to the heavier computational load. As before, the solid line indicates the new theoretical upper bound and the dashed line the RD performance of the dynamic algorithm. It is also interesting to look more in detail at the segmentation/allocation choices performed by the algorithm for different bit rate constraints; this is displayed in figure 6a–c with respect to the RD points A, B and C

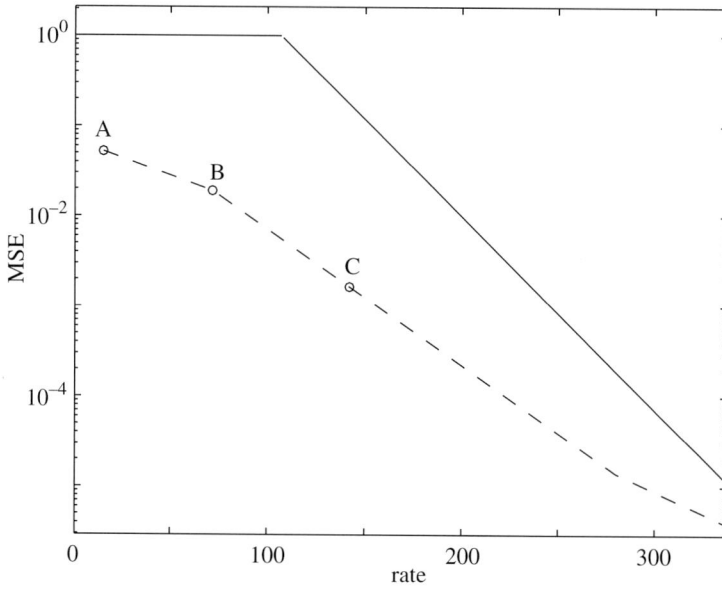

Fig. 5: Theoretical (solid) and experimental (dashed) RD curves for the dynamic segmentation algorithm.

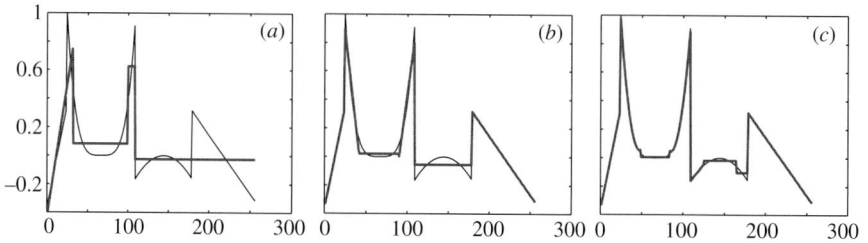

Fig. 6: Approximations provided by the dynamic segmentation algorithm corresponding to points A, B and C on the RD curve in figure 5.

marked by circles in figure 5. In figure 6 the thin line shows the original piecewise polynomial function while the thick lines show the algorithmic results; while not always very intuitive, these low-bit-rate approximations are nonetheless optimal in an MSE sense.

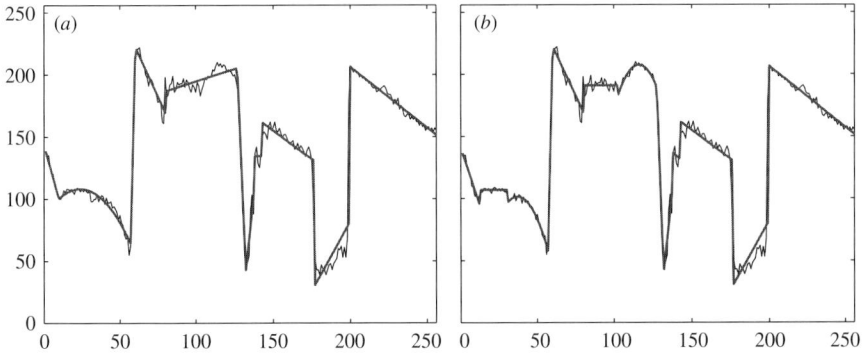

Fig. 7: Piecewise polynomial approximations of a line of Lena.

5 Conclusions

We have derived two operational upper bounds for the coding of piecewise polynomial functions. The results show the large superiority, especially at high bit rates, of a coding method based on direct polynomial modelling; this method is, however, much more expensive computationally, and its advantages are not immediately apparent in a low-rate fast compression framework. Yet, these bounds are concerned mainly with the operational way in which the coded data are represented rather than with the way the data are coded (wavelets versus polynomials) and the main lesson is that there is still room for gains once the peculiar properties of the expansion are taken into account; in particular, in a wavelet coding scenario, the vanishing moment property should influence the way the coefficients are coded. This is where our current research efforts are aimed at.

As a final note, we can ask ourselves two further questions: how does this framework extend to real-world signals, which are clearly not exactly piecewise polynomial? And more, can this framework be applied and compared to practical coding scenarios in which wavelets are known to perform very well, such as image compression? Unfortunately, dynamic programming techniques do not work for two-dimensional problems, and it is not clear how to fit polynomial surfaces in a globally optimal way. Yet, we can gain some intuition about both questions by looking at figure 7*a*, *b*. The thin line represents a single line of the 'Lena' image, for a total of 256 pixels; the thick lines are the piecewise polynomial approximations of the data, at increasing rates, obtained with the dynamic segmentation algorithm introduced above. We could argue that, for increasing bit rates, the algorithm captures more and more finely the local polynomial trends underlying the image surfaces, while the finer details can be represented as an additive noise-like residual. Whether this can lead to an efficient approximation scheme for images is, however, hard to say at present.

Appendix A. Local Legendre expansion

Legendre polynomials are usually defined over the $[-1, 1]$ interval by the recurrence relation

$$(n+1)L(n+1; t) = (2n+1)t\, L(n; t) - nL(n-1; t), \qquad (A1)$$

where $L(n; t)$ is the Legendre polynomial of degree n, and they constitute an orthogonal basis for $L[-1, 1]$.

A *local* Legendre expansion over the interval $I = [\alpha, \beta]$ can be obtained by defining a translated set of orthogonal polynomials:

$$L_I(n; t) = L\left(n; \frac{2}{\beta - \alpha}t - \frac{\alpha + \beta}{\beta - \alpha}\right); \qquad (A2)$$

the orthogonality relation becomes

$$\int_\alpha^\beta L_I(n; t)L_I(m; t)\, dt = \frac{\beta - \alpha}{2n + 1}\delta(n - m), \qquad (A3)$$

and for any polynomial $p(t)$ of degree N over $[\alpha, \beta]$, we can write

$$l_n = \int_\alpha^\beta L_I(n; t)p(t)\, dt, \quad n = 0, \ldots, N, \qquad (A4)$$

$$p(t) = \sum_{n=0}^{N} \frac{2n + 1}{\beta - \alpha} l_n L_I(n; t). \qquad (A5)$$

Appendix B. Estimate of the series truncation error

The estimate in (3.29) can be obtained as follows: assume that at scale j_0 the step discontinuity at t_0 falls within the interval $[h2^{-j_0}, (h+1)2^{-j_0})$ for some h. Then in the Haar wavelet series for $s(t)$ the indices of the non-zero coefficients $c_{j,k}$ for $j > j_0$ satisfy

$$h2^{j-j_0} \le k < (h+1)2^{j-j_0}. \qquad (B1)$$

The wavelet set $\{\psi_{j,k}(t)\}$ with j and k as above form an orthonormal basis for the $[h2^{-j_0}, (h+1)2^{-j_0})$ interval minus the addition of a scaling function $\varphi(t) = 2^{j_0/2}$ over the same interval. We can therefore write

$$D_t = \sum_{j=0}^{\infty} \sum_{k=h2^{j-j_0}}^{(h+1)2^{j-j_0}-1} c_{j,k}^2 = \int_{h2^{-j_0}}^{(h+1)2^{-j_0}} s^2(t) - \left[2^{-j_0/2}\int_{h2^{-j_0}}^{(h+1)2^{-j_0}} s(t)\right]^2,$$

$$(B2)$$

where we have used Parseval's identity. Now consider the location of the step (see figure 1*b*) and let $\tau = t_0 - h2^{-j_0}$; we can safely assume that $\tau \in \mathcal{U}[0, 2^{-j_0}]$ and we can rewrite (B2) as a combination of simple area measures involving τ and $x_{1,2}$, the values of $s(t)$ left and right of the jump, respectively, which are independent uniformly distributed variables over $[-\frac{1}{2}, \frac{1}{2}]$:

$$D_t = \tau x_1^2 + (2^{-j_0} - \tau)x_2^2 - 2^{j_0}(\tau^2 x_1^2 + (2^{-j_0} - \tau)^2)x_2^2 + 2\tau(2^{-j_0} - \tau)x_1 x_2. \tag{B3}$$

By taking expectations over these independent quantities we finally have

$$E[D_t] = (\tfrac{1}{3}6)2^{-j_0}. \tag{B4}$$

6   References

Bellman, R. 1957 *Dynamic programming*. Princeton University Press.

Berger, T. 1971 *Rate distortion theory*. Englewood Cliffs, NJ: Prentice Hall.

Cohen, A., Daubechies, I. & Vial, P. 1993 Wavelet bases on the interval and fast algorithms. *J. Appl. Comput. Harmonic Analysis* **1**, 54–81.

Cohen, A., Daubechies, I., Guleryuz, O. & Orchard, M. 1999 On the importance of combining wavelet-based non-linear approximation in coding strategies. (In preparation.)

Coifman, R. R. & Wickerhauser, M. V. 1992 Entropy-based algorithms for best basis selection. *IEEE Trans. Information Theory* **38**, 713–718.

Donoho, D. L., Vetterli, M., DeVore, R. A. & Daubechies, I. 1998 Data compression and harmonic analysis. *IEEE Trans. Information Theory* **44**, 2435–2476.

Gersho, A. & Gray, R. M. 1992 *Vector quantization and signal compression*. Dordrecht: Kluwer.

Mallat, S. 1997 *A wavelet tour of signal processing*. San Diego, CA: Academic.

Mallat, S. & Falzon, F. 1998 Analysis of low bit rate image transform coding. *IEEE Trans. Signal Processing* **46**, 1027–1042.

Ortega, A. & Ramchandran, K. 1998 Rate-distortion methods for image and video compression. *IEEE Signal Processing Magazine* November, pp. 23–50.

Prandoni, P. 1999 Optimal segmentation techniques for piecewise stationary signals. PhD thesis, École Polytechnique Fédérale de Lausanne, Switzerland.

Prandoni, P., Goodwin, M. & Vetterli, M. 1997 Optimal time segmentation for signal modeling and compression. In *Proc. ICASSP*, vol. 3, pp. 2029–2032.

Ramchandran, K. & Vetterli, M. 1993 Best wavelet packet bases in a rate-distortion sense. *IEEE Trans. Image Processing* **2**, 160–175.

Shannon, C. E. 1948 A mathematical theory of communication. *Bell Syst. Tech. Jl* **27**, 623–656.

Shannon, C. E. 1959 Coding theorems for a discrete source with a fidelity criterion. *IRE Nat. Conv. Rec.* 142–163.

Shapiro, J. M. 1993 Embedded image coding using zerotrees of wavelet coefficients. *IEEE Trans. Signal Processing*, **41**, 3445–3462.

Shoham, Y. & Gersho, A. 1988 Efficient bit allocation for an arbitrary set of quantizers. *IEEE Trans. Acoust. Speech Signal Processing* **36**, 1445–1453.

Vetterli, M. & Kovačević, J. 1995 *Wavelets and subband coding*. Englewood Cliffs, NJ: Prentice Hall.

Xiong, Z., Ramchandran, K., Herley, C. & Orchard, M. T. 1994 Flexible time segmentations for time-varying wavelet packets. *IEEE Proc. Int. Symp. on Time–Frequency and Time–Scale Analysis*, pp. 9–12. New York: IEEE.

12

The contribution of wavelets to the analysis of economic and financial data

James B. Ramsey

Department of Economics, New York University,
269 Mercer St., New York, NY 10003, USA

Abstract

After summarizing the properties of wavelets that are most likely to be useful in economic and financial analysis, the literature on the application of wavelet techniques in these fields is reviewed. Special attention is given to the potential for insights into the development of economic theory or the enhancement of our understanding of economic phenomena. The paper is concluded with a section containing speculations about the relevance of wavelet analysis to economic and financial time-series given the experience to date. This discussion includes some suggestions about improving our understanding and evaluation of forecasts using a wavelet approach.

Keywords: wavelets; economics; finance; time-scale; forecasting;
non-stationarity

1 Introduction

Imagine a dynamical system that is characterized by the decentralized interactions of a large number of participants. Participants make decisions about their future behaviour facing enormous degrees of freedom using only local information over both time and space. Imagine further that this system is constantly evolving, open and non-isolated so that its performance over time reflects its reactions to external stimuli and even the manipulation of the fundamental relationships of the system. Imagine yet again that attempts to control the system generate anticipatory reactions by the participants, each of whom is trying to optimize their position within the system. Imagine also that each agent is operating on several time-scales at once, so that not only current events affect behaviour, but the distant past as well as the agent's anticipation of the distant future.

If you can imagine this challenge to the analyst's art, you have imagined the difficulties that face the empirical economist. As a result of these the statistical and analytical tools that work so well in other fields have limited success in economic contexts, and then perform best only as short-term approximations to more complex phenomena. The econometrician, in order to achieve some limited success, must attempt to reconcile the irreconcilable. Even though the economy is constantly evolving, one must often assume in order to forecast that the system is at least locally constant. The econometrician must assume that only a few variables are, as a practical matter, involved in any relationship. Most often, it is the violation of this technically necessary assumption that produces unanticipated and sometimes dramatic shifts in observed economic relationships.

In such a context, the search for useful analytical tools is paramount. Key issues to be considered by the putative analyst are robustness of procedure to erroneous assumptions, flexibility of regression fit to deal with imprecise model formulations, the ability to handle complex relationships, efficiency of the estimators to be able to make useful distinctions on few data points, and simplicity of implementation. This, the econometrician's 'Holy Grail', may be impossible to achieve, but is nevertheless a worthwhile overall objective.

With this background, the potential promised by wavelets is readily apparent. While wavelets do not meet all the criteria required of the 'Holy Grail', they meet some of them sufficiently well to indicate that wavelets may well be able to provide new insights into the analysis of economic and financial data.

One of the first benefits of a wavelet approach is the flexibility in handling very irregular data series, as illustrated in Donoho *et al.*(1995). Given my remarks above, it is no surprise that the ability to represent highly complex structures without knowing the underlying functional form is of great benefit in economic and financial research. A corollary facility that wavelets possess is that of being able to precisely locate discontinuities, and isolated shocks to the dynamical system, in time regime shifts. Further, it is vital that the process of representation should be able to deal with the non-stationarity of the stochastic innovations that are inevitably involved with economic and financial time-series.

A critical innovation in estimation that was introduced by wavelets, although by no means necessarily restricted to wavelets, is the idea of shrinkage. Traditionally in economic analysis the assumption has universally been made that the signal, $f(t)$, is smooth and the innovations, $\varepsilon(t)$, are irregular. Consequently, it is a natural first step to consider extracting the signal $f(t)$ from the observed signal $y(t) = f(t) + \varepsilon(t)$ by locally smoothing $y(t)$. However, when the signal is as, or even more, irregular than the noise, such a procedure no longer provides a useful approximation to the signal. The process of smoothing to remove the contamination of noise distorts the appearance of the signal itself. When the noise is below a threshold and the signal variation is well above the threshold, one can isolate the signal from the noise component by selectively shrinking the wavelet coefficient estimates (Donoho & Johnstone 1995; Donoho *et al.*1995).

The most important property of wavelets for economic analysis is decomposition by time-scale. Economic and financial systems, like many other systems, contain variables that operate on a variety of time-scales simultaneously so that the relationships between variables may well differ across time-scales.

The remainder of this paper is in two sections. The first provides a selective review of the use of wavelets in the analysis of economic and financial data. The second section speculates on the potential for wavelet analysis in the context of economic and financial data.

2   A selective review of the literature

While applications in economics have not yet extensively used the special properties of wavelets, there have been a number of interesting applications and the list is sure to grow. We might fruitfully discuss the various applications in terms of four main categories. The first and largest category in terms of numbers of papers is that emphasizing the role of non-stationarity and the ability to handle complex functions in Besov space. The second category is those papers that are most concerned with structural change and the role of local phenomena. A third involves the use of time-scale decompositions and the last is directly concerned with forecasting. I will conclude this section with a few comments about some papers that do not easily fit into any of these categories.

2.1   Non-stationarity and complex functions

A first paper that warrants discussion uses applications in biology, but the statistical and methodological issues are quintessentially similar to those in economics. Von Sachs & MacGibbon (unpublished research) derived the bias and variance for wavelet coefficient estimators under the assumption of local stationarity; this paper is an extension of an earlier paper by Johnstone & Silverman (1997), who analysed the statistical properties of wavelet coefficient estimators under the assumption of stationary, but correlated, data of both short and long run. Further, von Sachs & MacGibbon (1997) incorporated into their analysis pulsatile components and simultaneously allowed for non-stationarity. An important assumption in their analysis is that while the non-stationarity is slowly changing over the entire sample period, it does so in such a manner that more observations per unit time would lead to locally asymptotic convergence of the estimators. The authors demonstrate the asymptotic properties of their estimators. In particular, they suggest under these circumstances the use of a local MAD estimator to estimate the variability of the wavelet coefficients in segments of quasi-stationarity within each level. The importance of this paper stems from the fact that the single most obvious variation in economic data, especially after first differencing, is the presence of second-order non-stationarity, which is sometimes modelled as an ARCH process; for a review in the context of financial data see Bollerslev *et al.*(1990).

A nearly contemporaneous article on a similar topic is that by Gao (unpublished research), who contemplated the model

$$y_i = f(t_i) + \sigma(t_i)z_i, \tag{2.1}$$

where $\{z_i\}$ is distributed as i.i.d. Gaussian noise. The model proposed here is a simplification of that considered by von Sachs & MacGibbon. The model is motivated by noting that heteroscedasticity arises from converting unequally spaced data to equally spaced data. When the variances are known the process of obtaining suitable estimators is straightforward, but when the variances are unknown, they must be estimated along with the wavelet coefficients. Gao's procedure extends Donoho & Johnstone (1997). Gao considers four alternative definitions of the shrinkage rule, the usual soft and hard rules, an intermediate rule that he labels 'firm', and the garrote. Under the usual assumptions supplemented by the knowledge that the variances are heteroscedastic, the wavelet coefficients w, have the distribution

$$w = Hy \sim N(\Theta, HD^2H') \tag{2.2}$$

where D is the covariance matrix for the model, H is the wavelet transform and $\Theta = Hf$. Under these conditions the near optimality in terms of expected mean-squared-error risk of the waveshrink estimators continues to hold.

Gao's solution when the variances are unknown is to use a non-decimated wavelet transform on y to obtain the variation at the finest level of detail. He next applies a running MAD estimator to obtain estimates for the standard deviation. These estimators are not distributed as Gaussian, but approximately as Student t with heavier tails. Near-optimal mean-squared-error expected risk also holds in this situation.

Pan & Wang (1998) introduce a novel approach when the data generating mechanism must be regarded as 'evolutionary', so that the wavelet coefficients are varying over time; the model is

$$y_t = W_t^T w_t + \epsilon_t. \tag{2.3}$$

The authors interpret the model in terms of a Kalman filter, so that the pair of equations that define the time path of coefficients is given by

$$\left. \begin{aligned} y_t &= W_t^T w_t + \epsilon_t, \\ w_t &= w_{t-1} + V_t, \end{aligned} \right\} \tag{2.4}$$

where V_t is the time-varying covariation for the wavelet coefficients. The model was applied to the monthly stock price index, y_t, that was regarded as some function $\tilde{f}(r_t)$, where r_t is the dividend yield. $\tilde{f}(r_t)$ was represented in the model by the wavelet transform $W_t^T w_t$. The empirical results were encouraging both as to the overall degree of approximation and to the extent that turning points were correctly indicated.

An article that represents research arising from earlier concerns in the analysis of the stock market is Ramsey *et al.*(1995). In this article, the chief topics of interest were the degree of statistical self-similarity of the daily stock return data and whether there was any evidence of quasi-periodicity. The wavelet analysis indicated that while there was little evidence of scaling in the data, there was surprisingly clear evidence for quasi-periodic sequences of shocks to the system. Thus, these results confirm what is by now a commonplace statement about this type of financial data (see, for example, Ramsey & Thomson 1998); they are very complex and are more structured than mere representations of Brownian motion.

The last two papers to be discussed in this section introduce a useful generalization to wavelet analysis in those cases where one wishes to begin the analysis in a purely exploratory phase. In Ramsey & Zhang (1996, 1997), highly redundant representations in terms of 'waveform dictionaries' were used (see Mallat & Zhang 1993). The approach taken in these papers is one that generalizes both wavelets and Fourier analysis. For a given choice of scaling function $g(t) \epsilon L^2(R)$, scale parameter s, position in time u and a given frequency modulation ξ, we define the time-frequency atom by

$$g_\gamma(t) = \frac{1}{\sqrt{s}} g\left(\frac{t-u}{s}\right) e^{e\xi t}, \tag{2.5}$$

$$\gamma = (s, u, \xi). \tag{2.6}$$

The function $g_\gamma(t)$ is normalized to have norm 1 by the component $1/\sqrt{s}$. The function $g_\gamma(t)$ is centred at the abscissa u, where its energy is concentrated in a region that is proportional to s. The Fourier transform of $g_\gamma(t)$ is centred at the frequency ξ and has its energy concentrated in a neighbourhood of ξ with size proportional to $1/s$. Matching pursuit algorithms are used to obtain the minimum squared error collection of atoms that will most parsimoniously represent the function.

Ramsey & Zhang (1996) examined the 16 384 daily observations of the Standard and Poor's 500 stock-price index on the New York Stock Exchange from January 3rd 1928 to November 18th 1988. The time-frequency distributions indicate virtually no power for any frequencies, although there is evidence that some frequencies wax and wane in strength. However, most of the power is in time-localized bursts of activity. The bursts do not appear to be isolated Dirac delta functions, but highly localized chirps that are characterized by a rapid build-up of amplitude of signal and rapid oscillation in frequency. A plot of the squared weights of the coefficients relative to those that would be obtained from random data indicate very strong approximations with relatively few coefficients. What is even more striking is that if we restrict our attention in the time-frequency plots to those scales that will produce the highest resolution of the frequencies we observe very clear evidence of quasi-periodicity and regularity in the data.

Ramsey & Zhang (1997) applied similar techniques to tic-by-tic foreign exchange rates worldwide for a year. The three exchange rates so examined were the Deutschmark–US dollar, the Yen–US dollar and the Yen–Deutschmark.

The results obtained were qualitatively similar for all three exchange rates. Both the levels and the first differenced data were examined, because there has been some controversy in the economics literature about the appropriate data-generating mechanism; the presence of a unit root in the data being of great concern. In analysing the levels data using the waveform-dictionary approach, some evidence of structure was discovered, but only with very low power. As discovered in the stock-market data, there was evidence for frequencies that waxed and waned over the year. However, most of the power seems to be in localized frequency bursts. The economic importance of this result is that instead of viewing the transmission of information in the foreign exchange market as being represented by a swift, almost effortless, adjustment, these results indicate that there is a period of adjustment of some noticeable length and that the implied frequencies of oscillation build up and decay as in a chirp. Another key insight provided by these data is that despite the relatively low number of atoms needed to provide a very good approximation to the data, about 100 is sufficient, there is little opportunity for improved forecasting. This is because, while relatively few structures are needed to represent the data, the bulk of the power is in chirps and there does not seem to be any way of predicting the occurrence of the chirps. In short, as most of the energy of the system is in randomly occurring local behaviour, there is little opportunity to improve one's forecasts.

2.2   Time-scale decompositions

For cognate disciplines such as economics and biology, one of the most useful properties of the wavelet approach is the ability to decompose any signal into its time-scale components. It is well known that physical and biological processes are phenomenalogically different across different time-scales. In economics as well, one must allow for quite different behaviour across time-scales. A simple example will illustrate the concept. In the market for securities there are traders who take a very long view (years in fact) and consequently concentrate on what are termed 'market fundamentals'; these traders ignore ephemeral phenomena. In contrast, other traders are trading on a much shorter time-scale and as such are interested in temporary deviations of the market from its long-term growth path; their decisions have a time horizon of a few months to a year. And yet other traders are in the market for whom a day is a long time.

An effort along these lines is illustrated in Davidson *et al.*(1997), who investigated US commodity prices. Even though the differences across scales were not pursued fully, the authors did consider the different properties of the wavelet coefficients across scales and calculated a measure of the relative importance of the coefficients between scales.

In two related papers (Ramsey & Lampart 1998*a, b*), a very different approach was taken to the use of scale in wavelet analysis. In both of those papers the concept was that the relationship between two variables may well vary across scales. For example, one may have a simple linear relationship between aggregate income and consumption, but that the coefficient relating consumption to income might well differ across scales. The objective in the two papers mentioned was to examine this issue in the context of two relationships; one was that between personal income and consumption and the other was that between money and gross domestic product, the so-called 'money velocity' relationship. Wavelets were used to provide both an orthogonal time-scale decomposition of the data and a non-parametric representation of each individual time-series. At each scale a regression was run between consumption and income, or between money and income, and the results were compared across scales. For example, using the consumption–income relationship, we have

$$\left.\begin{array}{l} C[S_J]_t = \alpha_J + \beta_J Y[S_J]_t + \varepsilon_t, \quad \text{or} \\ C[D_j]_t = \alpha_j + \beta_j Y[D_j]_t + \varepsilon_t, \quad j = 1, 2, \ldots, J - 1, \end{array}\right\} \quad (2.7)$$

where $C[S_J]_t$ and $Y[S_J]_t$ represent the components of consumption and income at the highest scale and $C[D_j]_t$ and $Y[D_j]_t$ represents consumption and income at intermediate scales.

The results were productive and informative. First, the relationship between economic variables does vary significantly across scales; that is, statistically significant differences were discovered in the estimated values of the coefficients β_j. Second, the decomposition resolved some empirical anomalies that existed in the literature. Third, the decomposition indicated that the delay in the relationship between two variables might well be a function of the state space; a result that heretofore had not been suspected. More specifically, the authors discovered that the slope coefficient relating consumption and income declines with scale, a theoretically pleasing result; and that the role of the real interest rate in the consumption–income relationship is strong and of the theoretically correct sign for the longest time-scales, but is insignificant for the shortest scales; which is another theoretically plausible result.

However, a more far-reaching result and one that has not been anticipated in the literature is that the phase of the relationship between two variables ordered in time, or between the scale components of two variables, may well vary with the state of the system. In economics a conventional assumption is that two variables, say consumption and income, are related contemporaneously; a secondary assumption that is occasionally invoked is that there is a delay between the two variables as there often is between stimulus and response. But what Ramsey & Lampart (1998a, b) discovered is that the delay between two variables might well be a function of the state of the system; that is

$$C[D_j]_t = \alpha + \beta Y[D_j]_{t-d[X_t]} + \varepsilon_t,$$

where X_t represents some component of the dynamical system within which the consumption–income relationship is embedded. An example of this in our current context is provided by recognizing that the 'timing of consumption, or certainly the timing of a purchase' is as much an economic decision as is the amount of the purchase.

A long-standing debate in the economics literature concerns the direction of causality between money and income; essentially does money cause income, or does income cause money? Prior empirical research had obtained conflicting results. Ramsey & Lampart (1998a, b) indicated that the decomposition of money and income into their corresponding time-scale components followed by performing

Wavelets in economics

Table 1 Granger causality tests: I (Results of Granger causality tests on individual crystals and log differenced data for M1 and nominal personal income (NPI): *P* values in parentheses.)

	results	null hypotheses M1 $\Rightarrow$ NPI	null hypotheses NPI $\Rightarrow$ M1
D6 (5 lags)	feedback	6.398 (0.000)	6.968 0.000)
D5 (20 lags)	feedback	4.491 (0.000)	5.242 (0.000)
D4 (19 lags)	M1 $\Longrightarrow$ NPI	3.334 (0.000)	0.809 (0.695)
D3 (17 lags)	M1 $\Longrightarrow$ NPI	1.838 (0.023)	1.294 (0.193)
D2 (23 lags)	M1 $\Longrightarrow$ NPI	5.620 (0.000)	1.146 (0.293)
D1 (14 lags)	M1 $\Longleftarrow$ NPI	1.558 (0.089)	5.194 (0.000)
log diff. (12 lags)	inconclusive	0.534 (0.892)	2.063 (0.186)

Table 2 Granger causality tests: II (F-tests of Granger causality between M1 and nominal personal income (NPI) at [D4] across phase shifts: *P* values in parentheses.)

	in–out 4/1962–8/1967 M1 $\Longrightarrow$ NPI	out–in 8/1967–5/1974 M1 $\Longleftarrow$ NPI4	in–in 4/1962–5/1974 inconclusive
phase shifts time period			
NPI $\Rightarrow$ M1	6.401 (0.00)	0.997 (0.434)	0.643 (0.696)
M1 $\Rightarrow$ NPI	1.465 (0.208)	2.714 (0.020)	0.933 (0.474)

a sequence of regressions between money and income at each scale level helped to resolve this debate. The main result is that the direction of the relationship depends on the level of the time-scale in a clear manner for each scale; but that the relationship between the aggregates over all time-scales is ambiguous. These results are illustrated in table 1.

Further, at the D4 scale level the 'causality tests' were ambiguous in that the results depended on the pattern of the phase relationship between money and income; that is, if money and income are moving into phase the causality goes one way, in the opposite direction if moving out of phase, and the results are anomalous if the mixture of phase relationships is random. These results are illustrated in table 2.

2.3 Forecasting

There are only two major papers in this section, notwithstanding the importance of the topic. The first approach is discussed in Ariño (unpublished research). The main idea is relatively simple and direct. Ariño's approach is to decompose the signal into its time-scale components and then to treat each approximation at each time-scale as a separate series. He provides forecasts for each component by using a regular ARIMA formulation. The final forecast for the complete series is obtained by adding up the component forecasts. Ariño restricts his attention to three components: trend, seasonal fluctuations and noise. The technique is applied to Spanish concrete production and car sales with interesting results that indicate the procedure has potential. Ariño's paper leaves numerous distributional issues unaddressed, but the experiment promises to be useful.

The next approach is presented in a series of papers by Alex Aussem, Fionn Murtagh and co-workers (see, for example, Aussem & Murtagh 1997; Aussem *et al.*1998). The major innovation introduced by these authors is to analyse the individual time decompositions by neural networks and to base the forecasts on the neural-network estimates. The wavelet transform used was the isotropic non-orthogonal linear à-trous wavelet.

As with Ariño, the forecast for the complete series is obtained by adding up the individual forecasts. The technique is applied to the famous sunspot series and to the daily Standard and Poor's stock-market series. For the latter the authors' provide five day ahead forecasts. As with the work reported by Ariño, the distributional properties of the combined forecasts obtained by adding up the individual time-scale-based forecasts is still an open issue, but the results to date are of interest and indicate substantial potential.

2.4 Some miscellaneous papers

This subsection contains a discussion of some miscellaneous, but important, examples of the use of wavelets in economic analysis.

2.4.1 *Density estimation*

The estimation of density functions has been a tradition in economics and the debate concerning the distribution of income and wealth in particular is of great topical concern. Further, because of such factors as 'minimum-wage' laws, 'poverty boundaries' or discontinuous tax relationships, income distributions frequently exhibit anomalous local behaviour that is best analysed in the context of wavelets. Härdle *et al.*(1998) illustrate the benefits of using a wavelet approach to the estimation of densities relative to the standard kernel smoothing procedures.

A recent and important line of research in economics is the analysis of fractionally integrated models; that is, models of the form

$$\Phi(L)(1 - L)^d (x(t) - \mu) = \theta(L)\epsilon(t), \qquad (2.8)$$

where $\Phi(L)$ and $\theta(L)$ are polynomials in the lag operator L, $x(t)$ and $\epsilon(t)$ are stochastic processes and d is the fractional differencing coefficient such that $|d| <$ 0.5. Jensen (1997a, b) in a recent series of articles has explored the use of wavelets in the estimation of the fractional differencing coefficient, d. Jensen has several reasons for choosing wavelets as a basis for the analysis of the long-memory processes defined by the ARFIMA models. The major reasons are that the wavelet-based calculations are far less intensive to calculate than the regular exact MLE, the estimates are much more robust to modelling errors (the orders of the polynomial lags in equation (2.8) are unknown), and the estimates are robust to not knowing the mean of the process. A key element in the computational gains achieved using wavelets is the relative sparsity of the wavelet coefficients, especially for the longer time-scale components of the process.

2.4.2   Structural change

Given the concern by economists about the potential effects of sudden regime changes and isolated shocks to the system it is surprising that more research on this subject has not been carried out so far. An exception is a paper by Gilbert (unpublished research), who looked for evidence of structural change using Haar wavelets as a basis. The data used were a set of macroeconomic indicators that are seasonally adjusted, some series were quarterly and some were monthly. Using these procedures, only the inflation rate index and the index of nominal GNP growth gave any indication of an abrupt structural change.

Another example that illustrates attempts to take advantage of the ability of wavelets to resolve local features is provided by the manuscript by Jungeilges (unpublished research). The major concern in this manuscript is to provide improvements on the specification error test procedures created by Ramsey (1969) and refined in Ramsey & Schmidt (1976) and the corresponding tests generated by White (1989). The latter tests used neural networks as a device for providing semiparametric representations of any time-series. The essential idea underlying Jungeilges's manuscript is that by using wavelets to provide a robust representation of the presumed structure of the data the residual variances should be uncorrelated and distributed identically with equal variances across the sample. Unfortunately, although the test designed by Jungeilges has power against many alternative errors, it is not as powerful in general as the original tests provided by Ramsey (1969). This is one of the few examples where the use of wavelets has not, at least marginally, led to an improvement in test results or in the efficiency of estimators. The difficulty resides in the author's design of the test in that by relying mainly on the effect of the specification errors on the *heteroscedasticity* of the residuals, rather than on the *structure* of the time varying conditional mean, the author lost both power and robustness.

2.4.3 Miscellaneous papers

Spectral techniques have long been a useful, but relatively neglected, tool in economic analysis. Three important papers on this topic are Neumann (1996), Chiann & Morettin (unpublished research) and Lee & Hong (unpublished research). Neumann (1996) derived the asymptotic normal distribution of empirical wavelet coefficients used to estimate the spectral density function of a non-Gaussian process. The wavelet estimator is shown to be superior to kernel estimators in terms of mean-squared-error criteria, especially when there are spatially inhomogeneous features in the spectral density. Lee & Hong advanced this literature by proposing a consistent test for serial correlation of unknown form by wavelet methods. The simulation results confirm the theoretical prediction that when there are spatial inhomogeneities the wavelet estimator is to be preferred and when there are no spatial inhomogeneities the kernel estimator is preferred.

Finally, mention should be made of a nearly unique approach to the use of wavelets to analyse economic data. In Morehart *et al.*(1999) wavelets are used to model the *geographic distribution* of economic and financial measures used in agriculture. The idea is to provide a non-parametric framework within which multivariate spatial relationships can be described. The authors base their analysis on the non-decimated à trous wavelet. As indicated by the authors, the use of non-decimated wavelets enhances the detection of local features and anomalies. There is one other paper that uses wavelets to analyse 'spatial' differences, namely Chen & Conley (unpublished research). In this paper, the statistical context is one involving panel data where the number of observations over time is relatively short. The authors' approach is to compensate for the lack of time-series data by using information on the spatial distribution of agents. Agents that are spatially close together can be assumed to behave in a similar manner, so that a 'cluster' of agents will almost provide a set of repeated time-series experiments. The empirical results are summarized in terms of estimates of the spectral density function.

3 The research potential

In this section I speculate on the potential for wavelets in the analysis of economic and financial data. The comments can be grouped into four categories of applications.

3.1 Exploratory analysis: time-scale versus frequency

Waveform dictionaries provide an excellent exploratory tool where one is particularly interested in determining the relative importance of frequency components to time-scale components. In economics and finance a preliminary examination of the data in order to assess the presence and the ebb and flow of frequency components is useful. In particular, little attention has been paid so far to the idea that frequency components may appear for a while, disappear and then reappear. Conventional spectral tools would most likely miss such frequency components altogether. Further, as will be argued below, in economics and finance, a matter of considerable concern is to separate the local from the global, the ephemeral from the permanent; wavelets provide the best tool for separating out these effects.

3.2   Density estimation and local inhomogeneity

We have seen in the discussion above that in general wavelets are superior, say to kernel estimators, whenever there are local inhomogeneities. I argued in the introduction that a prominent characteristic of economic data is the presence of spatial inhomogeneities, so that one quickly concludes that wavelets are a natural choice for analysing such data. At the very least, wavelets should be used as a first effort in the exploratory stages of analysis.

As a particular example of the argument above, the estimation of density and spectral density functions in economics and finance requires the recognition of the presence of spatial inhomogeneities. Further, these inhomogeneities are of theoretical and policy importance, so that one wants to do much more than merely note their presence. One will want to describe the inhomogeneity effects in detail. For example, the effects of minimum-wage legislation are debated frequently and at length. In order to assess the effects of minimum-wage legislation on the income distribution one requires to be able to isolate and describe analytically the spatial inhomogeneity that such legislation creates. Other inhomogeneities are created by tax legislation, by rigidities in trading rules on exchanges and commodity markets and perhaps by the process of innovation itself.

3.3   Time-scale decomposition

I now come to one of the most promising opportunities for the use of wavelets in economics, decomposition by time-scale. I demonstrated above how insights were gained and the resolution of apparent statistical anomalies were removed by recognizing the potential for relationships between variables to be at the scale level, not the aggregate level. That is, one should recognize that the relationship between consumption and income varies depending on the time-scale that is involved. Much work remains to be done in this context. Gains can be made if interest is restricted to a specific scale by using non-decimated wavelets, but if one wants to observe the data at a sequence of alternative scales and to preserve orthogonality, or at least near orthogonality, of the time-scale decompositions, alternative formulations will be needed. One thought is to pursue the work of Chui *et al.*(1995), in which expansions are in terms of the scaling functions

$$\left.\begin{aligned} \phi_k^{j,n}(t) &= (2^j \alpha_n)^{1/2} \phi(2^j \alpha_n t - k), \\ \psi_k^{j,n}(t) &= (2^j \alpha_n)^{1/2} \psi(2^j \alpha_n t - k), \end{aligned}\right\} \tag{3.1}$$

and

$$\alpha_n = \frac{2^N}{n + 2^N}, \quad N > 0, \quad n = 1, 2, \ldots, 2^N - 1$$

provides $2^N - 1$ additional levels between any two consecutive octave scales. For a given application, there may be a preferred value for α_n that provides the most useful time-scale decomposition.

In this context, the investigation of the potentially time-varying phase relationships between the variables of interest is an important and until now neglected aspect of empirical analysis. Wavelet decompositions are of course uniquely well suited to answering such questions. It is likely that many examples of 'complex relationships between variables' may well on further examination be found to be caused by shifting phase relationships. In short, this discovery has opened up a substantial area for future research.

3.4 Aspects for forecasting

The last and equally important future use of wavelets is in the role of improving forecasts. There are in fact several characteristics of wavelets that may well enhance the forecasting of economic and even financial time-series. First, as we have already seen, gains can be made by decomposing the series to be forecasted into its time-scale components and devising appropriate forecasting strategies for each. In fact, this approach has a long history in forecasting; the wavelet approach has merely formalized old notions of forecasting trend, seasonal and business cycle components. What is of far greater importance, I believe, is that wavelets enable one to isolate the local from the global. In forecasting, one attempts to forecast the permanent components of the series, not the strictly local events. In so far as wavelets enable one to separate out the two effects, one will be able to improve forecasts by eliminating the contamination from strictly local non-recurrent events.

A particular suggestion for the enhancement of time-series forecasts in the context of wavelet analysis is to capitalize on the time-scale decomposition by devising 'test' strategies to check for variations in the structure at the time-scale analysed. At the level of the overall trend using the father wavelet, one might well speculate that if there is to be a structural change away from the predicted long-term path, that change can be characterized initially in terms of an exponential growth path either above or below the current forecast path. Consequently, one can devise a specific test strategy to test for exponential deviations away from the current path. In principle this strategy should provide a more powerful test for deviations from the trend and provide an 'earlier warning' system than is currently possible. For the higher-frequency oscillations, techniques based on detecting changes in frequency as illustrated by the Doppler example of Donoho & Johnstone (1995) is a likely beginning. In any event, one major task facing all such suggestions is to evaluate the distributional properties of the recombined forecast estimates from the individual time-scale forecasts.

The wavelet approach also facilitates an important, but frequently neglected, aspect of forecasting; the limits to forecasting. Many of us forget that not all series can be forecasted, even in principle, and that often there may well be severe limitations on the extent of the forecast horizon. Using wavelets we can resolve these issues at each scale level. It is likely, for example, at the lowest detail level that noise will predominate so that there is no opportunity for forecasting at all. At a higher level we may discover that the maximum period of forecastability is commensurate with the time-scale and not beyond. At even higher scales we may recognize that we need to separate the long-term behaviour from the episodic occurrences of local phenomena. In short, the task of forecasting is considerably more complex than has been generally recognized.

4   Conclusion

In conclusion, it is apparent that wavelets are particularly well adapted to the vagaries of the statistical analysis of economic and financial data. Only the surface has been explored so far and the future indicates that some exciting and revealing analysis will be generated by the application of wavelets. The potential benefits are far greater than the mere application of new techniques. As I have tried to indicate, wavelet-based analysis should lead to new insights into, and novel theories about, economic and financial phenomena.

5   References

Aussem, A. & Murtagh, F. 1997 Combining neural network forecasts on wavelet-transformed time series. *Connection Sci.* **9**, 113–121.

Aussem, A., Campbell, J. & Murtagh, F. 1998 Wavelet-based feature extraction and decomposition strategies for financial forecasting. *J. Computat. Intell. Finance* **6**(2), 5–12.

Bollerslev, T., Chou, R. Y., Jayaraman, J. & Kroner, K. F. 1990 ARCH modeling in finance: a review of the theory and empirical evidence. *J. Econometrics* **52**, 5–60.

Chui, C. K., Goswami, J. C. & Chan, A. K. 1995 Fast integral wavelet transform on a dense set of the time-scale domain. *Numerische Mathematik* **70**, 283–302.

Davidson, R., Labys, W. C. & Lesourd, J.-B. 1997 Wavelet analysis of commodity price behavior. *Computat. Economics* **11**, 103–128.

Donoho, D. L. & Johnstone, I. M. 1995 Adapting to unknown smoothness via wavelet shrinkage. *J. Am. Statist. Ass.* **90**, 1200–1224.

Donoho, D. L. & Johnstone, I. M. 1998 Minimax estimation by wavelet shrinkage. *Ann. Statist.* **26**, 879–921.

Donoho, D., Johnstone, I., Kerkyacharian, G. & Picard, D. 1995 Wavelet shrinkage: asymptopia? (with discussion). *J. R. Statist. Soc.* **57**, 301–369.

Härdle, W., Kerkyacharian, G., Picard, D. & Tsybakov, A. 1998 *Wavelets, approximation, and statistical applications.* Springer.

Jensen, M. J. 1997*a* An alternative maximum likelihood estimator of long-memory processes using compactly supported wavelets. Working paper no. 97-05. Department of Economics, University of Missouri–Columbia.

Jensen, M. J. 1997*b* Using wavelets to obtain a consistent ordinary least squares estimator of the long memory parameter. Working paper no. 97-09. Department of Economics, University of Missouri–Columbia.

Johnstone, I. M. & Silverman, B. W. 1997 Wavelet threshold estimators for data with correlated noise. *J. R. Statist. Soc.* B **59**, 319–351.

Mallat, S. & Zhang, Z. 1993 Matching pursuit with time-frequency dictionaries. *IEEE Trans. Signal Process.* **41**, 3397–3415.

Morehart, M., Murtagh, F. & Starck, J.-L. 1999 Spatial representation of economic and financial measures used in agriculture via wavelet analysis. *Int. Jl Geographic Information Science.* **13**(6), 557–575.

Neumann, M. H. 1996 Spectral density estimation via nonlinear wavelet methods for stationary non-Gaussian time series. *J. Time Series Analysis* **17**, 601–633.

Pan, Z. & Wang, X. 1998 A stochastic nonlinear regression estimator using wavelets. *Computat. Economics* **11**, 89–102.

Ramsey, J. B. 1969 Tests for specification errors in classical linear least-squares regression analysis. *J. R. Statist. Soc.* B **31**, 350–317.

Ramsey, J. B. & Lampart, C. 1998*a* Decomposition of economic relationships by time scale using wavelets: money and income. *Macroecon. Dynam.* **2**, 49–71.

Ramsey, J. B. & Lampart, C. 1998*b* The decomposition of economic relationships by time scale using wavelets: expenditure and income. *Stud. Nonlinear Dynam. Econometrics* **3**, 23–42.

Ramsey, J. B. & Schmidt, P. 1976 Some further results on the use of OLS and BLUS residuals in specification error tests. *J. Am. Statist. Ass.* **71**, 389–390.

Ramsey, J. B. & Thomson, D. J. 1998 A reanalysis of the spectral properties of some economic and financial time series. In *Nonlinear time series analysis of economic and financial data* (ed. P. Rothman), pp. 45–85. Boston, MA: Kluwer.

Ramsey, J. B. & Zhang, Z. 1996 The application of waveform dictionaries to stock market data. *Predictability of complex dynamical systems* (ed. Y. A. Kravtsov & J. B. Kadtke), pp. 189–205. Springer.

Ramsey, J. B. & Zhang, Z. 1997 The analysis of foreign exchange rate data using waveform dictionaries. *J. Empirical Finance* **4**, 341–372.

Ramsey, J. B., Usikov, D. & Zaslavsky, G. 1995 An analysis of U.S. stock price behavior using wavelets. *Fractals* **3**, 377–389.

White, H. 1989 Some asymptotic results for learning in single hidden-layer feed-forward models. *J. Am. Statist. Ass.* **84**, 1003–1013.

13

Harmonic wavelets in vibrations and acoustics

David E. Newland

*Department of Engineering, University of Cambridge, Trumpington Street,
Cambridge CB2 1PZ, UK*
den@eng.cam.ac.uk

Abstract

Four practical examples from mechanical engineering illustrate how wavelet theory
has improved procedures for the spectral analysis of transient signals. New
wavelet-based algorithms generate better time–frequency maps, which trace how
the spectral content of a signal changes with time. The methods are applicable to
multi-channel data, and time-varying cross-spectra can be computed efficiently.

Keywords: wavelet; harmonic; transient; time–frequency; cross-spectrum

1 Introduction

The author is interested in measuring and characterizing the dynamical behaviour
of materials and structures. New techniques for the interpretation of transient
and intermittent data, using wavelets, allow system and excitation properties to be
deduced from measured data with more precision and greater speed than before.

The applications described in this paper come from the field of mechanical
engineering. Four examples will be considered, three giving results that have not
been published before. They involve the analysis of transient vibration signals.
The practical objective is to extract as much information as possible from measured
results. The signals may be short in duration because the phenomenon they represent
happens quickly. Or their characteristics may change with time because of changes
in the signals' underlying physical cause.

One example is the analysis of vibration data recorded during the transmission
of bending waves in a beam subjected to impact loading. This is an essentially
intermittent phenomenon as wave reflections occur and energy is transmitted
backwards and forwards along the beam. Another similar, but more complicated,
example uses data for pressure fluctuations recorded in an acoustic waveguide.
Here there are many different waves interfering with each other. A third example
uses data for ground vibration recorded near an underground train in London.
Disturbance from the rumble of underground trains is becoming increasingly
intrusive but it is very hard to predict. Finally, the fourth example computes
time-varying cross-spectra for multi-channel measurements of soil vibration in a

centrifuge test designed to model earthquake response. Simultaneously measured acceleration signals at different points allow the changing soil properties that occur under dynamic loading to be explored. The first two examples are laboratory demonstrations used as student experiments in the author's department. The second two examples are taken from research in progress at Cambridge for which wavelet analysis now provides an investigative tool of considerable importance.

2   Harmonic wavelet theory

The theory of harmonic wavelets has been described in previous papers by the author (Newland 1993, 1994*a*, *b*, 1998, 1999). In their simplest form, orthogonal harmonic wavelets provide a complete set of complex exponential functions whose spectrum is confined to adjacent (non-overlapping) bands of frequency. Their real part is an even function which is identical with the Shannon wavelet. Their imaginary part is a similar but odd function. Their equal spacing along the time axis is twice that of the corresponding set of Shannon wavelets. In their practical application, the boxcar spectrum of harmonic wavelets is smoothed (to improve localization in the time domain) and the spectra of adjacent wavelet levels are overlapped to give oversampling in order to improve time–frequency map definition.

These wavelets have been found to be particularly suitable for vibration and acoustic analysis because their harmonic structure is similar to naturally occurring signal structures and therefore they correlate well with experimental signals. They can also be computed by a numerically efficient algorithm based on the fast Fourier transform (FFT).

For time–frequency mapping, there are similarities between the harmonic wavelet transform (HWT) and the short-time Fourier transform (STFT). The advantage of the HWT over the STFT is that the HWT is a computationally efficient variable-bandwidth transform. Therefore, the time–frequency map it generates can have a variable-bandwidth basis, with the analysing wavelet's bandwidth altered from one frequency to another to suit the problem being studied. In contrast, a time–frequency map constructed by the STFT always has a constant-bandwidth basis, giving the same frequency resolution at high frequencies as it gives at low frequencies. This means that the STFT is less flexible and may lead to a requirement for (much) more computation than is required by the harmonic wavelet transform. A detailed discussion of the merits of the two related methods is given in Newland (1998).

3   Numerical algorithm

A practical algorithm for time–frequency analysis is illustrated diagrammatically in figure 1 (Newland 1998). The correlation calculation at the heart of the wavelet

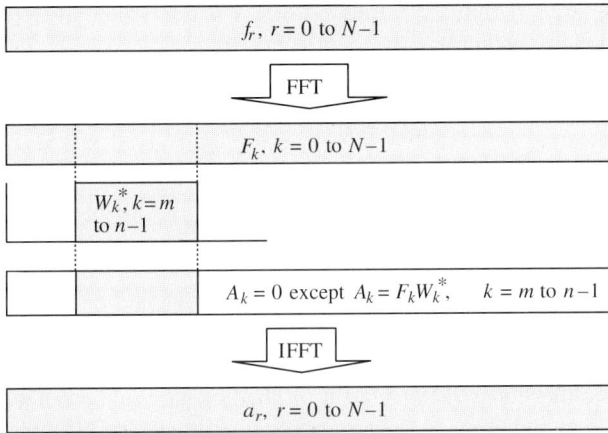

Fig. 1: FFT algorithm to compute harmonic wavelet coefficients for wavelets in the frequency band $m2\pi \leq \omega < n2\pi$ (for a record of unit length).

method is carried out in the frequency domain, where it becomes a multiplication operation rather than a convolution. The input signal $f(t)$ is represented by the N-term series $f_0, f_1, f_2, \ldots, f_{N-1}$ in the top box. It is transformed from time to frequency by the FFT to give the Fourier coefficients $F_0, F_1, F_2, \ldots, F_{N-1}$ in the second box. The record length is assumed to be unit time, so that the sampling interval is $1/N$ and the Nyquist frequency $N\pi$. By definition, the (complex) harmonic wavelet has a Fourier transform W_k, $k = m$ to $n-1$, which is zero everywhere except in the range

$$m2\pi \leq \omega < n2\pi, \tag{3.1}$$

where $m < n$ (Newland 1994b). Because the wavelets are complex, their Fourier transform is one-sided, so that W_k remains zero for all negative frequencies.

On carrying out the multiplication operation, the F_k and W_k terms are multiplied to generate the new series

$$A_k = F_k W_k^*, \qquad k = 0 \text{ to } N-1, \tag{3.2}$$

in the next box in figure 1. These A_k are the Fourier transforms of the wavelet coefficients. Computing their inverse (IFFT) reverts to the time domain to give the series of wavelet coefficients a_r, $r = 0$ to $N-1$, which are shown in the bottom box.

The time-scale runs from $0 \leq t < 1$ and $a_r = a(t = r/N)$ gives the result of calculating the wavelet coefficient for the wavelet centred at the chosen position $t = r/N$ on the time axis. The usual circularity property of the discrete Fourier transform method applies and when a wavelet runs off the end of the unit time-scale, it wraps round and reappears at the opposite end.

The computation in figure 1 therefore gives N wavelet coefficients for reference wavelets in all the possible N positions along the time axis. Only $n - m$ of these are needed to form an orthogonal set (Newland 1994b), and usually less than the (large) number N is needed to produce adequate resolution along the time axis. This is achieved by selecting $N1$ equally spaced values from the total available. If $N1$ is not a factor of N, appropriate methods of interpolation can be used. An efficient method of doing this is very important. The method used here is described in Newland (1999).

Instead of computing the IFFT of the N-term series A_k, $k = 0$ to $N - 1$, in the lowest but one box in figure 1, this interpolation method computes the IFFT of a shorter $N1$-term series B_k, $k = 0$ to $N1 - 1$, whose first $n - m$ terms are the non-zero A_k, $k = m$ to $n - 1$, and whose remaining terms are all zeros. It is shown in the reference that this generates a set of coefficients b_s, $s = 0$ to $N1 - 1$ that correspond to selected terms in the longer series a_r, $r = 0$ to $N - 1$, provided that $sN/N1$ is an integer. If it is not, then the b_s interpolate between the nearest two values of a_r. The magnitudes of corresponding terms are the same. Therefore, a time–frequency amplitude map drawn by computing the shortened $N1$-term series b_s defined above will faithfully represent an amplitude map computed from the full-length N-term series a_r. The phase angles of corresponding terms will generally be different according to (Newland 1999)

$$b_s = \exp(-e2\pi ms/N1)a_{r=sN/N1}, \qquad s = 0 \text{ to } N1 - 1, \qquad (3.3)$$

but allowance can be made for these differences.

The centre frequency of the wavelet Fourier transform in figure 1 is $(m + n)\pi$ and its bandwidth is $(n - m)2\pi$. By changing the centre frequency, or bandwidth, or both, and repeating the calculation, a new series of wavelet coefficients, a_j, $j = 0$ to $N - 1$, is generated. If this process is carried out for $N4$ different centre frequencies and each output series a_j downsampled to give $N1$ terms, the resulting $N4 \times N1$ array $A(N4, N1)$ is generated. This array is used to draw time–frequency maps to show how the amplitude and phase of the wavelet coefficients change over time and frequency. In the author's programs using these principles, the parameters $N1$ and $N4$ have the above meaning. Wavelet bandwidth is allowed to change linearly from $n - m = N2$ to $n - m = N3$ over the full frequency range of the calculation.

The algorithm in figure 1 applies for all harmonic wavelets, namely wavelets defined in the frequency domain with a compact spectrum such that $W(\omega) = 0$ outside a defined (generally narrow) band of frequencies. This no longer defines an orthogonal family of wavelets, but since reconstruction of the signal being analysed is not required, that does not matter. For the results given below, the boxcar spectrum of orthogonal harmonic wavelets has been windowed by a Hanning function, so that the function in the third box in figure 1 is given by

$$W_k = \frac{1}{2\pi(n - m)}\left(1 - \cos\frac{2\pi(k - m)}{n - m}\right), \qquad m \leq k < n. \qquad (3.4)$$

This has been found to give good localization in the time domain.

4 Phase interpretation

This calculation procedure generates complex wavelet coefficients, a_r (figure 1). Their phase depends on the relative position of the signal and its analysing wavelet. This defines the ratio of the imaginary part of a_r (correlation with the odd part of the harmonic wavelet) to that of the real part of a_r (correlation with the even harmonic wavelet). When, for a constant harmonic signal, the wavelet is moved to a new position, its phase will be different. Therefore, absolute phase is not a useful indicator because it depends on wavelet location. But phase gradient, defined as the rate of change of phase with time for wavelets in the same frequency band, is an interesting parameter because it is constant when $f(t)$ is a harmonic of fixed frequency and phase. It is shown in Newland (1999) that, for a single harmonic of frequency ω_0, the phase gradient

$$\frac{\partial \phi}{\partial t} = \omega_0 - \Omega + \pi B, \qquad \Omega - \pi B \le \omega_0 < \Omega + \pi B, \qquad (4.1)$$

where Ω is the centre frequency and $2\pi B$ the bandwidth of the analysing wavelets.

The essential property is that the rate of change of phase with time is constant for a harmonic of fixed frequency and phase so that a two-dimensional map of phase gradient, with $\partial\phi/\partial t$ on a frequency–time base, is sensitive to phase changes in the signal being analysed. This will be illustrated in one of the examples of wave propagation given below, for which sudden changes in phase gradient occur in between successive reflections of energy in local frequency bands.

A different example in which absolute phase can be used helpfully is the last example in Section 11, which is the analysis of simultaneous multi-channel recordings of ground movement after shock loading. Corresponding time records are analysed by the same wavelet arrays, with the wavelets in the same time location for the different channels. This enables differences in the phase of the channels to be detected and mapped as a function of time and position. It will be shown how changes in system properties (caused, for example, by soil slippage) can be detected by harmonic wavelet analysis from corresponding changes in the measured phase response.

5 Bending wave transmission in a beam

The first example comes from a laboratory experiment in the Department of Engineering at the University of Cambridge. The experiment illustrates bending wave propagation in a thin steel beam. The beam is suspended on light cords with its long axis vertical and is hit gently at one end by a soft-ended impulse hammer. This generates lateral bending waves which travel to the other (far) end of the beam, where they are reflected and return to the point of impact, before undergoing successive reflections until eventually they are dissipated by damping after several seconds. A small accelerometer is mounted on the beam so as to detect lateral vibration. For

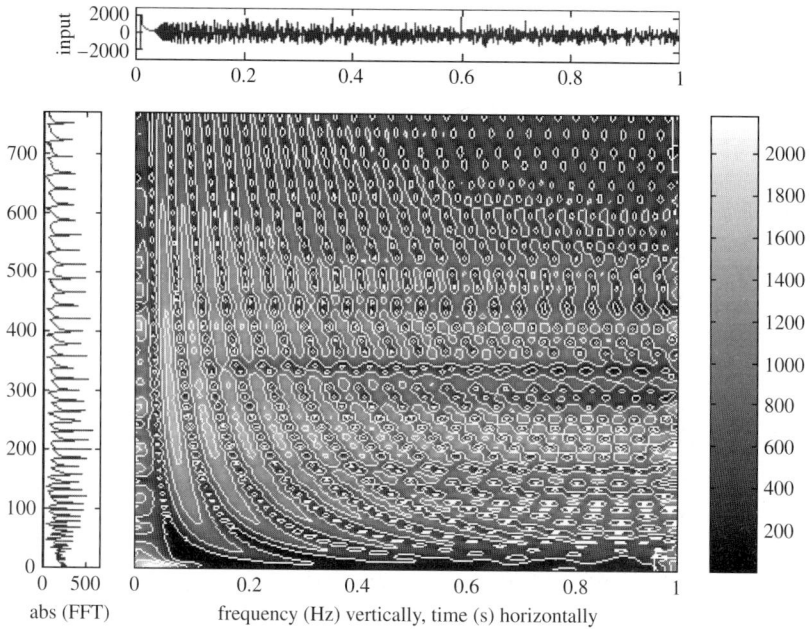

Fig. 2: Amplitude time–frequency map for the bending vibration of a thin steel beam subjected to impact loading at one end. The acceleration response is measured close to the point of impact. Top view, time history; left, spectrum; right, density scale for wavelet amplitude (all arbitrary units).

the results shown below, the accelerometer is positioned close to the first end of the beam, near its point of impact. The beam is 7.2 m long and has a rectangular cross-section 32.1 × 6.3 mm^2. The impulse hammer had a soft tip designed so that only low-frequency vibrations were generated (up to *ca.* 1 kHz). The sampling frequency was 4096 Hz.

Because the group velocity of bending waves depends on frequency (velocity proportional to frequency$^{1/2}$), groups of high frequency waves travel faster than low-frequency waves. Therefore a time–frequency map should show more frequent reflections for high frequencies than for low frequencies. This behaviour is, of course, not at all evident from the recorded time-domain response, which is shown for one second duration in the top view in figure 2, or from the spectrum which is drawn at the left-hand side of figure 2 with the frequency scale running from 0 to 750 Hz, approximately. A short length of this signal has been included in Newland (1998), but the full record has not been considered previously.

The map in figure 2 is a contour map of the three-dimensional surface obtained by plotting the magnitude of the wavelet coefficients against time and frequency. To

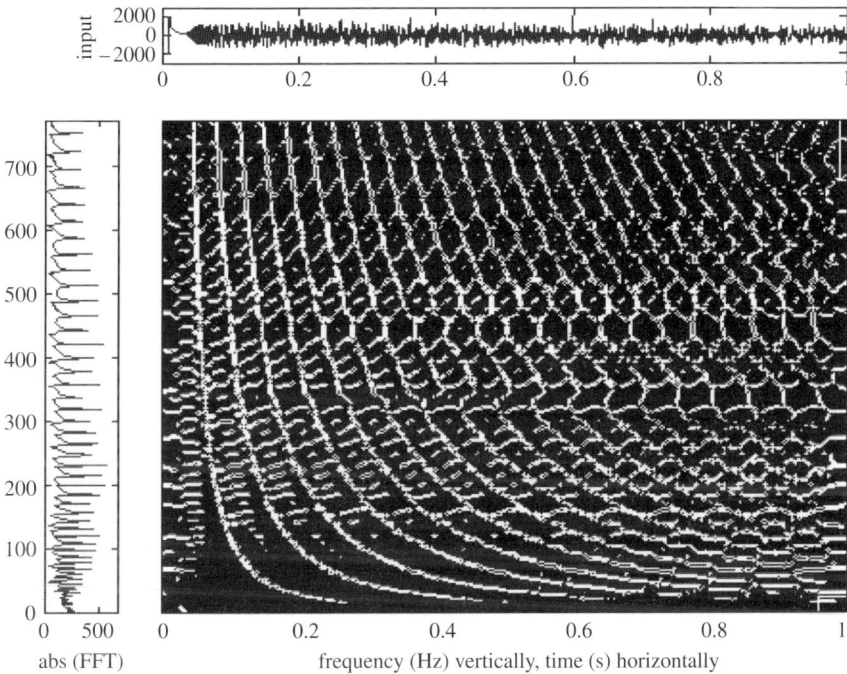

Fig. 3: Ridges of the amplitude time–frequency map in figure 2.

plot the diagram, the bandwidth of the harmonic wavelets has to be chosen and, in the example given, bandwidth is changed in proportion to centre frequency. That chosen is indicated on the map by the small rectangular 'tile' in the top and bottom right corners. The height of this tile shows the bandwidth B Hz of the analysing wavelet at that frequency; the width of the tile shows the (mean-square) width of the wavelet T s, satisfying the uncertainty limit $BT = 1$.

As time passes, the regular pattern of curved ridges in figure 2 is interrupted by some transverse 'valleys' that run from left to right in the figure. These appear to be caused by non-bending modes into which vibrational energy 'leaks' as the wave propagation process continues. Within the frequency range of figure 2, there are about 50 bending modes which are excited. There are also about nine twisting modes and two longitudinal modes whose frequencies lie in range. Some of these may be unintentionally excited by the impulsive input being slightly away from the geometrical centre of the beam or slightly off-line in direction, or they may be coupled to the bending modes, for example by the action of the supporting elastic cords or the mass loading of the accelerometer.

When a harmonic wavelet with a narrower bandwidth is used for the analysis, reduced definition along the time axis is achieved; for a harmonic wavelet with a wider bandwidth, reduced definition along the frequency axis is obtained (specific examples are given in Newland (1998)). The 'optimum' is determined by trial and error, guided by the shape of the uncertainty tile.

The unavoidable smearing of spectral features that occurs in figure 2 can be reduced by plotting only the ridges of the three-dimensional surface whose contours generate the figure. The exact identification of ridges is difficult (Eberly 1996) and identifying their precise position is complicated. The approach used by the author (Newland 1999) is to seek the height maxima of sections cut in the direction of the (smoothed) surface's greatest curvature. When this strategy is applied to the surface plotted in figure 2, the result is that shown in figure 3. Each ridge marks the arrival at the measurement point of successive groups of bending waves. At high frequencies the group velocity is higher, so successive reflections arrive more quickly than at low frequencies, when the ridges are further apart. Knowing the length of the beam, by measuring the time between successive reflections, the group velocity can be estimated as a function of frequency.

6   Response of an acoustic waveguide

A similar, but more complicated, example is provided by the reflection of pressure waves within an acoustic waveguide. This is also a laboratory experiment at Cambridge. Internal air pressure perturbations are generated in a closed circular duct of approximately 12 m length and 0.75 m diameter. These perturbations are caused by a pulse-like electrical input to a small loudspeaker mounted near the edge of one of the rigid ends of the duct. This excites several different families of acoustic waves, which travel backwards and forwards within the duct. A microphone mounted at the centre of the end with the loudspeaker records the resulting pressure fluctuations and this signal has been used to generate the diagrams in figures 4 and 5.

Figure 4 shows the ridges of an amplitude time–frequency map, computed as described above. In addition to the main ridges, there are numerous small, generally horizontal ridges which arise from local fluctuations in surface height. They can be eliminated by introducing more smoothing before ridge detection and it is a matter of judgement to generate the ridge map which is the 'best' for a required purpose. As for figure 3, figure 4 has the input time history shown for comparison along the top, and the modulus of this signal's Fourier transform plotted along the left-hand side (using arbitrary units). For convenient scaling, the square root of the Fourier transform is plotted.

The underlying physical processes represented in this map are quite complicated. Axial, plane waves travel at constant velocity (independent of frequency) and bursts of energy from these plane waves arrive periodically at the microphone. They show as equally spaced vertical lines in figure 4. Knowing the dimensions of the duct and the acoustic properties of air, the position of these lines can be calculated and their theoretical position is superimposed on their experimental position in figure 4. Within the frequency range of these maps there are two other families of non-axial

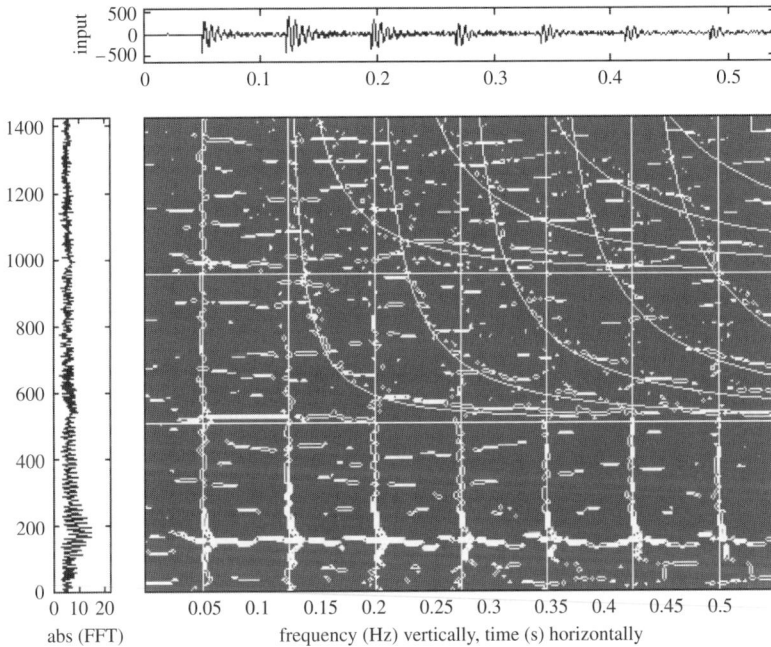

Fig. 4: Time–frequency ridge map of acoustic reflections in a closed duct: comparison of theory and experiment (from Newland 1999).

waves which are dispersive (their group velocity depends on frequency). Their passage time between reflections is given by

$$T(\omega) = \frac{2L}{c\sqrt{1 - \Omega_0^2/\omega^2}},\qquad(6.1)$$

where $L = 12.16$ m is the length of the duct, $c = 334$ m s^{-1} is the speed of sound, ω is the wave frequency, and Ω_0 is the cut-off frequency. For plane waves, the cut-off frequency is zero, and for the first two families of non-axial waves which are detected by a microphone at the centre of the duct it is $\Omega_0 = 3.83c/a$ and $7.02c/a$, where $a = 0.386$ m is the duct's radius (see, for example, Skudrzyk 1971, p. 431). These results have been used to plot the theoretical lines on figure 4, measuring time forward in steps of $T(\omega)$ from the instant of impulsive excitation.

The calculated cut-off frequencies are plotted as the horizontal lines in figure 4. Successive reflections of the dispersive waves appear as the two families of curved lines that are asymptotic to the cut-off lines. The horizontal ridge at *ca.* 200 Hz is due to ringing of the loudspeaker's diaphragm and is not associated with the travelling wave acoustic phenomena.

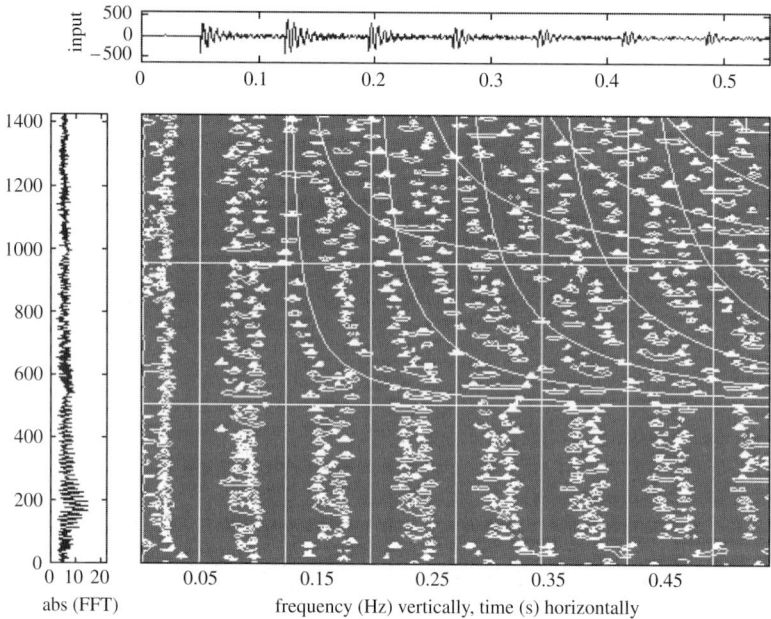

Fig. 5: Time–frequency differential phase map corresponding to figure 4 (from Newland 1999).

Figure 5 shows a differential phase map of the same data. It is a contour plot of the three-dimensional surface obtained by plotting the modulus of phase gradient on a base of frequency versus time. It can be seen that phase perturbations occur generally between ridge positions indicating phase changes at every reflection of the travelling wave energy. A characteristic of this presentation is that the vertical distribution (distribution over frequency) of the phase perturbations is correlated with the position of the peaks in the Fourier transform of the input signal (plotted along the left-hand side). By placing a straight edge across a similar diagram drawn to larger scale, it can be shown that the phase perturbations align quite closely with the positions of the troughs in the spectral data plotted on the left-hand side (Newland 1999). This is not evident so clearly in corresponding graphs of wavelet amplitude.

7   Underground train vibration

The transmission and attenuation of ground-borne vibration is an extremely difficult computational problem because of the geometric complexity of ground and building structures and because of lack of understanding of the dynamic response properties of soils and foundations (Newland & Hunt 1991, 1992, 1996). For example, for the

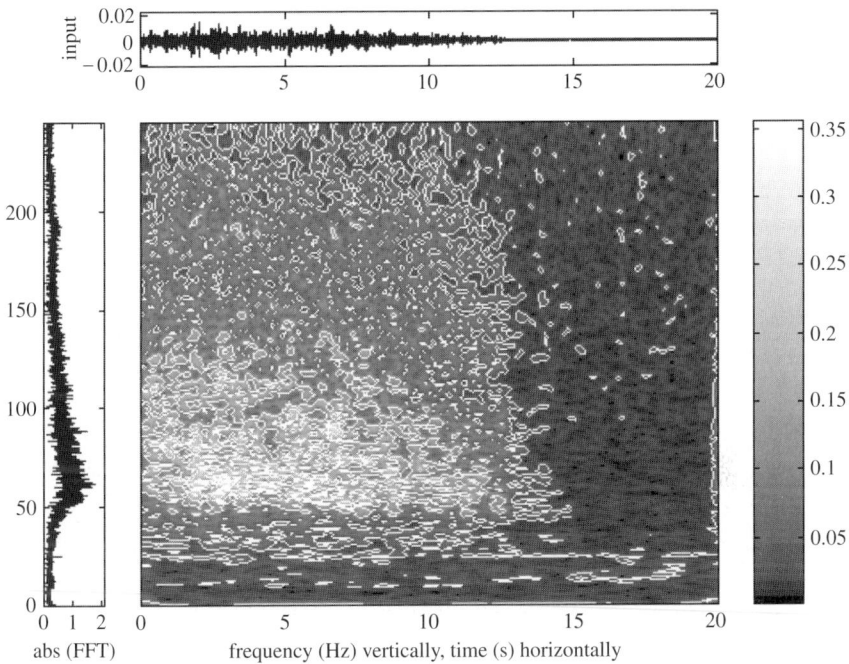

Fig. 6: Time–frequency analysis of the recorded ground acceleration during the passage of a train on a nearby underground railway line.

new Folkestone to London high-speed railway line, there are currently no agreed protocols to compute groundside vibration or the level of anticipated disturbance in local buildings, old or new. This is a serious, and currently unsolved, design problem.

Recently, ground acceleration data have been measured near to a curved section of the Piccadilly underground line in London. For the results given here, an accelerometer was secured to a stone step in an adjacent building and 20 s of the passage of a train recorded. The time history of this process is shown in the upper view in figure 6, the units being gs. The recording begins with the train already passing, and continues until it has passed out of hearing in 20 s.

Figure 6 shows a harmonic wavelet amplitude map for this vibration, covering the frequency range from 0 to 250 Hz (half the Nyquist frequency of 500 Hz). It is evident that there is a broadband response as the train passes, with ground vibrational energy in a wide range of frequencies as a result of wheel and rail surface irregularities, wheel flange-to-rail contact, mechanical train noise and electrical collector noise.

Examination of the higher-frequency content of the recorded signal for the first 5 s of the data (figure 7) shows a marked local variation in vibrational intensity. This

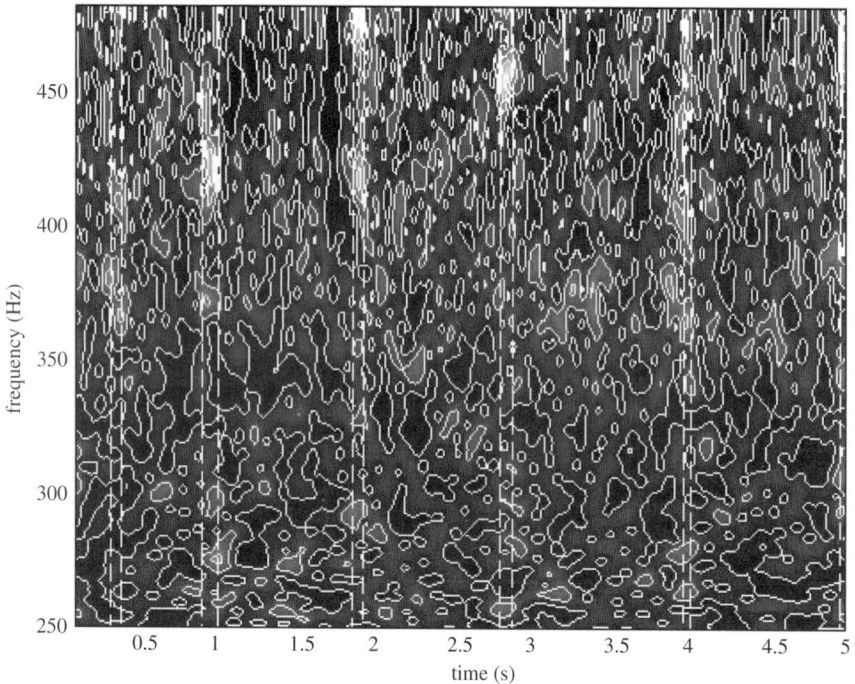

Fig. 7: Analysis of part of the same data as in figure 6, with segmentation markers to identify locally high-amplitude response.

is apparent from the mean-square graph plotted in figure 8 (bottom), which is the energy represented by the appropriate summation of wavelet amplitudes squared, for the frequency band of figure 7 (only). Above it in figure 8, the amplitude discriminator $d(j)$ is plotted, where $j = 1$ to 300 is the index of columns in figure 7. This function is a measure of the amplitude difference of all the wavelet coefficients (in the frequency band considered) between one column of the array plotted in figure 7 and the immediately adjacent column (Tait & Findlay 1996; Newland 1998). The segmentation markers in figures 7 and 8 are chosen to coincide with local peaks in $d(j)$ and provide a means of identifying the extent of the mean-square peaks in the bottom view in figure 8. These local peaks of high-intensity vibration appear to result from intermittent impact between wheels and rails. It can be seen from the top view in figure 8 that there is a degree of arbitrariness in the selection of the appropriate peaks of the amplitude discriminator $d(j)$ which denote a sudden change in vibrational spectral content. The likely explanation is that the onset and termination of the wheel flange to rail interaction process is variable and that the response to this process is confused by the vibration generated by other sources, in particular by rail joints and irregularities.

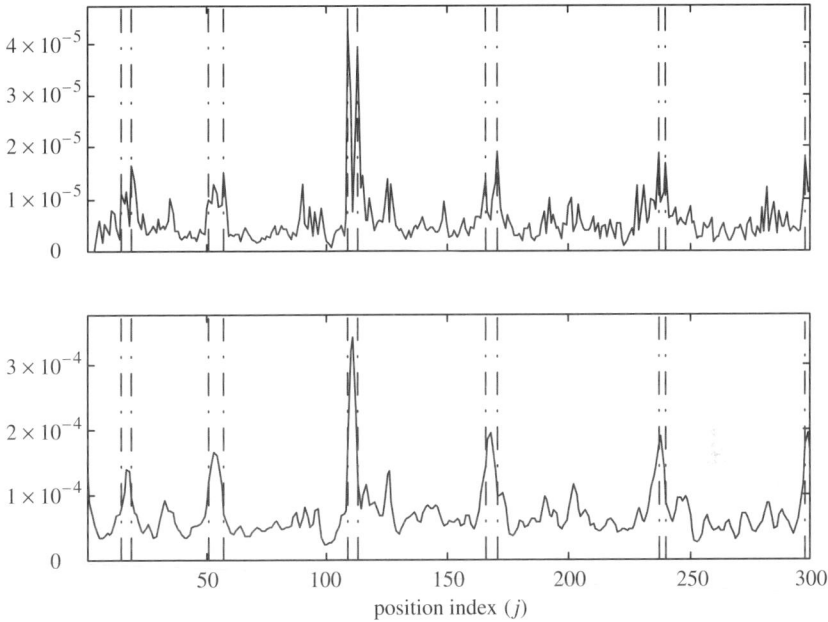

Fig. 8: Bottom: mean-square response for the frequency band shown in figure 7. Top: amplitude discriminator $d(j) = \sum_k (|a(k,j)| - |a(k, j-1)|)^2$, where $a(k,j)$ is an element in the two-dimensional array plotted in figure 7. The index j runs from 1 to 300 because the map in figure 7 is plotted from an array which has 300 columns.

8   Geotechnical centrifuge testing

Fundamental knowledge of the (large amplitude) dynamic behaviour of soil under earthquake excitation is meagre. Studies at Cambridge in our Geotechnical Centrifuge Centre and elsewhere (Lee & Schofield 1988; Taylor 1995; Butler 1999; and others) have obtained good data on the transient vibration of soil models under earthquake conditions. The levels of excitation cause large deflection intergranular movements which lead to so-called soil liquefaction effects when the soil's response is closer to that of a fluid than a solid. Because excitation lasts only for a second or two with excitation frequencies ranging up to *ca.* 200 Hz, data analysis can only be done if there are good methods of transient vibration analysis. Wavelet methods make this possible and good preliminary results have already been achieved using harmonic wavelets (Newland & Butler 1998). New research is concentrating on developing these methods to estimate time-varying cross-spectra between adjacent measuring points. This is seen as a very important area of further theoretical and experimental research.

9 Experimental data

The test data used below are those published in Newland & Butler (1998) and were obtained from geotechnical centrifuge tests. The experimental system represented a saturated sand model poured at two relative densities and mounted within a flexible container. The container is shown in figure 9. It is rectangular in shape with its side walls made of a series of flat rings, each mounted to the next by a rubber gasket. The intention is that the loaded container functions as an equivalent shear beam, whose shear modulus matches approximately that of the enclosed soil medium. Sand was poured at a density of 1576 kg m^{-3} in the lower 160 mm of the container and at 1670 kg m^{-3} for the remaining 365 mm to the top of the container.

The container and its contents were centrifuged to apply an acceleration vertically downwards (in figure 9) of $50g$ in order to simulate the response of a large ground volume in a small model. Horizontal shear excitation to the base of the container was supplied by a device called a stored angular momentum actuator. This consists of a flywheel which is connected to a reciprocating rack by a clutch assembly. When the clutch is engaged, there is a sudden burst of oscillatory energy which shakes the container and its contents while this is being centrifuged.

Sand movement was detected by miniature piezoelectric accelerometers. As seen in figure 9, these were stacked vertically within the test specimen with four in the bottom layer of sand and ten in the top layer. Previous tests have shown that they have an accuracy of ±5% within the frequency range 20 Hz to 2 kHz. Their natural frequency when embedded in sand is estimated to be *ca.* 4 kHz compared with frequencies of interest up to *ca.* 400 Hz. Each transducer was carefully oriented in the sand to record the resulting horizontal motion (the x-direction in figure 9) within the saturated model.

Data were stored in a digital data-acquisition system developed in Cambridge as part of the centrifuge's instrumentation, from which they are retrieved for detailed computer analysis. The input motion to the base of the model container had a fundamental frequency of 27 Hz with a displacement amplitude of ±1.5 mm. The duration of the shaking excitation was set to 1.2 s.

For purely harmonic movement, these displacements correspond to a lateral acceleration amplitude of *ca.* $4.4g$. However, loose-play and nonlinearities in the mechanism introduce a harmonic content to the excitation, as will be apparent from the measured results below. The measurement points are shown in figure 9. The signals f_1 to f_6 were recorded at the following six locations: f_1 at 7726, f_2 at 7828, f_3 at 7319, f_4 at 7709, f_5 at 12 611 and f_6 at 12 612, but only two of these measurements, f_1 and f_4, are used for the results given below.

10 Power spectral densities

All signal processing computations have been done by the harmonic wavelet method using the algorithm described above. Results are shown as before as two-dimensional

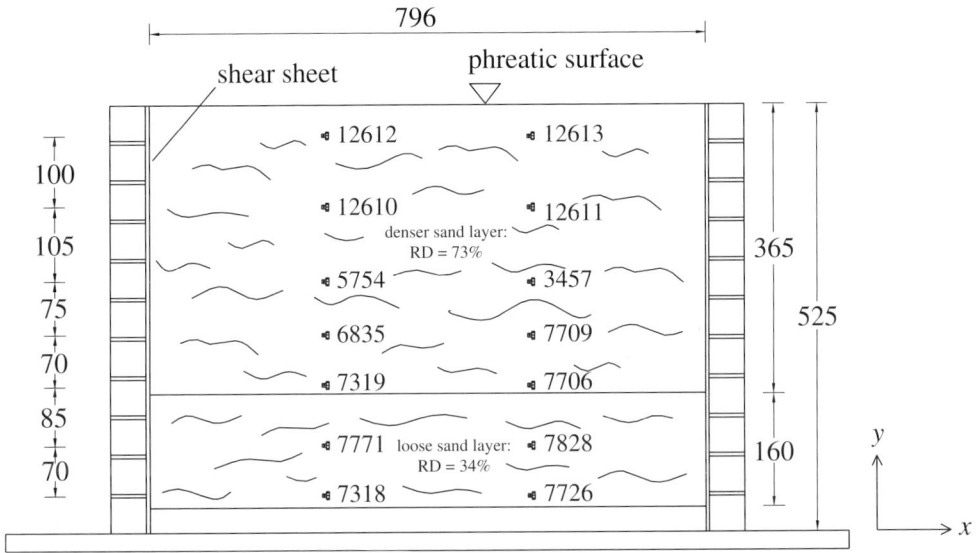

Fig. 9: Instrumentation layout. All accelerometers have their sensitive axis in the *x*-direction (Newland & Butler 1998).

maps of three-dimensional surfaces plotted for the relevant parameter. For ease of identification for a multi-channel system, it is convenient to refer to (i) the power spectral density of a measurement, (ii) the amplitude of the cross-spectral density between two measurements, and (iii) the phase of the cross-spectral density. These terms are not strictly correct because they are defined for stationary random processes, whereas we are concerned with transient and non-stationary processes. However, the amplitude squared of the wavelet coefficient is called a power spectral density since, for an orthogonal set of harmonic wavelets, the mean-square signal is equal to the sum of the (weighted) wavelet amplitudes squared (see, for example, Newland 1993). When the signal is oversampled to generate extra wavelet coefficients, the same analogy may be used. Similarly, the product of two wavelet amplitudes, when computed for the same wavelet at the same instant of time for two signals, represents the amplitude of the cross-spectral density between these signals (for that time instant and frequency band). Also the phase difference between the same two wavelet coefficients gives the phase of the cross-spectral density between these signals.

Time-varying auto-spectral densities calculated this way are plotted for records f_1 and f_4 (positions 7726 and 7709 in figure 9) in figures 10 and 11.

The forced displacement excitation of the base of the model consists of the fundamental component at *ca.* 27 Hz with unavoidable superimposed harmonics of all orders. These can be seen in figure 10. The time history f_1 is plotted along the

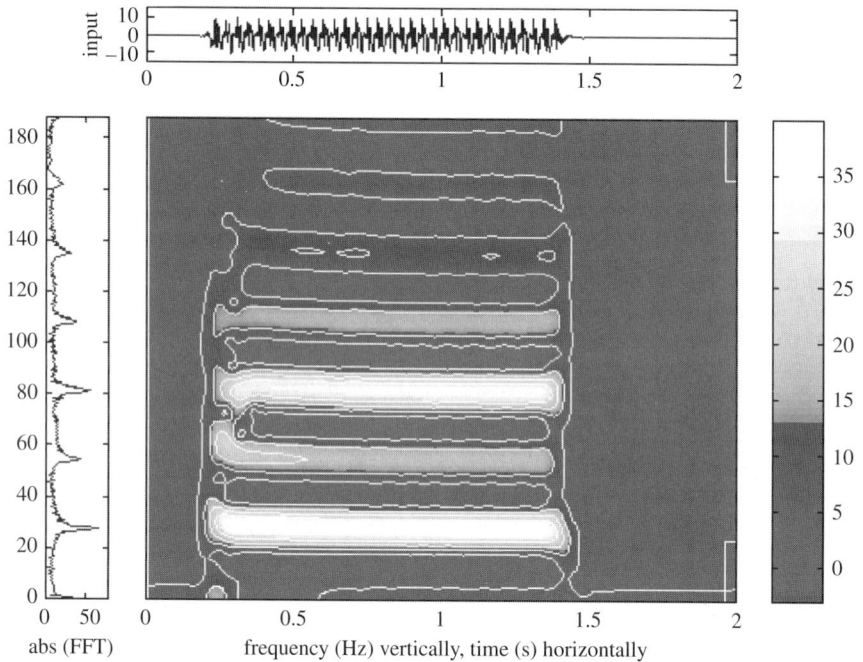

Fig. 10: Experimental data: power spectral density for signal f_1 measured at position 7726 (see figure 9).

top of the map and its power spectral density (also referred to as the auto-spectral density) is plotted along the left-hand side.

The amplitude map in figure 10 shows that the vibration close to the bottom of the box remains approximately constant as shaking continues because the horizontal stripes have approximately constant width and continue for the full duration of the shaking process. In contrast, all the other power spectral densities, for example f_4 in figure 11, show obvious changes with time. This must be due to the changing physical properties of the soil model as a result of its changing dynamic properties with time.

11   Cross-spectral densities

Power spectral density data indicate the total energy in a signal and its distribution over frequency and time. Relative changes in two signals are described by the cross-spectral density. This provides a measure of the local correlation between two signals. The cross-spectral density's amplitude is measured by the product of the wavelet amplitudes of both signals; its phase is measured by the wavelets' relative

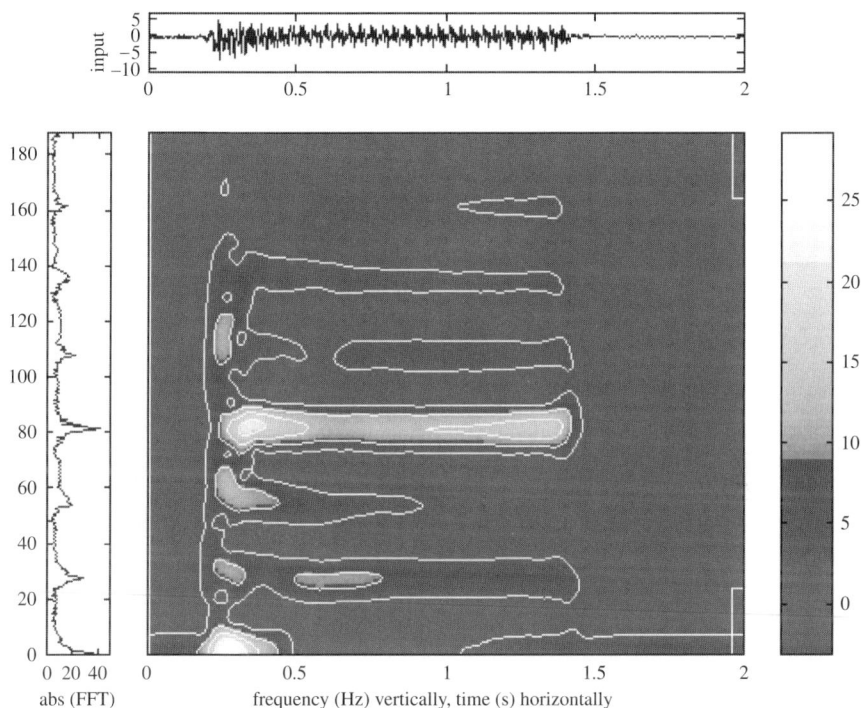

Fig. 11: Experimental data: power spectral density for signal f_4 measured at position 7709 (see figure 9).

phase. As for all time–frequency analysis, all spectral calculations are estimates for the chosen frequency band and time window considered. For the harmonic wavelet method, this is defined by the bandwidth and time duration of the chosen wavelet, which is under the control of the investigator and is indicated by the shape of the rectangular 'tile' shown in the top right and lower right corners of the maps (vertical height equals the bandwidth, horizontal width equals the time duration).

For the experimental data, plate 1 shows the phase of the time-varying cross-spectral density for f_4 with f_1 These results have not been published before, and some explanation of their interpretation is needed. The density legend is shown on the right-hand side of plate 1. It runs from $\pi = 180°$ at one extreme to $-\pi = -180°$ at the other extreme. Before the forced motion has been applied, and after it has finished, there are residual noise signals and these give rise to the haphazard phase represented by 'marbling' on the right- and left-hand sides of the map. During the forced motion, the map has obvious horizontal stripes, each corresponding to one harmonic of the motion (identified on the left-hand side spectrum, which is for f_1). Around the fundamental frequency of 27 Hz, the density of the stripe indicates that

f_4 and f_1 are approximately in phase with each other. The same is true for the second harmonic initially, but as motion continues there is a phase change from approximately zero through minus 90° and then through −180° to approach zero again. For higher harmonics, the density of each stripe changes as time passes, with the transition sometimes being quite sudden.

In between each horizontal stripe there is a thinner marbled stripe where there is uncorrelated response between two harmonics and phase coherence is not maintained.

12   Simulated results

For comparison with the experimental results, the behaviour of a linear model has been simulated. This model consists of a box in three horizontal sections (identified as 1:bottom, 2:middle and 3:top) with parameters chosen to give natural frequencies and damping ratios of 12 Hz, 0.13 and 31.5 Hz, 0.33. The fundamental frequency of the displacement excitation was set at 25 Hz and there are harmonics up to the 8th (200 Hz) included. The acceleration response of the model has been computed by numerically integrating the equations of motion (with the system's parameters constant). To model the experimental system, the deterministic response was supplemented by a low-amplitude random signal to represent noise.

Plate 2 is the phase of the cross-spectral density between f_2 and f_1 for the model (middle and bottom). The random noise causes the 'marbling' and the slight variations in colour density along each horizontal stripe. However, it can be seen from plate 2 that the cross-spectral phases remain approximately constant for each harmonic during the shaking phase.

By comparing plates 1 and 2, it is clear that significant phase changes occur during the duration of shaking in the experimental case (plate 1), which are not duplicated in the simulated comparison (plate 2).

By introducing time-varying parameters into a linear model, it is possible that the experimental results could be simulated. However gradual, progressive changes in parameters would not account for the observed behaviour. It is more likely that nonlinear effects caused by sudden slippages, or liquefaction, or, as a result, abrupt contacts with the side walls of the model container may account for the observed behaviour. To properly simulate these effects is a challenging task!

13   Acknowledgements

The laboratory experiments which generated the beam bending and acoustic duct data were devised by my colleague Dr Jim Woodhouse. I am grateful to him for making the data available for analysis and for numerous discussions about the results. Dr Woodhouse and two of his colleagues first applied time–frequency analysis to similar problems some years ago, using the short-time Fourier transform method

(Hodges *et al.*1985). Dr Hugh Hunt has worked with me on problems of ground vibration transmission and I thank him for providing the measured underground train data included above. We hope that analysis of this and similar data will lead eventually to better means of alleviating traffic noise problems. The work of my colleague, Professor Andrew Schofield, who was responsible for the design and development of the Cambridge geotechnical centrifuge and its derivatives elsewhere in the world, is well known to foundation engineers (Schofield 1980; Schofield & Steedman 1988). I am grateful to Professor Schofield and his PhD student, Gary Butler, for providing the centrifuge data for which the wavelet analysis method has been able to illuminate transient dynamic behaviour in a way that had not previously been possible. Only some illustrative results are given above from the extensive data that are now being analysed by those working in this field.

14   References

Butler, G. D. 1999 A dynamic analysis of the stored energy angular momentum actuator used with the equivalent shear beam container. PhD thesis, Department of Engineering, University of Cambridge, UK.

Eberly, D. 1996 *Ridges in image and data analysis.* Englewood Cliffs, NJ: Prentice Hall.

Hodges, C. H., Power, J. & Woodhouse, J. 1985 The use of the sonogram in structural acoustics and an application to the vibrations of cylindrical shalls. *J. Sound Vib.* **101**, 203–218.

Lee, F. H. & Schofield, A. N. 1988 Centrifuge modelling of sand embankments and islands in earthquakes. *Geotechnique* **38**, 45–58.

Newland, D. E. 1993 Harmonic wavelet analysis. *Proc. R. Soc. Lond.* A **443**, 203–225.

Newland, D. E. 1994*a* Wavelet analysis of vibration. Part 1. Theory. Part II. Wavelet maps. *ASME J. Vib. Acoustics* **116**, 409–425.

Newland, D. E. 1994*b* Harmonic and musical wavelets. *Proc. R. Soc. Lond.* A **444**, 605–620.

Newland, D. E. 1998 Time–frequency and time-scale signal analysis by harmonic wavelets. In *Signal analysis and prediction* (ed. A. Procházka, J. Uhlír, P. J. W. Rayner & N. G. Kingsbury), ch. 1. Boston: Birkhäuser.

Newland, D. E. 1999 Ridge and phase identification in the frequency analysis of transient signals by harmonic wavelets. *ASME J. Vib. Acoustics* **121**, 149–155.

Newland, D. E. & Butler, G. D. 1998 Application of time–frequency analysis to strong motion data with damage. *Proc. 69th Shock and Vibration Symp., Session HB1, SAVIAC (Shock and Vibration Information Analysis Center, US Dept of Defense), Minneapolis.*

Newland, D. E. & Hunt, H. E. M. 1991 Isolation of buildings from ground vibration: a review of recent progress. *Proc. IMechE* C **205**, 39–52.

Newland, D. E. & Hunt, H. E. M. 1992 Isolating buildings from vibration. *Proc. 2nd. Int. Congress on Sound and Vibration, Auburn, USA*, pp. 779–786. AL: Int. Science Publications.

Newland, D. E. & Hunt, H. E. M. 1996 The effect of variable foundation properties on railway-track vibration. *Proc. 4th Int. Congress on Sound and Vibration, St. Petersburg, Russia*, vol. 2, pp. 1065–1072. AL: Int. Science Publications.

Schofield, A. N. 1980 Cambridge geotechnical centrifuge operations. *Geotechnique* **25**, 743–761.

Schofield, A. N. & Steedman, R. S. 1988 Recent development of dynamic model testing in geotechnical engineering. *Proc. 9th World Conf. on Earthquake Engineering, Japan Association for Earthquake Disaster Prevention, Kyoto–Tokyo*, vol. 8, pp. 813–824.

Skudrzyk, E. 1971 *The foundations of acoustics*. Springer.

Tait, C. & Findlay, W. 1996 Wavelet analysis for onset detection. *Proc. Int. Computer Music Conf., ICMA, Hong Kong*, pp. 500–503.

Taylor, R. N. (ed.) 1995 *Geotechnical centrifuge technology*. London: Blackie Academic & Professional and Chapman & Hall.

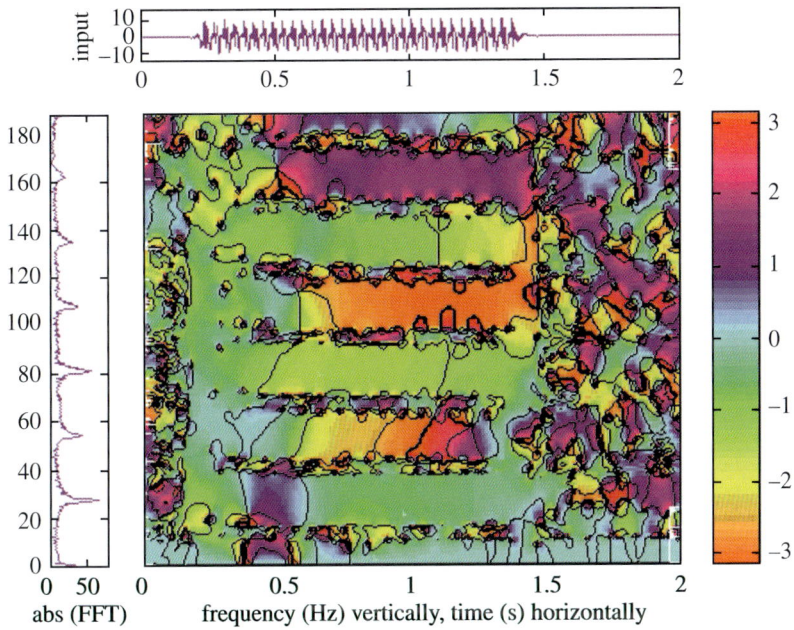

Plate 1: Experimental data: phase of cross-spectral density for signals f_4 and f_1 (measured at positions 7709 and 7726, see figure 9).

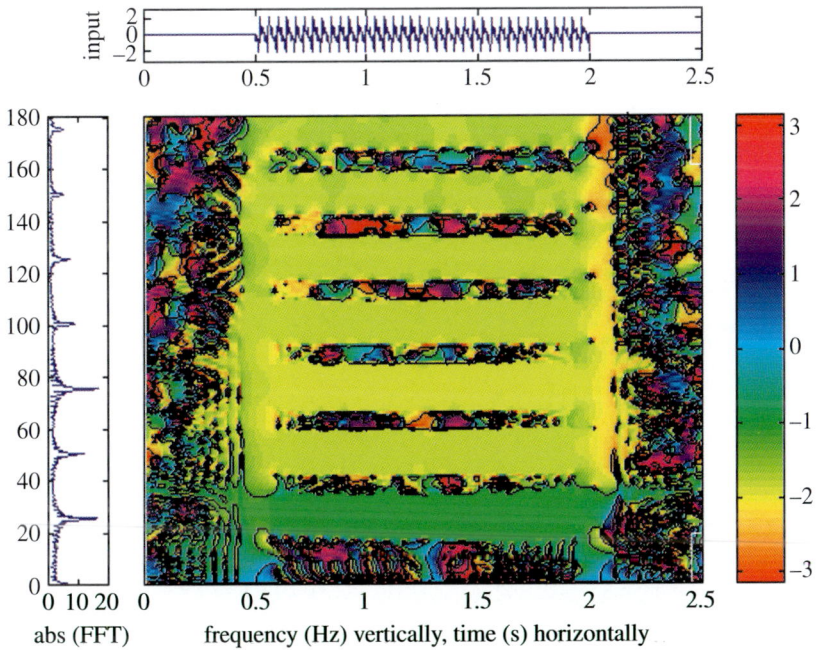

Plate 2: Simulated data: phase of cross-spectral density for signals f_2 and f_1 (at the middle and bottom of the model).